Advanced Information and Knowledge Processing

Series Editors

Lakhmi C. Jain
University of Canberra and University of South Australia

Xindong Wu
University of Vermont

Information systems and intelligent knowledge processing are playing an increasing role in business, science and technology. Recently, advanced information systems have evolved to facilitate the co-evolution of human and information networks within communities. These advanced information systems use various paradigms including artificial intelligence, knowledge management, and neural science as well as conventional information processing paradigms. The aim of this series is to publish books on new designs and applications of advanced information and knowledge processing paradigms in areas including but not limited to aviation, business, security, education, engineering, health, management, and science. Books in the series should have a strong focus on information processing - preferably combined with, or extended by, new results from adjacent sciences. Proposals for research monographs, reference books, coherently integrated multi-author edited books, and handbooks will be considered for the series and each proposal will be reviewed by the Series Editors, with additional reviews from the editorial board and independent reviewers where appropriate. Titles published within the Advanced Information and Knowledge Processing series are included in Thomson Reuters' Book Citation Index.

More information about this series at http://www.springer.com/series/4738

Longbing Cao

Metasynthetic Computing and Engineering of Complex Systems

Longbing Cao
Advanced Analytics Institute
University of Technology, Sydney
Sydney, New South Wales
Australia

ISSN 1610-3947
Advanced Information and Knowledge Processing
ISBN 978-1-4471-7062-4 ISBN 978-1-4471-6551-4 (eBook)
DOI 10.1007/978-1-4471-6551-4

Springer London Heidelberg New York Dordrecht

Softcover reprint of the hardcover 1st edition 2015

Printed on acid-free paper

Springer-Verlag London Ltd. is part of Springer Science+Business Media (www.springer.com)

Preface

We are in the era of complex systems.

An increasingly critical research issue facing researchers in different disciplines is how to compute and engineer complex systems and how to build corresponding problem-solving systems.

Various system metaphors have accordingly been proposed to address the intricacies of complex systems. This effort has become increasingly focused in recent decades as complex systems have become omnipresent in our everyday, social, business, and web worlds.

Reductionism has proved to be a very effective philosophy for tackling complexities in physical and chemical systems, in particular by dedicating priority to the parts of a system rather than the whole. Its principal ideas have widely influenced the research and engineering of complex scientific problems in many domains and business applications.

In recent decades, holism has emerged, bringing with it a focus on systems thinking and an emphasis on understanding complexities in the whole rather than merely in part. System sciences and cognitive sciences have a particular interest in holism. The science of complexity urges the exploration of holism for understanding complex natural and social systems.

Complementing the challenges of applying reductionism and holism for complex systems, systematology combines top-down holistic methodologies with bottom-up reductionistic approaches to consider the complexities in the whole system and its parts, as well as their respective connections. Systematology appears to be really important for addressing the complexities of open complex giant systems, in which it may not be effective to apply only reductionism or holism.

Faced with open complex giant systems like the Internet, we often do not know what we do not know. A qualitative-to-quantitative metasynthesis which explores the synergy of human intelligence and machine intelligence in a human-centered, human-machine-cooperative way may be helpful in understanding system complexities and building problem-solving systems.

Metasynthetic computing and engineering take human-centered, human-machine-cooperative, qualitative-to-quantitative metasynthesis as the main guide for understanding open complex systems. In this book, we outline the corresponding aspects: Chap. 1 presents an overview of complex systems, Chap. 2 considers ubiquitous intelligence in complex systems, Chap. 3 examines system methodologies, Chap. 4 explores computing paradigms, Chap. 5 focuses on metasynthesis, Chap. 6 looks at organization- and service-oriented methodology for engineering complex systems, Chaps. 7, 8, and 9 investigate visual modeling, formal modeling, and integrative modeling of complex systems, Chaps. 10 and 11 discuss the architecture and detailed design of complex problem-solving systems, Chap. 12 addresses ontological engineering, and Chaps. 13, 14, and 15 detail several case studies in building problem-solving systems for actionable knowledge discovery and understanding complex behavior and social data.

The overarching goal of compiling this book is to encourage and inspire discussion and reflection, and the implementation of effective methodologies and tools for computing and engineering open complex systems and problems, while synergizing ubiquitous intelligence, including human, domain, social, network, data, and machine intelligence, during the respective problem-solving processes and in problem-solving systems.

Sydney, Australia Longbing Cao

Contents

Chapter 1
Complex Systems

1.1 Introduction

Complex systems are ubiquitous and have become increasingly focused on scientific and business domains, since they are part of our daily life, business, and environment. Deeply understanding the intricacies of complex systems is thus a basic task in the scientific domain. In this chapter, we explore:

- The system complexities of complex systems, to summarize the main characteristics of complex systems
- System transparency, to outline general categories of complex (as well as simple) systems in terms of the transparency of their content and complexity to users
- System classification, to create multiple dimensions for categorizing complex systems
- Open complex systems, to discuss their characteristics and challenges
- Large-scale systems, to discuss those complex systems that have a huge number of components
- Hybrid intelligent systems, to show those systems that hybridize different techniques, methods, and tools
- Computing and engineering complex systems, to summarize the main computing paradigms, system analysis and design, and the objectives of metasynthetic computing and engineering

1.2 System Complexities

The term *system complexities* refers to the system features and built-in nature that drive the evolution and dynamics of a system, and the emergence and dynamics of system intelligence and intelligent behavior. System complexity can be described as having the

L. Cao, *Metasynthetic Computing and Engineering of Complex Systems*,
Advanced Information and Knowledge Processing,
DOI 10.1007/978-1-4471-6551-4_1

following characteristics: *autonomy, flexibility, reactivity,* and *proactivity*. In these aspects, open complex systems share the same features as normal intelligent systems. On the other hand, open complex systems also present specific complexities that define them as open and complex. In complex systems, the following features play a particularly important role in system evolution and emergent behavior: *openness, hierarchy, sociality, evolution, human–machine cooperation,* and *metasynthesis of intelligence*.

1. Openness
   Openness refers to the exchange of energy, information, and/or material between the system and its external environment and other outside systems. This exchange may take the following forms:

   - The system and its subsystems share different types or amounts of energy exchange, information, and/or material with the external environment and/or systems.
   - The subsystems in the system acquire knowledge or evolve by learning from each other.

   In some cases, *environment* is also open. *Open environment* refers to an environment that is inaccessible, uncertain, and dynamically continuous. Open systems differ substantially from closed systems. The characteristic of openness enables systems to work in a complex mode and to interact with the environment.

2. Hierarchy
   The hierarchy of open complex systems may present the following characteristics:

   - There may be many levels of hierarchy from those known to us at the different subsystem levels compared to the system as a whole; sometimes it is unknown how many layers of hierarchy exist in the system.
   - There may be many modes of subsystem structures, such as parallel structure, serial structure, master–slave structure, hierarchical structure, matrix-based structure, ring structure, and linear structure; sometimes it is not clear to us what kinds of structures exist in the system or subsystems.
   - Multiple structures often coexist in the system or subsystems, forming hierarchical structures; different distributed structures may exist in a system.
   - Compound structures may appear in a system or subsystem through the combination of multiple simple structures, which are coupled in different relationships.

   Complex systems are embedded with implicit or explicit hierarchical structures. In hybrid systems, two or more types of structure are combined and presented on different layers.

3. Sociality (Social Characteristics)
   Social (or societal) characteristics in open complex systems may be presented in terms of either single or combined forms of the following aspects:

   - A system is composed of subsystems or components (system members) which are coupled in terms of different relationships, such as a spatiotemporal relationship.

- The subsystems or components are distributed, flexible, and/or automated.
- The human as a social agent plays a built-in role in the system or subsystems; in some cases, the human serves as a system constituent.
- Subsystems or components taking different roles interact with each other in diverse social interaction modes (such as cooperation, negotiation, collaboration, and/or coordination) and communication languages (such as logical and/or probabilistic) and certain rules and policies (such as social rationality, norm, law). They jointly solve a problem by taking social responsibility while influencing each other and following certain social constraints.
- Subsystems or components behave according to different behavioral rules, norms, social rationality, law, protocols, or social policies but collaborate with each other to achieve certain social objectives and solve problems in a mutually influenced manner.
- A system component may take different roles within a system. Subsystems interact with each other in terms of interactive mechanisms (such as the social interaction modes of cooperation, coordination, negotiation, etc.). Sociality is presented as concurrency in terms of the temporal aspect, while the spatiotemporal characteristics are embedded in the data, control of the system, domain knowledge, and information distribution.
- The interaction and collaboration between subsystems or components create certain forms and levels of trust between one another, and contribute to joint problem-solving.
- Each subsystem or component owns a certain level of privacy within the subsystem level and in the system; privacy may be presented in terms of information, energy, and/or material.
- The temporal coupling may take place in serial or parallel form, while the spatiotemporal distribution may involve the distribution of data, control, domain knowledge, operations, and resources.

4. Evolution
   Evolution is a typical feature of complex systems, especially those that interact with the environment. Evolution in a complex system may be embodied as follows:

   - There may be continuous, sometimes substantial, changes taking place in the system structures, composition and formation, types of components, status of components, interactions between and within system elements, and the behaviors of the system or its subsystems; changes may occur at the design time or at run time.
   - There are changes taking place in system components, subsystems, and system states at different time points; changes in the interactions between system components and subsystems drive a system's dynamics and the evolution from one state to another.
   - System states and the underlying drivers—interactions between subsystems—are not determined at the design time. A system may have the

self-organizing ability to evolve from one state to another, enabled by simple or complex interaction rules embedded in a system.
- Evolution may take place at different levels within the system, which further contributes to the emergence of certain patterns, structures, architectures, relations, or new system elements at the local and/or global level; such emergent behaviors contribute to the intelligence and problem-solving capabilities of the system.
- The local interactions between and within the system members or subsystems may contribute to the emergence of unique new system features and behaviors, for instance, patterns emerging from self-organization within the system, and the so-called emergence of swarm intelligence. Such local-to-global emergent behaviors and intelligence form unique capabilities for understanding global system behaviors, their dynamics and problem-solving processes. Evolution-based emergence is a typical problem-solving ability embedded in complex systems.

5. Human–Machine-Cooperative Problem-Solving
The cooperation between human and machine in open complex systems is different from what is usually called human–machine interaction. We are interested in the problem-solving roles shared through cooperation between human and machine in open complex systems. This may take one of the following forms:
   - A typical (and substantial) characteristic in open complex systems is that the human forms a key system element, actually a senior intelligent entity, in the composition of the system, and is involved in the problem-solving process.
   - In terms of the problem-solving process and shared workload, the human and the problem-solving machine collaborate as a team to form solutions and achieve the combination of human cognition and recognition with machine reasoning and computation; a joint human–machine problem-solving system features both human cognitive intelligence and a machine's reasoning intelligence, as well as their synthesis.
   - The cooperation between human and machine produces intelligent behaviors which form the core drivers of problem-solving; a problem cannot be solved well if only machine intelligence is involved, and humans play a necessary role in the problem-solving in terms of intelligence, domain knowledge, and creative thinking (especially imaginative and inspired thinking). Cooperation may involve a group of domain experts rather than individuals.

Human–machine cooperation is widely seen in artificial systems, enterprise systems, and social systems. For complex problem-solving, it is not possible to rely only on machine intelligence, which is quantitative in nature. Human intelligence is likely to be more qualitative, and cooperation with the machine enables the problem-solving process to be both qualitative and quantitative.

6. Metasynthetic Feature in Open Complex Systems
Metasynthesis is a methodology proposed by Qian [1, 2] to understand and solve *open complex giant systems* (see Chap. 5). Here we discuss the metasynthetic complexities of complex systems.

- There are different types of intelligence embedded in an open complex system: human intelligence, machine intelligence, data intelligence, and the surrounding environment which forms network (web) intelligence, domain intelligence, social intelligence, and organizational intelligence. Different types of intelligence play respective and irreplaceable roles in a system and its problem-solving. For instance, a domain expert can provide domain knowledge and inspiration that cannot be delivered by machine. Data intelligence is increasingly recognized in the process of evidence-driven problem-solving through descriptive and deep analytics and knowledge discovery. Network intelligence serves such needs as information retrieval, search, and distributed computing.
- Different types of intelligence complement each other during the problem-solving process and system evolution. Qualitative human intelligence combines with quantitative data and machine intelligence to deepen our understanding of a complex system and its problem-solving.
- Social intelligence behavior and problem-solving capability in an open complex system are the results of interaction and cooperation between human, system, and other contextual components.
- The challenge of understanding a complex system lies in the extraction and definition of different types of intelligence, their roles, constraints, interactions, and complementation in problem formation and problem solving. An accurate and deep understanding will ease the challenge of analyzing, designing, implementing, and controlling a complex system.

1.3 System Transparency

The system classification criteria used here include system contents and their transparency to users. For example, a system is either a trading system or a data mining system according to its objectives, content, and degree of transparency in content disclosure during the interaction between users and the system. In light of this benchmark, we classify systems into four basic categories: *black boxes*, *white boxes*, *grey boxes*, and *glass boxes*.

1.3.1 Black Boxes

A *black box* system is specified entirely in terms of its function and interface. A user has no knowledge of its internal structure or method of implementation, hence all the knowledge to be used must be made explicit by the definition of the system itself and the environment in which it is to interact. The system is used in exactly the form in which it is provided. In human–computer interaction, only the externally visible behavior of a black box is considered but not its implementation or "inner workings."

Most commercial trading and data mining systems are in a black box. Such systems are generally described as "trading systems" or "data mining systems" because they run on proprietary algorithms that are not disclosed. These algorithms are run over a database and give users trading/data mining recommendations or suggestions, but they do not usually tell users "why." The provider of the system is responsible for the maintenance and quality of the algorithm and the system components.

1.3.2 White Boxes

A white box system is provided with all the source codes so that all details of the structure and implementation of the component are transparent to the user. A white box system looks more like a *toolbox* system. The system can be modified and adapted to suit the exact needs of the user. This is probably the most widespread type of reuse, and it operates in an unstructured and ad hoc way.

A *toolbox* is a box for holding hand tools. To be classified as a toolbox, the software must either fully disclose algorithms or let users decide what the box does by calling the tools available in the box. The user will be entirely responsible for the modifications, deployment, quality, and maintenance of the system. Usually, this type of system does not create trading suggestions. The user conducts the analysis and makes his/her own decisions.

1.3.3 Glass Boxes

A *glass box* is a white box component which is used unmodified. Using a system this way brings most of the benefits of a black box without the disadvantages. Not all the knowledge needed to make successful use of the component is made explicit in the interface definition. By examining the internal structure and implementation, the user can glean additional knowledge that is not made explicit in the formal definition. This knowledge can help the user to trade or mine more flexibly and consciously.

1.3.4 Grey Boxes

With only minor modifications, a white box can take the form of a grey box. A grey box is a box that occupies an intermediate position between the black box and the white box, with some degree of customization by users. Trading/data mining systems in grey boxes can generate suggestions from proprietary algorithms. However, they provide a general idea of how the formula works and sometimes allow the user to modify the settings or parameters. They may also have an associated toolbox component.

1.4 System Classification

There are different methods for classifying artificial or natural systems. Here, we discuss some fundamental aspects that may be involved in classifying systems.

1. *Openness*

   The *openness* of a system determines its exchange of energy, information and/or materials (mass) with its environment (including contextual systems). Accordingly, a system may be classified as:

   - *Closed system*: A system does not transfer energy, information, and/or mass to its environment.
   - *Semi-closed system*: A semi-closed system consists of social factors or components. It is an automated system, in which the problem-solving is automated.
   - *Semi-open system*: A semi-open system is a social system in which some individuals or components involve a societal process to achieve solutions. Intelligence emerges as a fact of social interaction, while conflict and negotiation involve the interaction between individuals and components.
   - *Open system*: If a system is open, it exchanges energy, information, and/or materials with its environment. In such a system, system formation is dynamic, and humans often play an important role in the system formation and/or problem-solving. Open systems present system complexities in terms of hierarchy, socialization, evolution, human–machine cooperation, and intelligence metasynthesis.

   The main differences between the above systems are discussed below:

   - The difference between closed and open systems lies in whether a system is social, that is, whether the system's intelligent behavior is social or not. Accordingly, a closed system is automated, while an open system is social. In terms of coupling relationships between system components, closed systems are tightly coupled, while open systems are loosely coupled.
   - The difference between closed and semi-closed systems lies in whether a system is completely automated. A semi-closed system partially consists of social constituents represented in different forms.
   - The difference between open and semi-open systems lies in whether the human is part of the system, and the degree of societal and hierarchical characteristics in the system.

2. *Scale*
   The *scale* of a system is often used as a primary factor in the classification of systems. The scale refers to the number of individuals, components, or subsystems embedded in a system. Systems are thus often classified into different scales according to largely empirical and subjective judgment, and relative position, as follows:

   - *Small-scale systems* refer to systems composed of dozens of system elements and components.
   - *Middle-scale systems* refer to systems composed of hundreds and thousands of elements and components.
   - *Large-scale systems* refer to millions of elements and components embedded in one system.
   - *Ultra-large-scale systems*, or *giant systems*, may consist of billions of objects in a system.

   In data systems, scale often refers to the size of the data. Typically, one might find data of kilobyte (10^3), megabyte (10^6), or gigabyte (10^9). Big data may involve terabyte (10^{12}), petabyte (10^{15}), extabyte (10^{18}), or zettabyte (10^{21}).

3. *Heterogeneity*
   The complexity of a system also depends on the *heterogeneity* of the system elements, subsystems, or components it contains. Homogeneous systems consist of one type of element only. Heterogeneous systems consist of different types of objects.

   The heterogeneity of a system is often related to its scale, thus affecting system complexity and classification.

4. *Interaction*
   Coupling relationships exist between system constituents and cause various levels and forms of interaction. There may be different types of coupling relationships between system constituents or subsystems. These may present as semantic relations, syntactic relations, hierarchical relations, or social relations. Interaction is embodied in terms of collaboration, cooperation, communication, negotiation, or coordination.

   The other aspect is the level of coupling relationships. Some systems may consist of strong couplings, while others may incorporate weak couplings. A system may consist of different interaction modes which present different levels of interaction frequency, strength, parties, and effects.

5. *System Classification*
   Openness, scale, heterogeneity, interaction, and other possible aspects form the foundation of classification systems. Broadly speaking, a system is either simple or complex:

   - *Simple systems*: If a system only consists of homogeneous objects or subsystems, and has a weak and low level of interaction and/or social components, it is a simple system, even if its scale is very large, or giant.

- *Complex systems*: A system that is composed of a large scale of heterogeneous objects, especially human objects, and/or has particularly strong interaction with its environment is a complex system.

1.5 Complex Agent Systems

1.5.1 Multiagent Systems

1.5.1.1 What Are Multiagent Systems

Multiagent systems (MAS) [3, 4] are systems composed of multiple interacting computing elements, known as *agents*. Agents are typically capable of cooperating to solve problems that are beyond the abilities of any individual member [4]. An agent is a physical or virtual computing system which can act, perceive its environment (in a partial way), and communicate with others. In this process, it *interacts* with other agents and the agent environment, and presents *autonomy,* deciding by itself "what to do" to achieve its ultimate goals. Typical MAS include the Internet, virtual ecosystems, e-markets and trading systems.

The studies on the two interwoven strands of autonomous agents and multiagent systems, or AAMAS [5] have, since the 1980s, formed a newly and fast-developing science which has been widely recognized since the mid-1990s. From its inception, AAMAS has been expected to be an effective and efficient tool for dealing with tough residual problems in artificial intelligence (AI), such as hybrid intelligent systems, and multiple, inter- and cross-disciplinary problems, such as IT-enabled electronic markets and trading systems.

A huge number of branches of MAS persist, all from different theoretical backgrounds. They are, for instance, the Distributed Artificial Intelligence by Lesser, Gasser, Feber, and Sycara; the Rational Agents branch by Rao and Georgeff, Shoham, and Castelfranchi; the Reactive Agents branch by Brooks, Steels, Drogoul, Ferber, and Demazeau; Artificial Social Systems [6, 7] and Artificial Life by Newell, Langton, and Mitchell [6, 8, 9], and so on. These studies, although maintaining separate points of view and dealing with differing problems, as well as having their own foundations and applications, are actually very complementary.

MAS are important primarily because they have been found (i) to be able to deal with some of the open problems which cannot be handled using existing methods or techniques, and (ii) to have very wide applicability, in areas as diverse as complex software systems, ecosystems, artificial life, virtual environment and simulation, industrial process control and electronic commerce, knowledge discovery and data mining.

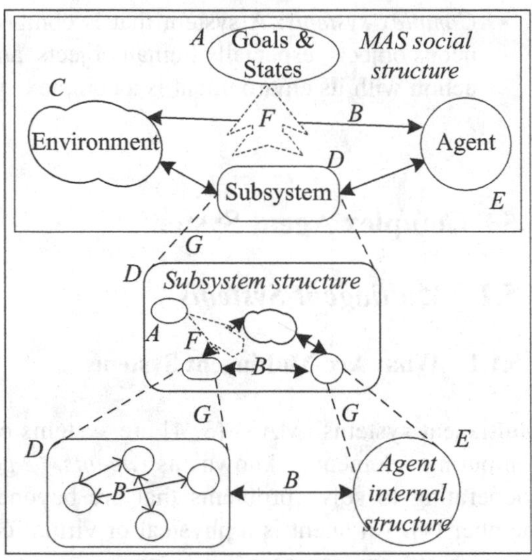

Fig. 1.1 MAS research map

1.5.1.2 Multiagent System Research Map

The science on MAS has been advocated not only as a technique for building high-quality intelligent information systems, but also as a methodology for dealing with a number of demanding problems in AI and multiple other disciplines. In reality, the research on MAS has been extended to a huge number of classes. Fig. 1.1 is an attempt to capture the main research scopes in the MAS domain, which are categorized in terms of organizational metaphor and are marked in symbols from A to G. The individual research scopes are described as follows.

A. *Goals and States*

The research on goals investigates how agents accomplish their personal goals and also how multiple agents interact and cooperate with each other to reach common or superordinated (super) goals. This also concerns how a super goal is decomposed into or fulfilled by multiple subgoals.

Cognitive and reactive agents normally express inner and outer states in achieving their goals, whether they are personal goals or common goals. States are affected in the process of perceiving the environment. The research on states investigates which agents or agent organizations maintain and adjust their personal or global states in their perception of the world, and how. Specifically, mental or cognitive states are embodied through cognitive activities. Cognition obtained from cognitive activities is further embodied through cognitive elements such as belief, desire, and intention.

B. *Interaction and Rules*

The research on interaction has become a significant branch of the MAS domain. In this work, we classify all the following interactive patterns—coordination, cooperation, communication, conversation, negotiation, auction, brokerage, wrapping, embassy, mediation, matchmaking, teamwork, and coalition [10–12]—into the family of interaction. In an interaction activity, agreements are reached, and certain relationships and rules are normally adhered to by interactive agents. Research on interaction therefore involves the definition and representation of interaction and goal distribution, task planning, communication [13], interaction constraints and methods, conflict resolution, decision-making, etc. In addition, studies also concern the layers on which the interaction and its rules take place in an agent system.

The studies in [10–12] are of main interest in respect of rules and relationships, social order, laws, policies, distributed constraints, and so on.

C. *Environment*

An agent can interact with its environment in several ways. It observes its environment with its sensors and modifies it with the help of its actuators. Additionally, an agent is able to sense and act inside its environment. The research on environment involves the methods, activity, and types of agent–environment interaction [14]; studies on states and their dynamics of environment are also an interesting topic. Evolution, adaptation, learning, and load balancing [10–12] in MAS are closely related to agent environment. Physical, social, and spatial environments [15], in which principles, processes, and populations play important roles in organization formation and evolution, are also attracting attention. Modeling the agent environment using such theories as Markov decision process [16], dynamical systems theory [14], and others is another interesting topic.

D. *System Architecture and Organization*

Previous research on MAS architectures has focused on three types of structure. These are deliberative architectures based on physical symbol system hypothesis [17, 18], reactive architectures inspired by symbolic AI [19, 20], and hybrid architectures that integrate multiple structures such as reactive, planned, and social behaviors [21].

Currently, research on hybrid intelligent system architectures [22–24] is a popular area in AI and AAMAS, with more attention being transferred to tough problems that remain in AI. In this field, studies focus on how to integrate different computing paradigms such as fuzzy logic, neurocomputation, evolutionary computing, probabilistic computing, machine learning, and other computing frameworks in terms of multiagent theory.

From the perspective of artificial organizations, organizational architectures are sensitive to organizational rules and organizational interactive patterns. The following structures may be used for designing agent system structures: centralized, market-like, pluralistic community and community with rules of behavior [25]; fixed hierarchical predefined structure, variable peer-to-peer evolutionary structure, variable peer-to-peer predefined structure, and evolutionary structure [26]; and architectural styles such as flat structure, pyramid, joint venture, structure-in-5, takeover, arm's length, vertical integration, co-optation, and bidding [27, 28].

E. *Agent and Agent Internal Structure*
The studies on agents include aspects of methodology, architecture, organization, interaction, communication, reasoning, and constraint, as well as mental elements such as belief, desire, and intention.

More recently, research on synthetic agents, artificial social agents, and agencies from artificial organizations and sociality has attracted special attention. This is probably inspired by organizational metaphor and societal problem-solving mechanisms, which work very well in human and social organizations.

F. *System Dynamics*
Research on agent system dynamics has attracted little attention from the AAMAS domain because previous and existing work has mainly concentrated on developing mechanisms on individual agents and simple agent systems. However, system or organization dynamics is a critical issue for understanding, analyzing, and deploying open complex agent systems (OCAS). This also refers to traditional relevant work on cybernetics and dynamical systems theory [14], and the newly emergent science of complexity [7, 29–32, 66–67]. In the science of complexity, one interesting topic related to system dynamics is self-organization in artificial social systems.

G. *Agent-Oriented Software Engineering*
The main tasks in agent-oriented software engineering (AOSE) comprise the following aspects [10–12, 33–39]:

- Methodologies and methods for AOSE
- Formal and informal system modeling for requirements analysis, system analysis, and design
- Modeling methods
- Refinement and synthesis
- System/goal/task decomposition
- Agent system architectures
- Software development environment, implementation techniques including programming languages, and standard API
- Performance evaluation, verification and validation, testing, and so forth
- Organizational and social impacts on AOSE
- Organization-oriented software engineering

In this book, AOSE is the main area of interest. We will first sketch the AOSE in Sect. 4.5.2 "Agent-Oriented Software Engineering", and in later chapters, some associated problems will be explored in more detail.

H. *Applications*
Multiagent systems have been widely used in multiple disciplines outside the computer sciences by academia. Two aspects need to be mentioned here: one is hybrid applications in which MAS technology is integrated into other existing problem domains or computing paradigms; the other is the industry-strength application of MAS in the real world, particularly for solving complex problems.

1.5.2 Large-Scale Systems

Large-scale systems are increasingly the principal focus in the study of complex systems. *Large-scale multiagent systems*, or *massively multiagent systems*, have recently become the focus of the multiagent system community. Increased efforts have been made to study the complexities of these systems, the software engineering methodologies and tools for analyzing and designing them, and the implementation problems encountered, including communication, architecture design, application integration, cooperation, and coordination. Typical issues studied in the community of large-scale multiagent systems include: the theoretical foundation for software engineering, requirement engineering, software architecture design, openness, interaction between a system and its environment, cooperation, mobility, reusability, dependability, and the empirical evaluation of system quality and performance.

Attention is paid to large-scale systems to address typical issues in static multiagent systems, including susceptibility to error, poor scalability, design time-based predefined system design, or ad hoc formation. Such systems cannot adapt to dynamics at running time and decision-making at running time, and are often small scale. As a result, such systems cannot usually satisfy the real-life need for dependable, robust, trustful, reusable, and scalable design and implementation. Large-scale systems have been proposed to address these issues and contribute to intelligent large-scale systems.

Large-scale systems possess characteristics and present intelligent behaviors that cannot be interpreted, simulated, or predicted in small-scale systems. This is due to the large scale of their system constituents, the complex interactions between a significant number of agent and subsystem categories, and the interactions between a system and its environment. A typical situation is the Web and online systems such as social network systems and social media systems. Such systems usually involve large scale system or contextual constituents which interact to collaborate and operate the business. In such systems, multiple interaction mechanisms exist, including autonomy, cooperation, communication, negotiation, and coordination between system members. The interaction mechanisms may not be fixed, certain, and precise, as system members share a high level of autonomy by following certain constraints, rules, policies, protocols, and procedures defined in the system. This enables the system and its interactions to be dynamic, adaptive, and evolutionary. Accordingly, analyzing, designing, implementing, and operating such systems is a very challenging process. Unexpected and exceptional behavior and/or effects can sometimes be seen to emerge at an individual, subsystem, or system level.

1.5.3 Large-Scale Multiagent Systems

1.5.3.1 Concepts and Issues

Large-scale multiagent systems, ultra-large-scale multiagent systems, or massively large-scale systems refer to those multiagent systems which consist of hundreds or thousands of agents of different types. Every type of agent has its own characteristics. Such a system has certain system features, such as reliability, security, self-adaptation, interoperability, scalability, maintainability, and/or reusability. Large-scale multiagent systems can be used as a paradigm for designing complex intelligent systems or a software engineering method for building multiagent systems.

In general, multiagent systems research focuses on a small number of types of agents, and a small scale of agents, in tens or so. Such systems often have ad hoc features, designed to test design objectives. These lab-based prototype systems are not effective for handling real problems and building practical systems. A key issue here is that the true level of complexity will be much higher when they are tasked with real-world problem-solving. The problem of managing hundreds of agents may be essentially different from tackling thousands of agents.

According to [40, 41], ultra-large-scale (ULS) systems will be interdependent webs of software-intensive systems, people, policies, cultures, and economics. The scale and complexity of such systems is increasing dramatically. ULS systems are systems of unprecedented scale in some of these dimensions:

- Lines of code
- Amount of data stored, accessed, manipulated, and refined
- Number of connections and interdependencies
- Number of hardware elements
- Number of computational elements
- Number of system purposes and user perception of these purposes
- Number of routine processes, interactions, and "emergent behaviors"
- Number of (overlapping) policy domains and enforceable mechanisms
- Number of people involved in some way

1.5.3.2 How Are ULS Systems Different? [40]

The sheer scale of ULS systems changes everything. ULS systems will necessarily be decentralized in a variety of ways, developed and used by a wide variety of stakeholders with conflicting needs, evolving continuously, and constructed from heterogeneous parts. People will not just be users of a ULS system; they will be elements of the system. Software and hardware failures will be the norm rather than the exception. The acquisition of a ULS system and its operation will be simultaneous, and will require new methods of control.

These characteristics may appear in today's systems and systems of systems, but in ULS systems, they will dominate. Consequently, ULS systems will place unprecedented demands on software acquisition, production, deployment, management, documentation, usage, and evolution practices.

The ULS systems notion has inspired us to ask new questions about software-reliant systems:

- What new quality attributes arise due to scale?
- What types of analysis is required to understand and design (at all levels) systems at scale?

- Are new architecture design principles needed?
- What new strategies are needed to control, predict, and bound the behavior of systems at scale?

Although many software engineering methodologies such as GAIA [42] and TROPOS [43] have been proposed for engineering complex agent systems, none has been instantiated into business. There are still many practical as well as theoretical issues surrounding the proposed methodologies; for instance, they cannot support the whole software engineering process; there are no tools to support software evaluation and quality assurance; no support for system dynamics analysis; no working tools like Rational Rose to support the process computerization; there is a lack of support for organizational relationships, semantics, etc.; no support for engaging humans as system components; the issue of how interaction and cooperation between human and system modules can be supported; whether the organizational relationships appearing in small-scale systems cover the full picture in large-scale systems; and the transfer capability of achievements realized in studying small-scale systems to large systems.

The above discussions show that it is necessary to develop effective tools to engineer large-scale multiagent systems. It is especially important to propose more practical abstraction mechanisms, analysis and design tools for capturing organizational factors and relationships, system dynamics, workflow management, and tools and algorithms to map a complex real-world problem to a computerized system.

1.5.3.3 Major Research Issues

The fundamental issues lie in the gaps between different scales of systems and the significant system complexities in large-scale systems which may be fundamentally different from simpler systems. The corresponding research and development is necessary to address those unique or sophisticated complexities in large-scale systems.

The main issues studied in large-scale multiagent systems are comprehensive. Some are listed below:

- System analysis and design methodologies and tools
- System architecture and framework design
- Design factory, design patterns, and architectural styles
- Modeling language
- Reflective architecture
- Architecture and tools to support coordination and cooperation between large-scale agents
- Reliability and dependability of system design
- Mobility, security, and distribution
- Context awareness
- Exception and fault tolerance management

- Large group design and tuning and monitoring
- Testing and evaluating systems
- Verification and validation

With many agents in a system, agents need to know each other and the context. Context consists of the circumstances, objects, and constraints for supporting the interactions between agents, and between agents and the environment. Understanding the context properly is crucial for designing working, large-scale systems.

- Context may be physical, environmental, informative, individual vs. social, practical vs. system-based. Irrespective of the type, an agent needs to acquire, sense, and reason the context in which the agent is located.
- Different agents may have a different understanding of the context, and may use different description tools to describe the context, which engenders complexity in the interaction and cooperation between agents.
- Large-scale multiagent systems often involve heterogeneous agents which share knowledge about context and may be situated in different contexts. Multiple contexts may be involved due to the different roles played in the system.
- Agents in the community may have different capabilities and preferences; some may be selfish or unpredictable.

There is a need to develop context-comprehending tools to ensure that transparent, integrative, and manageable architectures are built for the agent society. It is expected that the system will be dependable, robust, trustworthy, and extensible.

Dependability is a critical issue in large-scale multiagent systems. Dependability refers to the capability of agents to provide dependable services. The following are typical issues to be studied:

- There may be conflict between dependability and the other features of agents. For instance, an automated agent may have a problem in managing the ineffectiveness assumption. The mental situation of an agent may limit its introspection and hence affect its monitoring and control of unexpected issues.
- Is role a fundamental property of an agent that has nothing to do with service?
- How can it be ensured that an agent will provide dependable services, what key of exception assumption could it have, and how could agents be enabled to manage exceptions instantly?
- How can the confidentiality, integrity, reliability, availability, and maintainability be ensured?

In large-scale systems, communication plays a very important role, especially for open systems. This involves many fundamental issues, such as:

- How to use, control, and manage autonomy
- How to balance autonomy and interoperability
- How to ensure behavior integrity by designing proper coordination mechanisms
- How to design exception and fault tolerance to tackle issues in large-scale agents or distributed communications.

The above issues are very much related to the dependability of multiagent systems. Taking them on board, we need to incorporate dependability into the design and be prepared for running time issues. This requires systems to be automated, but also adaptive and self-learning.

1.5.4 *Open Complex Agent Systems*

1.5.4.1 Multiagent System Classification

The classification of MAS is usually performed from one of the following perspectives: (1) theoretical foundations, (2) system functionalities, (3) agent architecture, and (4) application problem domains.

- Theoretical foundations such as symbolic AI and logic, physical symbolic systems, distributed and concurrent systems, social sciences, systems science, biology, economics, philosophy, game theory, dynamical systems theory, science of complexity, and so on.
- System functionalities such as the classification method in [44], in which there are three high-level agents such as biological agents, computational agents, and robotic agents. There are software agents under the computational agents; at a lower level are the information agents, of which there are two groups, noncooperative and cooperative, both of which comprise agent classes like adaptive, rational, and mobile.
- Agent architecture, which is dependent on different dimensions. In [25], agent structures are classified into centralized, market-like, pluralistic communities, and communities that have rules of behavior, while in [26], Ferber classifies agent structures as fixed hierarchical predefined, variable peer-to-peer evolutionary, variable peer-to-peer predefined, and evolutionary.
- Application problem domains such as workflow, business process management, social simulation, virtual environments, e-markets [45], e-commerce, auction, information retrieval and management, Web services, human–computer interaction, knowledge management [46], computer-supported cooperative work, decision-support systems, and so on.

Here we give another classification based on the perspective of systems sciences, or more specifically the *system complexity* of agent systems. The system complexity of agent systems is evaluated based on the following attributes [47–49].

- Openness: whether there is live interaction and exchange of energy, information, and material between agents and the environment.
- Population: the number of system components, namely, agents.
- Hierarchy: many different levels exist in a MAS, from components which are clearly recognizable at an individual level, to the system in terms of the macrolevel. In some cases, it is not clear how many levels there are in a system.

- Uncertainty: whether there are many uncertain situations, for instance, the system is not decomposable.
- Sociality: whether temporal, spatial, and interwoven spatiotemporal interaction, activities, behaviors, and constraints exist within an agent system.
- Human orientation or human–computer coexistence: whether human beings are required or more directly involved as computational constituents of an agent-based problem-solving system.
- Evolution: whether agents and the agent organization have the adaptive capability to sense and learn from the live environment with or without cognition, so that states can be adjusted to find a collective solution to the problem.

Based on these attributes describing system complexity, we classify all agent systems into four categories as follows:

- Closed systems: There is no exchange or interaction between the agent and the agent environment; all agents are autonomous with tight coupling in the agent organization; agents self-interestedly control and determine their behaviors totally by themselves without reference to other agents and surroundings.
- Semi-closed systems: Similar to closed systems, but the tightly coupled agents determine their behaviors by reference to both themselves and other partners' reactions; in other words, the socially organized agents cope with problems in an autonomous way.
- Semi-open systems: The socially organized agents cope with problems in both an autonomous and a social way (a problem is solved through social interaction [47] with multiple relevant agents).
- Open systems: Both the organization and problem-solving within these agent systems take the form of social interaction.

Based on the degree of system complexity, we can further classify open agent systems into *open simple agent systems* and *open complex agent systems*. The definition and characteristics of open complex agent systems will be discussed in the next section.

1.5.4.2 Open Complex Agent Systems

Open complex agent systems (OCAS) are middle-size and large-scale open agent systems, for instance, a practical agent-based financial trading systems. In an OCAS, a great many of system components socially interact with each other and with the environment to achieve self-sufficient goals; a hierarchical and dynamic structure will emerge as a result of collective intelligence. OCAS are very critical because they are able to deal with those complex problems, which cannot be handled by traditional technologies or simple agent systems.

Compared with open simple agent systems, in which small populations of agents coexist, an OCAS is much more complicated. A best instance for understanding OCAS is the Internet. Analyzing the structure, behaviors, and dynamics of the Internet [50] would be very helpful for understanding the complexity of an OCAS.

The complexity of an OCAS, like the Internet, is pervasively embodied in a great many of agents or groups of agents, social interaction, hierarchical and dynamic structure, open niche, and surmisable and self-sufficient goals.

With the increase of the population of agents, the complexity of an OCAS also increases dramatically. More importantly, the complexity associated with an OCAS is normally not straightforward. The evolution process and the resulting phenomena of huge number of agents interacting with each other and with environment are not expectable, uneasy to be comprehended, and even hard to be explained.

The environment of an OCAS is open, networked, heterogeneous, and uncertain. The complexity of agent environment makes interaction among agents and management of them more difficult and increases the likelihood of the occurrence of exceptional situations, emergent behaviors, and unexpected outcomes. The situation may get much more beyond control if human users or software developers delegate more autonomy and trust on agents.

The strong sociality makes an OCAS much more complicated. In an OCAS, the interactions, agent activities, and behaviors take place temporally, spatially, or spatiotemporally. The constraints and the rules on the social interactions may result from temporal, spatial, or spatiotemporal dimensions.

In an OCAS, goals would be multiform and hierarchical. Low-level subgoals could be comprehensive; however, the global goals would be determinable. Decomposition and fulfillment of goals would not be an easy job. The achievement of goals would be as a result of intelligence emergence either with cognitions or not.

Because of social interactions, the system structure would be rather dynamic and complicated. The structure is evolutionary and might be presented as a collective emergence of a great number of interacting agents with partners and environment.

An OCAS is complicated. As the life cycle of any existing sciences have shown, there is also a long way for OCAS to go toward a more complete and scientific specification. This situation cannot be changed without hard work in both theoretical and technical exploration. More specifically, engineering an OCAS would also not be an easy work. In fact, software engineering of an OCAS would be a challenging issue. The development of a generally operable and applicable methodology and technique would be critical and hard work for engineering an OCAS. Even now, we still expect that the following qualities such as reliable, robust, scalable, usable, and effective and deployable are some basic criteria for the AOSE of OCAS. The next section will sketch the engineering of open complex agent systems.

1.6 Hybrid Intelligent Systems

1.6.1 Concept

The research on hybrid intelligent systems (HIS) [51] has been a popular topic in the AI and relevant communities in recent years. Why do we need hybrid intelligent systems? *Hybrid intelligent systems* (HIS) have been taken as a general approach for handling complex intelligent phenomena and building complex intelligent systems. There are many reasons or driving forces; a simple and straightforward one is because any single solution cannot handle the complexities in an underlying system, for instance, imprecision, incompletion, uncertainty, vagueness, dynamics, heterogeneity, and complementation. The involvement of multiple computing paradigms forms combined or hybrid approaches, which can collectively address different challenges. This thus provides a solution to handle the complexities beyond any single solution. The goal of hybridization is to increase problem-solving capability of a hybrid system while maintaining its automation, adaptation, and robustness and, at the same time, addressing the constraints from physical, logical, data-driven, and evolutionary aspects.

Hybridization can happen at any aspects or fields, such as statistics and machine learning. More specifically, HIS aim to integrate or synthesize two or more types of intelligent processing technologies to produce more intelligent, adaptive, and powerful intelligent systems. The resulting systems are then expected to handle physical, logical, data-driven, evolutionary constraints, etc. HIS researchers intend to create novel hybrid methods, methodologies, and frameworks, which can be used to develop automatic, adaptive, and evolutionary HIS systems and applications.

The main tasks of HIS are to address how soft computing can be integrated with AI technologies, such as fuzzy logic, neurocomputation, evolutionary computing, machine learning, data mining, and multiagent systems. This will create hybrid methods, algorithms, architectures, techniques, and systems for studying intelligence. A typical example is evolutionary fuzzy neural network, which combines techniques of genetic algorithms, fuzzy systems, and neural networks. Hybridization may also take the form of integrating AI technologies with others, such as ontological engineering and data mining, to form super-intelligent systems.

The hybridization may be built on top of two techniques, three techniques, or multiple techniques. In a hybrid system, either one or multiple techniques take the leading roles to contribute to the hybrid intelligence. For instance, hybrid neural computing systems are formed on the basis of neural systems, incorporated with other necessary techniques in AI and soft computing, to form hybrid methods, models, and systems.

1.6.2 Hybridization Strategies

Generally speaking, there are different ways to hybridize (integrate) two to more intelligent systems, techniques, or disciplinary tools. Different computing paradigms, combined, or hybridization approaches are studied to integrate multiple intelligent tools and leverage the weakness in one system, in order to address issues that cannot be handled by any single systems. The HIS community is dedicated to invent new and effective methodologies, methods, techniques, frameworks, and platforms.

In recent years, increasing efforts have been made to create new theories, techniques, and applications especially through combining soft computing, computational intelligence, statistics, machine learning, and data mining with artificial intelligence. This forms new approaches and tools such as fuzzy logic, neurocomputing, evolutionary computing, probabilistic machine learning, nature-inspired computing, and agent mining. Such combinations are embodied in terms of creating new aspects such as new intelligent methods, algorithms, tools, and systems. This may be through combining two or more kinds of techniques, such as genetic fuzzy neurosystems by integrating genetic algorithms and fuzzy set theory with neural system. In addition to the integration of artificial intelligence with other intelligent techniques, artificial intelligence is often combined with methods in other areas such as ontological engineering, multiagent systems, and data mining to form hybrid techniques such as agent mining.

The idea of hybrid intelligent systems can be described as follows: Assume there are N individual techniques, methods, and/or models $I = \{I_j | j \epsilon N\}$, where I_j represents a specific constituent technique and a hybrid system $⋒$ is an integrative outcome of n constituent techniques or systems $\cap_j$:

$$⋒ = \{\cap_1, \cap_2, \ldots\} = I_{j=1}^{n} I_j \tag{1.1}$$

The hybridization of multiple constituent techniques is built on certain conditions and constraints. Assume C_s is a constraint set for the target problem, $C_s = \{c_{s1}, c_{s2}, \ldots\}$, where $c_{sj}(j\epsilon N)$ is a specific constraint. Accordingly, we can build the following abstraction to explain the purpose and process of hybridization:

$$⋒' = \{(\cap, c) | \cap \epsilon ⋒ \, c \in C_s\} \tag{1.2}$$

HIS are increasingly used in different domains and applications in recent years. Below, we list some relevant techniques and applications.

- Machine learning integrating multiple learning methods and models
- Data mining combined different knowledge discovery methods and models
- Multiagent systems consisting of multiple agent technologies and agent-based models
- Agent mining synthesizing multiagent technology with data mining and machine learning
- Visual information processing combining image processing with intelligent information processing
- Human–machine-cooperated system embedded human intelligence and machine intelligence
- Automatic control systems integrating automation with cybernetics
- Bioinformatics combining biologics with informatics

More specific hybrid intelligent systems include:

- Integrative decision-support systems
- Intelligent information retrieval system integrating text mining with information retrieval
- Big data analytics system involving resources of business, social, mobile, and Web data

Hybrid intelligent systems research presents some typical characteristics:

- Their main tasks are to explore techniques for integrating two to multiple soft computing techniques and integrate soft computing with machine learning methods and models.
- The hybridization focuses on low-level technical solutions and applications.
- Very limited research on building effective methodologies for synthesizing different technologies and methodologies at a higher level; as a result, HIS research outcomes often present specific, individual, and operational method, process, and systems.

1.6.3 Design Strategies

There are many different strategies proposed for designing and implementing hybrid intelligent systems. In summary, we can categorize them into bi-hybridization, tri-hybridization, and multi-hybridization methods, according to the main computing methods, models, and techniques used in the hybridization. The way to hybridize different intelligent techniques is multifold. In binary, triple, or multiple combinations, either one or two methods may play a major role in the hybridization, or multiple methods take an equal responsibility in the hybrid systems. Typical hybrid neurosystems are built on top of neurosystems, by incorporating other soft computing techniques.

1. *Bi-hybridization*
   Bi-hybridization integrates two types of technologies into one unified tool, which may be presented in terms of bi-methods, bi-models, bi-systems, or hybrid modes such as one method with another system. Bi-hybridization is a very typical and widely used approach, for instance, evolutionary neural computing. Typical bi-hybridization methods include:
   - Two methods based machine learning, for example, supervised reinforcement learning
   - Based on one method, one of other methods is involved, for instance, neural computing has been integrated with many different methods and formed typical methods including fuzzy neural computing and evolutionary neural computing
   - Combining two applications, for instance, image signal processing to address both images and signals

2. *Tri-hybridization*
   Tri-hybridization integrates three types of methods, models, systems, and/or applications, to form a unified tool which is a tri-method, tri-model, tri-system, or in a hybrid mode such as a model integrated into two systems.
   - Three learning methods are combined, for example, the combination of supervised learning with unsupervised learning and reinforcement learning.
   - Three intelligent techniques are combined, for instance, an evolutionary fuzzy neural system combining genetic algorithms, fuzzy set system, and neural network.
   - Combining one (or two) type of intelligent technique with two (or one) other types of techniques, for example, agent-based multi-data source association rule mining combining multiagent systems with database and association rule mining.
3. *Multi-hybridization*
   Multi-hybridization involves more than three types of models, methods, systems, or applications into one solution or application. Although multiple components may provide substantial additional values to problem-solving, it also often brings about additional challenges and costs to effectively take advantage of the power of every constituent. Typical modes include:
   - Multi-learner combination, for instance, the combination of supervised learning, unsupervised learning, reinforcement learning, and cooperative learning
   - Multi-technique combination, for instance, the integration of ontological engineering with multiagent systems, data mining, and statistics to form super-intelligent methods, models, or systems

1.6.4 Typical Hybrid Applications

Many new methods, techniques, and applications have been generated through different design strategies. Below, we summarize a few.

1. *Hybrid Neural Systems*
   Hybrid neural systems are centered on neural computing technique to form different hybrid systems by integrating other techniques in artificial intelligence and soft computing, for instance:
   - Neuro-fuzzy systems combining neural networks with fuzzy set systems
   - Neuro-expert systems combining neural networks with expert systems
   - Integration of neural networks with global optimization algorithms
   - Integration of rule-based system with neural networks
   - Combination of neural networks with evolutionary computing

2. *Hybrid Fuzzy Systems*
   Hybrid fuzzy systems combine fuzzy set systems with other artificial intelligence and soft computing methods, such as:
   - Neuro-fuzzy systems
   - Fuzzy expert systems combining fuzzy set systems with expert systems
   - Fuzzy optimization combining fuzzy set systems with optimization
   - Evolutionary fuzzy systems combining fuzzy set systems with evolutionary computing
   - Rough fuzzy methods combining rough set with fuzzy set systems
3. *Hybrid Evolutionary Systems*
   Hybrid evolutionary systems combine evolutionary computing and genetic algorithms with other artificial intelligence and soft computing methods to form new evolutionary models, methods, structures, or systems, for instance:
   - Fuzzy evolutionary systems
   - Neuro-evolutionary systems
   - Rough evolutionary systems combining rough set with evolutionary computing
   - Evolutionary optimization combining particle swarm optimization and simulated annealing
   - Integrating evolutionary computing with bacterial foraging
   - Integrating evolutionary computing with molecular computing
   - Integrating evolutionary computing with optical computing
4. *Hybrid Learning and Reasoning Systems*
   Hybrid learning systems integrate multiple learning or reasoning methods, models, algorithms, structures, systems, or applications, for instance:
   - Combination of supervised learning, unsupervised learning, reinforcement learning, and cooperative learning
   - Integrating soft computing with machine learning, such as support vector machine, rough set, Bayesian network, probabilistic reasoning, and statistical relational learning
   - Integrating case-based reasoning and inductive logic programming with grammatical inference
5. *Hybrid Computing Systems*
   Hybrid computing systems incorporate multiple computing techniques, for instance:
   - Agent service-oriented computing combining agent-based computing with service-oriented computing
   - Agent organization-oriented computing combining agent-based computing with organizational computing
   - Agent-based social computing involving agent-based computing with social computing
   - The integration of agent-based computing, service-oriented computing, organizational computing, and social computing

6. *Hybrid Optimization*
   Multiple optimization methods are combined, for instance:
   - Integrating local optimization with global optimization techniques
   - Integrating fuzzy set systems with global optimization techniques
   - Integrating neural network with global optimization techniques
7. *Agent Mining*
   Agent mining is a recent technique proposed to integrate multiagent systems with data mining and machine learning to address challenges that cannot be handled by respective technique only.
   - Agent-based data mining, for instance, data mining-driven intelligence enhancement, user modeling based on data mining, agent recommender driven by data mining, agent-based distributed learning, agent-based reinforcement learning, and self-learning agents
   - Data mining-driven agents, for instance, agent-based activity mining, agent-based cross-database mining, agent-based distributed data mining, agent-based linkage analysis, agent-based Web mining, agent-based data processing, and mobile agent-based data mining

1.7 Evolution of Intelligent Systems

The evolution of *artificial intelligence* (AI) and *intelligent systems* has experienced a number of stages characterized by diverse focuses and perspectives. This is embodied in selected abstract objects, concepts, and/or mechanisms to study intelligence and intelligent behavior appearing at different layers or aspects. The abstraction of intelligence can be categorized into the following major categories:

- *Symbolic AI*: representing intelligence in terms of symbols
- *Connectionist AI*: representing intelligence in terms of connectionism and networking, especially artificial neural networks
- *Situated AI*: representing intelligence in terms of multiagent systems and the interactions within a system and between agents and environment
- *Social AI*: representing intelligence in terms of social interactions and collective intelligence in problem-solving

Table 1.1 illustrates the main characteristics and difference between the four categories in terms of underlying objects, research issues, representative work, and main approaches.

At the stage of symbolic intelligence and connectionist intelligence, the focus was on the study of the intelligence formation process and mechanisms in human cognitive systems, and the findings were then applied to simulate human intelligent behaviors in building machine intelligence. In the study of situated intelligence and social intelligence, human and artificial or social entities were taken as being the

Table 1.1 Evolution of artificial intelligence research

Stage	Research object	Target issues	Typical work	Features of the work
Symbolic intelligence	Symbolic mechanisms for representing and processing intelligence	Symbolic representation, heuristic programming, logic modeling, logic reasoning	Physical symbolic system hypothesis [53]	Logic-centered, top-down design, logical thinking
Connectionist intelligence	Neural network connectionism for representing and processing intelligence	Backpropagation algorithm, hierarchical neural network, computational intelligence	Artificial neural network theories [54]	Networks formed by nodes and connections, imaginary thinking
Situated intelligence	Interaction mechanisms between system and environment to represent and process intelligence	Intelligence emergence through situated interactions between intelligent behaviors, sensation–perception, and feedback mechanisms	Subsumption architecture [55], swarm intelligence, Web intelligence, domain intelligence, [56] etc.	Taking environment (context) as a part of an intelligent system, the emergence of intelligence through interactions between a system and its environment, intelligence without representation and reasoning, bottom-up design
Social intelligence	Mechanism design for organization, interaction, cooperation, collaboration, coordination, and communication between social entities and between social entities and environment	Social interactions, organizational factors, norms and policies, emergence of social intelligence	System modeling, multiagent system design and software engineering, negotiation, trust, privacy, law, norm, etc.	Agent society with human as system components if necessary, social interactions, swarm intelligence, interdisciplinary, top-down design
Nature-inspired intelligence	Objects and their dynamics in the natural world, the projection from natural world to problem-solving systems	Working mechanisms in the nature, theories and tools for computational intelligence	Biologically and linguistically motivated computational paradigms, evolutionary computing, neural networks	Neural networks, connectionist systems, genetic algorithms, evolutionary programming, fuzzy systems, and hybrid intelligent systems

(continued)

Table 1.1 (continued)

Stage	Research object	Target issues	Typical work	Features of the work
Metasynthetic intelligence	Open complex systems with large-scale objects, complex interactions between entities and with environments	Theories and tools to synthesize human intelligence with other ubiquitous intelligence and to implement the qualitative-to-quantitative decision-support systems	Working mechanisms in social cognitive systems, theories and tools for metasynthetic engineering, Hall for Workshop of Metasynthetic Engineering [1, 2, 15, 57, 58]	Systemism combining holism with reductionism, top-down plus bottom-up design

carriers of intelligence, with the target of exploring the interactions between agents (including human agents). *Agent-based modeling and simulation* were widely used in this stage. When nature-inspired intelligence became the focus of research, working mechanisms in natural systems, including animals and plants, were emphasized, such as biologically motivated computing theoretical systems and tools.

Recent highlights in intelligent systems are reflected through the paradigm shift from *situated intelligence* to *social and nature-inspired intelligence*. In a situated intelligence study, individuals (agents), sensation, and perception within and between agents, as well as between agents and environments, play key roles. This results in simple abstraction models that include a limited number of agents, very few agent types, and low-level and small-scale systems with simple interactions.

In a social intelligence study, a large scale of agents of many different types collocate within a system and interact with the environment via sophisticated interaction modes and protocols, following social and organizational mechanisms and policies. In this research, behavior is a very important concept. Behaviors refer to the activities, interactions, and their properties of multiagents and the environment. These may be embodied through different modes of cooperation, collaboration, coordination, communication, and negotiation.

In categorizing the major methodologies adopted in intelligent system studies from cognitive science and intelligence simulation, we are concerned with three main methodologies:

- Symbolism
- Connectionism
- Behaviorism

Table 1.2 provides a summary of the main components incorporated in the above three respective methodologies.

Table 1.2 Three methodologies in AI and intelligent systems

	Symbolism	Connectionism	Behaviorism
Representation	Symbols	Connections	Behaviors
Variables	Discrete	Continuous	Continuous
Information processing	Serial	Parallel	Parallel
Operations	Reasoning	Mapping	Interactions
Architecture	Local	Distributed networks	Distributed systems
Problem-solving	Top-down	Bottom-up	Bottom-up
Cognitive process	Logical abstraction	Neural network simulation	Perception and feedback
Computational process	Reasoning–searching	Connection–self-organization	Interaction–feedback
Technical fundamental	Physical symbolic system hypothesis, limited ration	Neural network models, neural learning	Agent-based modeling

Refer to social interactions including cooperations, collaborations, and coordinations

The community of AI and intelligent systems has invested much effort in hybrid intelligent systems by integrating the above methodologies. This has formed and generated new theories and tools, for example:

- The combination of symbolic intelligence with connectionism, such as fuzzy neural networks and hybrid expert systems incorporating multiple models.
- Behaviorism involves situated intelligence as well as the study of social intelligence.
- Logical reasoning and connectionist mechanisms are still used in behavioristic systems and tools.

With the involvement of domain-specific, organizational, social, and natural intelligence, the study of complex intelligent systems increasingly relies on key mechanisms: integration, synthesis, and complementation.

To address the challenges in complex systems, there is a strong need to incorporate and synthesize different methodologies and technical approaches, especially the combination of behaviorism with symbolism and connectionism as needed. This involves the need of combining reasoning with representation, knowledge management, social computing, behavior informatics, machine learning, and knowledge discovery from data. This is very helpful for handling challenges resulting from qualitative intelligence, uncertain intelligence, and system complexities beyond the capability of each individual solution.

1.8 Open Giant Intelligent Systems

Open giant intelligent systems are those with particular complexities:

- They are ill-structured which cannot be directly converted to structured problems in terms of the current problem-solving methods available in AI and intelligent systems; the presentation of intelligence is closely relevant to the spatiotemporal characteristics and social interactions in the systems.
- The complexities in such systems cannot be handled by existing methodologies, paradigms, and theoretical foundation built in the mainstream AI system family for intelligence and cognition representation, reasoning, and evolution.
- The existing methodologies, theories, and paradigms and the corresponding systems and tools available in the AI and intelligent system family were built mainly on top of the reductionism assumption. We cannot rely on the software engineering architectures, methodologies, and tools available in the classic intelligent systems to analyze and design the problem-solving systems for them.
- Necessarily, we may have to consider the synthesis of empirical and rational, explicit and implicit, qualitative and quantitative, common sense and evidence-based, and local and global knowledge, data and information in order to have a complete understanding of the problems.
- Logical thinking and imaginary thinking play mutual roles in designing the system problem-solving methodologies, methods, and tools.

As the AI community has been suffering from the missing mainstream focuses, which was not the case in the era of logic-based design, those methodologies and tools aiming for building automated intelligent systems are deterministically not suitable for open giant intelligent systems with the above characteristics and complexities. This may be embodied in aspects including intelligence abstraction, working mechanisms, system architectures, interactions, and software engineering tools. The recent progress made in knowledge science and data science, as well as systems science, makes it possible to explore open giant intelligent systems by an interdisciplinary methodology.

As per the system classification discussed in [1, 2, 58], an open intelligent system and its family can be described as follows.

Definition 1.1 (Giant Systems) Whether a system is a giant system is very much dependent on the number of subsystems and the number of subsystem types and the complex extent of those interactions and coupling relationships built in the subsystems:

- Giant systems: the number of subsystems is enormous, namely, hundreds of thousands, millions, or even billions.
- Simple giant systems: those giant systems with tens of subsystem types and simple interactions and coupling relationships.
- Complex giant systems: those giant systems with many subsystem types (in hundreds or more), hierarchical structures, and complex interactions and coupling relationships.
- Open complex giant systems: if those complex giant systems are also open, namely, having interactions and coupling relationships with the environments.

Table 1.3 Major counterintuitive working mechanisms

Working mechanism	Counterintuitive effect
Paradox	Inconsistent phenomena, generating inconsistency and contradictions
Uncertainty and Instability	Small local variations may trigger significant (global) effects and unexpected major effect
In-computability	Behaviors work beyond rules
Coupling	The summation of local parts does not produce the global function; a system cannot be interpreted by decomposed parts
Emergence	Unexpected intelligence (outcomes and effect) generated through self-organization and interactions between local parts

Open complex systems may present some counterintuitive features and status [55, 59] which are different from simple systems or do not appear in simple systems. These may include key working mechanisms: paradox and self-reference, uncertainty, in-computability, couplings, and emergence. Table 1.3 shows the summary of these mechanisms. They may contribute to the formation of complex intelligent phenomena, behaviors, and systems. The diversity of working mechanisms and contributions results in respective counterintuitive effects.

For giant intelligent systems, they can be classified into the following four types as per their design strategies (system analysis and design) and system structure (problem-solving process) [60]:

- Closed giant intelligent systems (for short, closed systems)
- Semi-closed giant intelligent systems (for short, semi-closed systems)
- Semi-open giant intelligent systems (for short, semi-open systems)
- Open giant intelligent systems (for short, open systems)

We further describe the main characteristics of the above four types of giant intelligent systems in Table 1.4.

The relationships between the above four types of giant intelligent systems are described below:

- Closed vs. open systems: the main difference lies in whether the system structure is social, namely, whether the intelligent behaviors are social, which determines whether the system is automated or social; they have different coupling degrees, closed systems are tightly coupled, while open systems are loosely coupled.
- Closed vs. semi-closed systems: the main difference is on whether a system is fully automated; semi-closed systems rely on a social process of system construction to integrate multiple relevant techniques.
- Semi-open vs. open systems: they are different in terms of whether the human is a key system module and in terms of the aspects of system societal characteristics and hierarchy.

Table 1.4 Family of giant intelligent systems

System type	Representative systems	Main characteristics	Contributions to AI
Closed giant intelligent systems	The Cyc Project [62]	Knowledgist; completion and integrity are built in the system; open Cyc system; automated system	Knowledge is power; system vulnerability is relevant to the completion of common knowledge
Semi-closed giant intelligent systems	Knowledge bus PACT systems [63]	System is built in a social way (relying on synthesis and integration); automated problem-solving system	Hybrid techniques are used in building such an intelligent system
Semi-open giant intelligent systems	Open semantic framework [64]	A system is social; some individuals in the system are social too in terms of problem-solving	Intelligence is embodied through social behavior; subsystems negotiate to resolve conflict
Open giant intelligent systems	Hall for Workshop Metasynthetic Engineering [1, 2]	The system structure and constituents are dynamic; human–system interaction and cooperation contribute to the system operations; metasynthetic engineering as a guide for system engineering	Human forms a key system module; system is hierarchical and dynamic; the evolution of the system involves interactions with its environment

1.9 Computing and Engineering Complex Systems

The complexities of complex systems require new methodologies and tools to be produced in order to effectively understand, analyze, design, implement, evaluate, and operate a complex system. The computing and engineering of complex systems aims to develop such theories and techniques. This forms the motivation and purpose of this book.

1. *Computing Paradigms*

   As system complexities are embodied in different aspects, different computing paradigms are needed to address the respective complexities. Such *computing paradigms* consist of:

   - Object-oriented methodology
   - Component-based methodology
   - Agent-oriented methodology
   - Service-oriented computing
   - Agent service-oriented methodology
   - Organizational computing
   - Social computing
   - Distributed computing
   - Cloud computing

Often several computing paradigms are required to engineer and compute a complex problem.

2. *System Analysis and Design*
The above computing paradigms provide different working mechanisms and building blocks for analyzing and designing a complex problem. Early-proposed computing paradigms such as object-oriented methodology can only cater for certain system complexities. Typical system complexities and challenges in complex systems cannot be well handled by directly using any single paradigm or existing systems.

    - None of the existing systems provide systematic and complete support to address critical system complexities especially openness, sociality, and metasynthesis of intelligence. Typical aspects include social norm, organizational structure, open environment, system dynamics, social interactions, and human involvement.
    - The handling of system complexities requires the level of large scale of problem-solving systems and the human–machine-cooperated manner in order to address the critical challenges in open complex systems.
    - A paradigm needs to provide complete support of the whole process from early to late requirement analysis, running time-based system design, workspace for human and machine cooperation, involvement of ubiquitous intelligence, and dynamic and systematic evaluation and validation.
    - Most of existing computing paradigms were proposed to address specific system complexities, such as different methodologies proposed for agent-based software engineering. Some of them do not address such complexities as sociality and human–machine cooperation. Such resulting theories and tools are often incomplete, inoperable, and unusable for real-life complex systems.

3. *Metasynthetic Computing and Engineering*
Based on massive working experience in many different domains and large national engineering projects, Qian and his colleagues proposed the methodology *qualitative-to-quantitative metasynthesis* through building a *Hall for Workshop of Metasynthetic Engineering*, or in short *metasynthetic engineering* [1, 2, 55, 58, 64, 65], to address the system complexities in open complex giant systems. The principal idea is as follows: since an open complex giant system is too difficult to be understood and handled at the beginning, a group of relevant domain experts are invited to share the understanding of the system complexities, and the relevant data, information, and domain knowledge are acquired and analyzed in high-performance computing power to assist in the experts' discussions (through approaches including seminars and workshops), to deepen the understanding and problem-solving.

    As you can see, metasynthetic computing and engineering involves the following key factors and principles in solving a complex problem:

    - The methodology for handling open complex giant systems is the metasynthesis of ubiquitous intelligence, so as to implement the process of qualitative-to-quantitative problem-solving.

- Following the principle of "*human-centered human–machine symbiosis,*" a problem-solving system is a human–machine symbiont, namely, a Hall for Workshop of Metasynthetic Engineering, where the human plays a major role.
- Both reductionism and holism are needed in understanding such a complex system. The systematology aims to integrate reductionism with holism, which addresses both bottom-up and top-down challenges and advantages, both reductionism with holism are needed in handling very complex systems.
- The implementation of *human-centered, qualitative-to-quantitative metasynthetic engineering* relies on different disciplinary knowledge, expertise, experts, and their seamless cooperation and co-inspiration. This requires knowledge and expertise from domains and disciplines including system engineering, cognitive science, artificial intelligence, computer science, statistics and mathematics, management science, and social science. The latest technological developments including Web technology, social network and media, social computing, data science, and advanced analytics are critical for synergizing their respective advantage and strengths toward maximizing the value of cooperation, interaction, and complementation of different intelligence and knowledge.

1.10 Summary

In this chapter, we present an overall picture of complex systems in terms of system complexities, transparency, and classification. More specifically, we discuss complex agent systems, hybrid intelligent systems, evolution of intelligent systems, and open giant intelligent systems. The challenges of open complex systems require the development of metasynthetic computing and engineering, which forms the main focus of this book.

In the remainder of this book, we will introduce the ubiquitous intelligence surrounding open complex systems, system methodologies, and computing paradigms for handling the challenges in complex systems. In particular, we focus on discussing computing and engineering methodologies and tools for complex systems.

References

1. Qian, X., Yu, J., Dai, R.: A new scientific field–open complex giant systems and the methodology (in Chinese). Chin. J. Nat. **13**(1), 3–10 (1990)
2. Qian, X.: Revisiting issues on open complex giant systems (in Chinese). Pattern. Recogn. Artif. Intell. **4**(1), 5–8 (1991)
3. Wooldridge, M., Jennings, N.R.: Intelligent agents: theory and practice. Knowl. Eng. Rev. **10** (2), 115–152 (1995)
4. Wooldridge, M.: An Introduction to Multiagent Systems. Wiley (2002)
5. http://www.aamas-conference.org/

6. Miller, J.H., Page, S.E.: Complex Adaptive Systems: An Introduction to Computational Models of Social Life. Princeton University Press, Princeton (2007)
7. Holland, J.H.: Signals and Boundaries: Building Blocks for Complex Adaptive Systems. The MIT Press, Cambridge (2014)
8. Langton, C. (ed.): Artificial Life. Santa Fe Institute Studies in the Sciences of Complexity. Addison-Wesley, Reading (1992)
9. Mitchell, M.: An Introduction to Genetic Algorithms. MIT Press, Cambridge, MA, London (1999)
10. Castelfranchi, C., Johnson, W.L.: Proceedings of AAMAS2002. ACM Press, New York (2002)
11. Rosenschein, S.J., Sandholm, T., Wooldridge, M., Yokoo, M.: Proceedings of AAMAS2003. ACM Press, New York (2003)
12. Jennings, N.R., Sierra, C., Sonenberg, L., Tambe, M.: Proceedings of AAMAS2004. ACM Press, New York (2004)
13. Huget, M.P. (eds.): Communication in Multiagent Systems. Springer, Heidelberg (2003)
14. Beer, R.D.: A dynamical systems perspective on agent-environment interaction. Artif. Intell. **72**, 173–215 (1995)
15. Odell, J.J., Parunak, H.V.D., Fleischer, M., Brueckner, S.: Modeling agents and their environment. In: AOSE 2002
16. Vasilyev, A.: Synergetic approach in adaptive systems. Master thesis, Transport and Telecommunication Institute, Riga, Latvia
17. Newell, A., Simon, H.A.: Computer science as empirical enquiry. Commun. ACM **19**, 113–126 (1976)
18. Chapman, D.: Planning for conjunctive goals. Artif. Intell. **32**(3), 333–377 (1987)
19. Brooks, R.A.: Intelligence without reason. In: Proceedings of the 12th International Joint Conference on Artificial Intelligence, IJCAI-91, pp. 569–595. Morgan Kaufmann, San Francisco (1991)
20. Wooldridge, M., Jennings, N.R.: Intelligent agents: theory and practice. Knowl. Eng. Rev. **10** (2),115–152 (1995)
21. Muller, J., Pischel, M.: The Agent Architecture InteRRaP: Concept and Application. Technical Report RR-93-26, DFKI Saarbrucken
22. Abraham, A., Koppen, M., Franke, K. (eds.): Design and Application of Hybrid Intelligent System. IOS Press, Amsterdam, The Netherlands, (2003)
23. Goonatilake, S., Khebbal, S. (eds.): Intelligent Hybrid Systems. Wiley, Chichester (1995)
24. Zhang, Z.L., Zhang, C.Q.: Agent-based Hybrid Intelligent Systems. LNAI-2938. Springer, Heidelberg (2004)
25. Gasser, L.: Social conceptions of knowledge and action: DAI foundations and open system semantics. Artif. Intell. **47**, 107–138 (1991)
26. Ferber, J.: Multi-agent Systems: An Introduction to Distributed Artificial Intelligence. Addison-Wesley, London (1999)
27. Castro, J., Kolp, M., Mylopulos, J.: Towards requirements-driven information systems engineering: the tropos project. Inf. Syst. **27**(6), 365–389 (2002)
28. Giorgini, P., Kolp, M., Mylopoulos, J.: Multi-agents architectures as organizational structures. Int. J. Autonom. Agents Multi-Agent Syst. **13**(1), 3–25 (2006)
29. Waldrop, M.M.: Complexity: The Emerging Science at the Edge of Order and Chaos. Simon & Schuster, New York (1992)
30. Meadows, D.H.: Thinking in Systems: A Primer. Chelsea Green Publishing, White River Junction (2008)
31. Page, S.E.: Diversity and Complexity (Primers in Complex Systems). Princeton University Press, Princeton (2010)
32. Mitchell, M.: Complexity: A Guided Tour. Oxford University Press, New York (2011)
33. Jennings, N.R.: On agent-based software engineering. Artif. Intell. **117**(2), 277–296 (2000)
34. Ciancarini, P., Wooldridge, M. (eds.): Agent-Oriented Software Engineering. Springer, Berlin (2001)

35. Garcia, A., Lucena, C.: Software engineering for large-scale multi-agent systems—SELMAS 2002 (workshop report.) ACM Software Eng. Notes **27**(5), 82–88
36. Garcia, A., et al. (eds.).: Software Engineering for Large-scale Multi-agent Systems. Springer, Heidelberg (2003)
37. Goos, G., et al.: Software Engineering for Large-scale Multi-agent System. Springer, Berlin (2002)
38. Mao, X.J., Yu, E.: Organizational and social concepts in agent oriented software engineering. In: AOSE04 (2004)
39. Cao, L.B., Zhang, C.Q., Luo, D., Chen, W.L., Zamani, N.: Integrative early requirements analysis for agent-based systems. The 4th International Conference on Hybrid Intelligent Systems. IEEE Computer Society Press, New York (2004)
40. www.sei.cmu.edu/uls/
41. Feiler, P.H., Sullivan, K., Wallnau, K.C., Gabriel, R.P., Goodenough, J.B., Linger, R.C., Longstaff, T.A., Kazman, R., Klein, M.H., Northrop, L.M., Schmidt, D.: Ultra-Large-Scale Systems: The Software Challenge of the Future. Software Engineering Institute, June (2006)
42. Zambonelli, F., Jennings, N.R., Wooldridge, M.: Developing multiagent systems: the GAIA methodology. ACM Trans. Softw. Eng. Meth. **12**(3), 317–370 (2003)
43. Giunchiglia, F., Mylopoulos, J., Perini, A.: The TROPOS software development methodology: processes, models and diagrams. In: AOSE02 (2002)
44. Klusch, M. (eds.): Intelligent Information Agents. Springer, Berlin (1999)
45. Veit, D.: Matchmaking in Electronic Markets: An Agent-Based Approach. LNAI-2882. Springer, Berlin/Heidelberg (2003)
46. Abecker, A. (eds.): Agent-mediated Knowledge Management. Springer, New York (2003)
47. Cao, L.B., Dai, R.W.: Social abstraction for agent-based open giant intelligent systems. In: Proceedings of International Conference on Intelligent Information Technology (ICIIT-02), 22–25 Sept, pp. 47–52. Beijing, China. ISBN 7-115-75100-5/0267
48. Cao, L.B., Luo, C., Li, C.S., Zhang, C.Q., Dai, R.W.: Open giant intelligent information systems and its agent-oriented abstraction mechanism. In: Proceedings of the Fifteenth International Conference on Software Engineering and Knowledge Engineering (SEKE 2003), pp. 85–89. ISBN: 1-891706-12-8
49. Cao, L.B., Li, C.S., Zhang, C.Q., Dai, R.W.: Open giant intelligent information systems and its agent-oriented analysis and design. In: Proceedings of the 2003 International Conference on Software Engineering Research and Practice (SERP'03), vol. 2, pp. 816–822. CSREA Press, Georgia (2003)
50. Dai, R.W., Cao, L.B.: Internet—an open complex giant system. Science in China (Series E). Sci. China Ser. E **33**(4), 289–296 (in Chinese) (2003)
51. Goonatilake, S., Khebbal, S. (eds.): Intelligent Hybrid Systems. Wiley (1995)
52. Nilsson, N.J.: The physical symbol system hypothesis: status and prospects. In: Lungarella, M., et al. (eds.) 50 Years of AI, Festschrift, LNAI 4850, pp. 9–17. Springer, Berlin (2007)
53. Bishop, C.M.: Neural Networks for Pattern Recognition. Oxford University Press, Oxford (1995)
54. https://www.princeton.edu/~achaney/. . ./Subsumption_architecture.html
55. Longbing, C., Ruwei, D.: Open Complex Intelligent Systems (in Chinese). Post & Telecom, Beijing, China (2008)
56. Cao, L., Zhang, C., Zhou, M.: Engineering open complex agent systems: a case study. IEEE T. Syst. Man Cybern. C: Appl. Rev. **28**(4), 483–496 (2008)
57. Cao, L., Dai, R., Zhou, M.: Metasynthesis: M-space, M-interaction and M-computing for open complex giant systems. IEEE Trans. Syst. Man Cybern. A **39**(5), 1007–1021 (2009)
58. Qian, X.: Building Systematology (in Chinese). Shanghai Jiaotong University Press, Taiyuan, China (2007)
59. en.wikipedia.org/wiki/Counterintuitive
60. Dai, R., et al.: Metasynthesis of Intelligent Systems. Zhejiang Science & Technology Press (in Chinese), Hangzhou, China (1995)
61. Lenat, D.: Hal's legacy: 2001's computer as dream and reality. From 2001 to 2001: common sense and the mind of HAL. Cycorp, Inc. 26 Sept 2009

62. www-ksl.stanford.edu/knowledge-sharing/papers/pact.tex
63. opensemanticframework.org/
64. Ruwei, D., Yaodong, L., Qiudan, L.: Social Intelligence and Metasynthetic System (in Chinese). Post & Telecom, Beijing, China (2013)
65. Dai, R.: Qualitative-to-quantitative metasynthetic engineering (in Chinese). Pattern. Recogn. Artif. Intell. **6**(2), 60–65 (1993)
66. Chu, D.: Complexity: Against Systems. Theory in Biosciences. Springer (2011)
67. Rocha, L.M.: Complex systems modeling: using metaphors from nature in simulation and scientific models. BITS: Computer and Communications News. Computing, Information, and Communications Division, Los Alamos National Laboratory, November (1999)

Chapter 2
Ubiquitous Intelligence

2.1 Introduction

Complex systems involve multiple aspects such as domain knowledge, constraints, human roles and interaction, life cycle and process management, and organizational and social factors. Many complex systems are dynamic and need to cater for online, run time, and ad hoc requests. With the involvement of social intelligence and complexities, such complex systems need to consider reliability, reputation, risk, privacy, security, trust, and actionability of problem-solving solutions. Research in one area can actually stimulate, complement, and enhance research in another. A typical example is agent mining technology [1–4], which synergizes the ubiquitous intelligence for handling complex intelligent problems and systems through the combined strengths of data mining, machine learning, and multiagent systems. Other typical examples that involve ubiquitous intelligence include open complex intelligent systems [5, 6], domain-driven actionable knowledge discovery [7], combined mining for discovering complex patterns [8], and ubiquitous computing [9].

A typical understanding of the above issues in complex systems comes from the angle of ubiquitous intelligence. Ubiquitous intelligence embodies a real-world complex problem that can be identified and categorized into the following types:

- Data intelligence
- Human intelligence
- Domain intelligence
- Network intelligence
- Organizational intelligence
- Social intelligence
- Metasynthesis of intelligence

L. Cao, *Metasynthetic Computing and Engineering of Complex Systems*,
Advanced Information and Knowledge Processing,
DOI 10.1007/978-1-4471-6551-4_2

In complex systems and problems, it is not only necessary to involve one of the above types of intelligence, but to also consolidate the relevant ubiquitous intelligence into modeling, evaluation, working process, and systems.

In this chapter, we discuss the concepts and aims of involving each type of intelligence and the corresponding techniques and case studies for incorporating them into complex systems. The listed ubiquitous intelligence and the consolidation in a system offer a new angle for observing key and mutual challenges in complex systems.

Consequently, we believe that many issues in complex systems can be addressed and solutions provided. As a result, many great opportunities will emerge with more advanced, effective, and efficient methodologies, techniques, means, tools and applications for dealing with complex problems and systems.

2.2 Data Intelligence

2.2.1 *What Is Data Intelligence?*

Definition 2.1 *Data intelligence* tells interesting stories and/or has hidden indicators in data about a business problem. The intelligence of data emerges in the form of interesting patterns and actionable knowledge.

There are two levels of data intelligence:

- *General level of data intelligence* refers to the knowledge identified from explicit data, presenting general knowledge about a business problem.
- *In-depth level of data intelligence* refers to the knowledge identified in more complex data, using more advanced techniques or disclosing much deeper information and knowledge about a problem.

Taking association rule mining as an example, a general level of data intelligence is found in the frequent patterns identified in basket transactions, while associative classifiers reflect a deeper level of data intelligence.

2.2.2 *Aims of Involving Data Intelligence*

We aim to disclose data intelligence from multiple perspectives. One angle for observing data intelligence is that of data explicitness or implicitness.

- *Explicit data intelligence* refers to the level of data intelligence that discloses explicit characteristics or that is exhibited explicitly. An example of explicit data intelligence is the trend of a stock market index or of stock price dynamics.
- *Implicit data intelligence* refers to the level of data intelligence that discloses implicit characteristics or that is exhibited implicitly. In stock markets, an example of implicit data intelligence is seen in the trading behavior patterns of a hidden group in which investors are associated with each other.

Both explicit data intelligence and implicit data intelligence may present intelligence at either a general level or an in-depth level.

Another angle for scrutinizing data intelligence is from either a syntactic or a semantic perspective.

- *Syntactic data intelligence* refers to the kind of data intelligence that discloses syntactic characteristics. An example of syntactic data intelligence is itemset associations.
- *Semantic data intelligence* refers to the kind of data intelligence that discloses semantic characteristics. An example of semantic data intelligence is the temporal trading behavior embedded in the temporal logic relationship between trading behaviors.

Similarly, both syntactic data intelligence and semantic data intelligence may present intelligence at either a general level or an in-depth level.

2.2.3 Aspects of Data Intelligence

Even though mainstream knowledge discovery systems focus on the substantial investigation of various types of data for hidden interesting patterns or knowledge, real-world data and its surroundings are usually much more complicated. Below is a list of aspects that may be associated with data intelligence:

- Data type such as numeric, categorical, XML, multimedia, and composite
- Data timing such as temporal and sequential
- Data spacing such as spatial and spatiotemporal
- Data speed and mobility such as high frequency, high density, dynamic, and mobile
- Data dimension such as multidimensional, high-dimensional data, and multiple sequences
- Data relation such as multi-relational, linkage record
- Data quality such as missing data, noise, uncertainty, and incompleteness
- Data sensitivity such as being mixed with sensitive information
- Data structure such as graph or sequence
- Data distribution such as Gaussian distribution or mixed distribution

Deeper and wider analysis is required to disclose in-depth data intelligence in complex data. Two kinds of approach, *data science* and *engineering*, need to be further developed for understanding data characteristics, structures, and distribution, and providing effective and efficient processing and management tools, as well as analyzing and modeling real-world large-scale data and various data complexities such as multidimensional, high-dimensional, mixed, and distributed data, and processing and mining unbalanced, noisy, uncertain, incomplete, dynamic, and stream data.

2.3 Domain Intelligence

2.3.1 What Is Domain Intelligence?

Definition 2.2 *Domain intelligence* refers to the intelligence that emerges from the involvement of domain factors and resources in complex systems, which encompasses not only a problem but its target data and environment. Domain intelligence is embodied through involvement in the modeling process, models, and systems.

Domain intelligence involves qualitative and quantitative aspects. They are instantiated in aspects such as domain knowledge, background information, prior knowledge, expert knowledge, constraints, organizational factors, business process, and workflow, as well as environmental aspects, business expectation, and interestingness.

2.3.2 Aims of Involving Domain Intelligence

Multiple types of domain intelligence may be engaged in complex systems:

- *Qualitative domain intelligence* refers to the type of domain intelligence that discloses qualitative characteristics or involves qualitative aspects. Taking stock data mining as an example, fund managers have qualitative domain intelligence such as "beating the market," when they evaluate the value of a trading pattern.
- *Quantitative domain intelligence* refers to the type of domain intelligence that discloses quantitative characteristics or involves quantitative aspects. An example of quantitative domain intelligence in stock data mining is whether a trading pattern can "beat VWAP[1]" or not.

The roles of involving domain intelligence in complex systems are multiform:

- They assist in the modeling and evaluation of the problem: An example is "my trading pattern can beat the market index return" when domain intelligence of "beat market index return" is applied to evaluate a trading pattern.
- They make mining realistic and business-friendly: By considering domain knowledge, we are able to work on a genuine business problem rather than an artificial one abstracted from an actual problem.

[1] VWAP is a trading acronym for volume-weighted average price, the ratio of the value traded to the total volume traded over a particular time horizon.

2.3.3 Aspects of Domain Intelligence

In the mainstream of complex systems, the consideration of domain intelligence is mainly embodied through the involvement of domain knowledge and prior knowledge, or by building the process and/or workflow associated with a business problem to incorporate domain intelligence into a solution.

In a specific domain problem, domain intelligence may be presented in multiple aspects. Several aspects of domain intelligence are listed below:

- Domain knowledge
- Background and prior information
- Meta-knowledge and metadata
- Understanding the underlying problems and challenges
- Constraints
- Business processes
- Workflows
- Rules, policies, and laws
- Predefined settings, patterns, templates, and protocols
- Seasonal, periodical, and cyclical factors
- Benchmarking and criteria definitions
- Business expectation and interest

2.4 Network Intelligence

2.4.1 What Is Network Intelligence?

Definition 2.3 *Network intelligence* refers to the intelligence that emerges from both Web- and broad-based network information, facilities, services, and processes surrounding a complex problem or system.

Network intelligence involves both *Web intelligence* and *broad-based network intelligence* such as information and resource distribution, linkages between distributed objects, hidden communities and groups, Web service techniques, messaging techniques, mobile and personal assistant agents for decision support, information, and resources from networks, in particular the Web, information retrieval, searching, and structuralization from distributed, cloud, and textual data. The information and facilities from the networks or cloud concerning the target business problem either consist of the problem constituents or can contribute to useful information for understanding the challenges and possible solutions that will lead to problem-solving.

2.4.2 Aims of Involving Network Intelligence

The aims of involving network intelligence in complex systems and problems have multiple aspects, for example, to:

- Involve data and information from a community or team
- Involve Web and online data and resources
- Involve networked software, applications, facilities, infrastructures, and platforms
- Support group-based problem-solving, crowdsourcing, and decision-making
- Support the sharing and transferral of knowledge and capability
- Support decision-making on top of mined patterns and knowledge
- Support social systems or solutions by providing facilities for social interaction in a team

In particular, we are concerned with:

- Retrieving, discovering, and utilizing the business intelligence in networked data related to a business problem; for instance, discovering market manipulation patterns across markets
- Discovering and utilizing networks and communities existing in a business problem and its data; for instance, discovering hidden communities in a market investor population
- Involving networked constituent information in problem-solving on target data; for example, mining blog opinion to verify abnormal market trading
- Involving public data and resources (software, platforms, and infrastructures) with private business data and facilities for more comprehensive problem-solving
- Utilizing networking facilities to pursue information and tools for actionable problem-solving; for example, involving cloud resources to support distributed and peer-to-peer problem-solving

2.4.3 Aspects of Network Intelligence

In respect of network intelligence, we expect on one hand to fulfill the power of the Web and network information and facilities for complex problem-solving in terms of many aspects, for instance:

- Information and resource distribution
- Linkages between distributed objects
- Hidden communities and groups
- Information and resources from networks and in particular the Web
- Information retrieval and query
- Structuralization and abstraction from distributed textual (blog) data
- Distributed and cloud computing
- Web network communication techniques
- Web-based decision-support techniques
- Dynamics of networks and the Web

- Cloudsourcing
- Networked communication, messaging, and conferences

On the other hand, we are specifically interested in disclosing and discovering Web and network intelligence. In this regard, there are many emergent topics to be studied. We list a few here:

- Social network mining
- Social media analysis
- Hidden group and community mining
- Context-based Web mining
- Opinion formation and evolution dynamics
- Sentiment analysis
- Distributed and multiple source mining
- Mining the changes and dynamics of networks
- Distributed data mining

2.5 Human Intelligence

2.5.1 What Is Human Intelligence?

Definition 2.4 *Human intelligence* refers to (1) the explicit or direct involvement of human knowledge or human person as a problem-solving constituent and (2) the implicit or indirect involvement of human knowledge or human person as a system component.

The explicit or direct involvement of *human intelligence* may consist of human empirical knowledge, belief, intention, expectation, run-time supervision, evaluation, and individual end users or expert groups. An example of *explicit human intelligence* is when a domain expert tunes parameters via user interfaces. The implicit or indirect involvement of human intelligence, in contrast, may present as imaginary thinking, emotional intelligence, inspiration, brainstorming, reasoning inputs, and embodied cognition such as convergent thinking through interaction with other members to assess identified patterns. Examples of involving *implicit human intelligence* are user modeling for game behavior design, collecting opinions from an expert group to guide model optimization, and utilizing embodied cognition for adaptive model adjustment.

2.5.2 Aims of Involving Human Intelligence

The importance of human involvement in complex systems and problem-solving has been widely recognized. With the systematic specification of human intelligence, we are able to convert complex systems toward more human centered, interactive, dynamic, and user-friendly systems, enhancing the capability of dealing with complex issues, forming closed-loop problem-solving systems, and strengthening the usability of system complexities.

- *Human-centered capability*: Human involvement, including individual and group knowledge, experience, preferences, cognition, thinking, reasoning, and more broad aspects linked to social and cultural factors (on which we will further expand in our discussion of social intelligence), makes it possible to utilize human intelligence to enhance complex problem-solving capability. Based on the depth and breadth of human involvement, the human cooperation with complex systems can be human centered or human assisted.
- *Interactive capability*: Human involvement takes place through interactive interfaces. This forms interactive problem-solving capabilities and systems that can effectively and sufficiently handle the integration of human intelligence into complex systems and problem-solving solutions.
- *Improving adaptive capability*: Real-life complex applications are often dynamic. Problem-solving models and solutions are often predefined and cannot adapt to these dynamics. The involvement of human intelligence can assist with the understanding and capture of such dynamics and can change and guide the corresponding adjustment and retraining of models.
- *User-friendliness*: Catering to user preferences, characteristics, and requests in complex systems will certainly make such systems more user-friendly.
- *Dealing with complex issues*: Many complex issues cannot be handled very well without the involvement of domain experts. Complex knowledge discovery from complex data can benefit from inheriting and learning from expert knowledge; enhancing the understanding of domain, organizational, and social factors through expert guidelines; embedding domain experts into complex systems; etc.
- *Closed-loop problem-solving systems*: In general, agent mining systems are open. As we learn from disciplines such as cybernetics, problem-solving systems are likely to be closed loop in order to deal with environmental complexities and to achieve robust and dependable performance. This is the same for actionable knowledge discovery and delivery systems. Human involvement can essentially contribute to closed-loop problem-solving systems and solutions.
- *Enhancing capability of problem-solving*: Driven by the involvement of human intelligence and the corresponding development and support, the ability of complex problem-solving systems can be greatly enhanced by the incorporation of human knowledge, inspiration, creativity, and teamwork.

2.5.3 Aspects of Human Intelligence

The aspects of human intelligence in complex systems are embodied in many ways, for instance:

- Human empirical knowledge
- Belief, intention, and expectation

- Sentiment and opinion
- Run-time supervision and evaluation
- Expert groups and teamwork
- Imaginary thinking
- Emotional intelligence
- Inspiration
- Creativity
- Brainstorming and collective intelligence
- Retrospection
- Reasoning inputs
- Embodied cognition such as convergent thinking through interaction with other members to assess identified patterns
- Conflict resolution capability
- Debating and agreement formation
- Consolidation and consensus building of opinions and inspiration

2.6 Organizational Intelligence

2.6.1 What Is Organizational Intelligence?

Definition 2.5 *Organizational intelligence* refers to the intelligence that emerges from the involvement of organization-oriented factors and resources in complex systems and in problem-solving solutions. Organizational intelligence is embodied through its incorporation into areas including system constituents, structures, evolution, dynamics, governance and regulation, working processes, modeling, and solutions.

2.6.2 Aims of Involving Organizational Intelligence

In a complex organization, the involvement of organizational intelligence is essential in many aspects, for instance in:

- Reflecting organizational reality, its needs, and constraints in business modeling and the delivery of findings
- Satisfying organizational goals and norms, policies, and regulation and convention
- Considering the impact of organizational interaction and dynamics in the modeling and deliverable design
- Catering for organizational structure and its evolution in data extraction, preparation, modeling, and delivery

2.6.3 *Aspects of Organizational Intelligence*

Organizational intelligence consists of many aspects, for example:

- Organizational structures related to key issues such as where data comes from and who in which branch needs the findings
- Organizational behavior related to key issues such as understanding the business and data and determining how individuals and groups act in an organization
- Organizational evolution and dynamics related to key issues such as data and information change, affecting model/pattern/knowledge evolution and adaptability
- Organizational/business regulation and convention related to key issues such as business understanding and finding delivery, including rules, policies, protocols, norms, law, etc.
- Business process and workflow related to key issues such as data (reflecting process and workflow) and business understanding, goal and task definition, delivery of findings, etc.
- Organizational goals related to key issues such as problem definition, goal and task definition, performance evaluation, etc.
- Organizational actors and roles related to key issues such as system actor definition, user preferences, knowledge involvement, interaction, interface and service design, delivery, etc.
- Organizational interaction related to key issues such as data and information interaction between subsystems and components, data sensitivity and privacy, and interaction rules applied to organizational interaction that may affect data extraction, integration and processing, pattern delivery, and so on

2.7 Social Intelligence

2.7.1 *What Is Social Intelligence?*

Definition 2.6 *Social intelligence* refers to the intelligence that emerges from the group interactions, behaviors, and corresponding regulation surrounding a complex system or problem. Social intelligence covers both human social intelligence and animat/agent-based social intelligence.

Human social intelligence is related to aspects such as social cognition, emotional intelligence, consensus construction, and group decision. *Animat/agent-based social intelligence* involves swarm intelligence, action selection, and foraging procedure. Both sides also engage social network intelligence and collective interaction, as well as social regulation rules, law, trust, and reputation for governing the emergence and use of social intelligence.

2.7.2 Aims of Involving Social Intelligence

In designing complex problem-solving systems in a social environment, both human social intelligence and agent-based social intelligence may play an important role, for instance in:

- Understanding the inbuilt social components, working mechanisms, and complexities
- Enhancing the social computing capability of complex problem-solving methods and systems
- Implementing complex problem-solving systems and evaluation in a social and group-based manner, under supervised or semi-supervised conditions
- Utilizing social group thinking and intelligence emergence in complex problem-solving
- Building social software on the basis of software agents, to facilitate human–mining interaction, group decision-making, self-organization, and autonomous action selection
- Defining and evaluating social performance including trust and reputation in developing quality social software and socially inspired problem-solving and evaluation
- Enhancing the project management and decision-support capabilities of the identified findings in a social environment

2.7.3 Aspects of Social Intelligence

Aspects of social intelligence are in multiple forms. We illustrate them from the perspective of human social intelligence and animat/agent-based social intelligence, respectively. *Human social intelligence* aspects consist of aspects such as social cognition, emotional intelligence, consensus construction, and group decision-making.

- Social cognition: aspects related to how a group of people process and use social information, which can inform what information to involve in complex problem-solving solutions, and how
- Emotional intelligence: aspects related to a group of people's emotions and feelings, which can inform interface and interaction design, performance evaluation, and the delivery of findings for complex problems
- Consensus construction: aspects related to how a group of people think and how thinking evolves toward convergence in a group, in divergent thinking situations in particular, which can aid conflict resolution if people from different backgrounds value different aspects in pattern selection, or if there is a conflict between technical and business interests

- Group decision-making: aspects related to strategies and methods used by a group of people to make a decision, which inform the discussion between business modelers and end users

Animat/agent-based social intelligence aspects consist of:

- Swarm/collective intelligence: aspects related to the collaboration and competition of a group of agents in handling a social problem, which can assist in complex problem-solving through multiagent interaction, collaboration, coordination, negotiation, and competition
- Behavior/group dynamics: aspects related to group formation, change and evolution, and behavior dynamics, which can assist in simulating and understanding the structure, behavior, and impact of understanding a group/community

Both human and agent social intelligence also involve many common aspects, such as:

- Social network intelligence
- Collective interaction
- Social behavioral network
- Social interaction rules, protocols, norms, etc.
- Trust and reputation, etc.
- Privacy, risk, security, etc., in a social context

2.8 Metasynthesis of Ubiquitous Intelligence

The involvement of ubiquitous intelligence is very important for handling open complex problems, such as open complex intelligent systems [10] and open complex agent systems [11]. Besides the individual engagement of each type of intelligence, a more critical problem is their synthesis into a problem-solving system. How they can be integrated becomes an interesting but difficult problem. As with the studies on the five proposed types of intelligence, the methodologies and approaches for doing this are not yet mature.

We have conducted a few trials in this regard, which are essentially helpful for detailing ubiquitous intelligence in the construction of open complex systems:

- Intelligence metasynthesis [12–14]: This is a methodology proposed to deal with open complex giant systems [12, 13]. It advocates a general framework for integrating human intelligence and computer intelligence into problem-solving systems, notably by establishing a hall for workshop of metasynthesis from qualitative to quantitative intelligence. Agent-based prototypes have been built for applying metasynthesis to handle macroeconomic decision-making [15].
- Metasynthetic computing [11]: This is a computing technique proposed to construct systems, mainly open complex intelligent systems, by utilizing the methodology of intelligence metasynthesis. A typical solution is to integrate

agents, services, organizational, and social computing for handling open complex systems [16] by establishing an M-space powered by M-computing and M-interaction [17].

A typical example of involving ubiquitous intelligence in data mining is domain-driven data mining [18] for actionable knowledge discovery and delivery [19]. There are many open issues in the integration of ubiquitous intelligence into agents, data mining, and agent mining. We believe that the studies on this will lead to great opportunities for innovative methodologies, techniques, means, and tools, as well as applications in relevant areas including agents, data mining, and agent mining. This will consequently definitely promote the transfer of the relevant disciplines to a more advanced, effective, and efficient stage from methodological, technical, and practical aspects.

2.9 Summary

Complex systems involve ubiquitous intelligence and respective issues:

- Data intelligence refers to both the general level and in-depth level of data intelligence from both syntactic and semantic perspectives.
- Human intelligence refers to both explicit or direct and implicit or indirect involvement of human intelligence.
- Domain intelligence refers to both qualitative and quantitative domain intelligence.
- Network intelligence refers to both web intelligence and broad-based network intelligence.
- Organizational intelligence refers to organizational goals, structures, rules, and dynamics.
- Social intelligence refers to both human social intelligence and animat/agent-based social intelligence.

We have discussed the definition, aims, aspects, and techniques for involving these types of intelligence into complex systems or complex problem-solving solutions. In Chap. 4, intelligence metasynthesis and metasynthetic computing for synthesizing ubiquitous intelligence will be discussed. In Chap. 14, ubiquitous intelligence is incorporated into knowledge discovery and data mining to form domain-driven actionable knowledge discovery delivery.

References

1. Cao, L., Gorodetsky, V., Mitkas, P.: Agent mining: the synergy of agents and data mining. IEEE Intell. Syst. **24**(3), 64–72 (2009)
2. Cao, L., Gorodetsky, V., Mitkas, P.: Guest editors' introduction: agents and data mining. IEEE Intell. Syst. **24**(3), 14–15 (2009)

3. Cao, L. (ed.): Data Mining and Multiagent Integration. Springer, New York (2009)
4. Gorodetsky, V., et al. (eds.): Autonomous Intelligent Systems: Agents and Data Mining. LNAI-4476. Springer, Berlin/Heidelberg (2007)
5. Cao, L., Dai, R.: Open Complex Intelligent Systems. Post & Telecom Press, Beijing PRC (2008)
6. Qian, X., Yu, J., Dai, R.: A new scientific field–open complex giant systems and the methodology. Chin. J. Nat. **13**(1), 3–10 (1990)
7. Cao, L., et al.: Domain-driven actionable knowledge discovery. IEEE Intell. Syst. **22**(4), 78–89 (2007)
8. Cao, L., Zhang, H., Zhao, Y., Zhang, C.: General frameworks for combined mining: case studies in e-government services. Submitted to ACM TKDD (2008)
9. Poslad, S.: Ubiquitous Computing: Smart Devices, Environments and Interactions. Wiley, Hoboken (2009)
10. Cao, L.: Metasynthetic computing for solving open complex problems. In: IEEE International Workshop on Engineering Open Complex Systems: Metasynthesis of Computing Paradigms, joint with COMPSAC 2008, IEEE Computer Society, pp. 896–901. New York (2008)
11. Cao, L.: Integrating agent, service, and organizational computing. Int. J. Softw. Eng. & Knowl. Eng. **19**(5), 573–596 (2008)
12. Qian, X.: Revisiting issues on open complex giant systems. Pattern Recogn. Artif. Intell. **4**(1), 5–8 (1991)
13. Symeonidis, A., Mitkas, P.: Agent Intelligence Through Data Mining. Springer, New York (2005)
14. Gorodetsky, V., et al. (eds.): Autonomous Intelligent Systems: Agents and Data Mining. LNAI-3505. Springer, Berlin/Heidelberg (2005)
15. Dai, R.: Qualitative-to-quantitative metasynthetic engineering. Pattern Recogn. Artif. Intell. **6**(2), 60–65 (1993)
16. Cao, L., Luo, C., Zhang, C.: Agent-mining interaction: an emerging area. In: Gorodetsky, V., et al. (eds.): Autonomous Intelligent Systems: Agents and Data Mining. LNAI-4476, pp. 60–73. Springer, Berlin/Heidelberg (2007)
17. Cao, L., et al.: Editor's introduction: interaction between agents and data mining. Int. J. Intell. Inf. Database Syst. **2**(1), 15 (2008)
18. Cao, L., Yu, P., Zhang, C., Zhao, Y.: Domain Driven Data Mining. Springer, New York (2010)
19. Cao, L., Dai, R.: Agent-oriented metasynthetic engineering for decision making. Int. J. Inf. Technol. Decis. Making **2**(2), 197–215 (2003)

Chapter 3
System Methodologies

3.1 Introduction

In human history, many different methodologies, metaphors, and philosophies [1–9] have been proposed by scientists, philosophers, theologians, and thinkers. Scientific development has greatly benefited from them, as is shown in the history of the physics family. The appropriate understanding of complex systems, system complexities, and the ubiquitous intelligence surrounding and embedded in complex systems, as well as engineering such systems and complexities, require full acknowledgement of the functions, characteristics, and suitability of respective methodologies.

The major system methodologies which have been developed by philosophers to understand and explain complex phenomena and systems can be categorized into three types:

- Reductionism: taking the assumption that a whole is equal to the sum of all parts
- Holism: assuming that the whole goes beyond the collection of parts
- Systematology: arguing the need for both the whole and its parts, which play different but complementary roles

In this chapter, the main ideas of reductionism, holism, and systematology are summarized, and the limitations and strengths of each discussed. Connections for them are constructed to address system complexities in complex systems. The principles of these system methodologies provide high-level support for explaining complex systems, in particular, the formation of corresponding cognition, theories, tools, and systems for understanding, explaining, computing, and engineering complex systems.

L. Cao, *Metasynthetic Computing and Engineering of Complex Systems*, Advanced Information and Knowledge Processing,
DOI 10.1007/978-1-4471-6551-4_3

3.2 Reductionism

Reductionism [10–12] is a methodology that claims a system is equivalent to the sum of its parts, or that the whole is nothing more than the sum of its parts. Accordingly, understanding a system can be equivalently achieved by exploring individual constituents. Reductionism may be defined in terms of such perspectives as structural, theoretical, conceptual, substantive, methodological, and ontological. Accordingly, reductionist theories have been developed to explain the phenomena related to the respective perspective, for instance, the reduction of one theory to another (forming concepts such as sub-theory or super-theory), or the reduction of explanations to the smallest possible entities or particles (with hierarchical concepts including system, subsystem, member, object).

Reductionism has typically driven scientific development in physics, chemistry, and mathematics, in which the complexity of a system is decomposed into smaller complexities in subsystems or smaller (even atomic) organizational units. In addition, theoretical reductionism holds such claims in the scientific domain as the foundations of chemistry being based in physics, and molecular biology being rooted in chemistry.

Reductionism may be more applicable and workable for simple systems in which no significant changes exist; the interactions and coupling relationships between system constituents can be ignored. Such systems are more likely to be linear, or capable of being equivalently converted to linear, through a bottom-up approach. Reductionism targets the understanding of each individual unit at the very bottom and then sums the findings from each individual to build a complete understanding of its higher-level components.

In real life, however, most systems are not simple enough to be treated as linear and cannot be understood by reductionism alone [13, 14]. Large scale system constituents, various types and levels of interactions and coupling relationships, dynamic evolution driven by nonlinear and uncertain factors, and heterogeneous components exist in an open complex system.

Reductionism may overlook explicit and/or implicit coupling relationships between individual constituents, instead treating constituents as being independent. This assumption unavoidably creates a paradox or presents challenges in explaining complex systems in which interactions and coupling relationships, evolution, nonlinearity, uncertainty and system complexities cannot be ignored. In complex systems such as natural systems, ecosystems, and human society, typical phenomena such as emergence, interaction, and feedback loops across system components pose challenges to reductionism. For instance, a system may grow additional parts that do not belong to the original system under certain evolutionary conditions. In such systems, what emerges is more than the sum of the processes from which it has evolved. New components, processes, procedures, laws, concepts, generalizations, and intelligence emerge during the evolution of a system which cannot be explored based on previous assumptions; rather, they require inspiration and creativity, or human involvement, in problem understanding and solving.

3.3 Holism

Philosophers and researchers in social science and systems science increasingly recognize the limitations of reductionism for explaining complex social and system phenomena. Typically, in the latter half of the twentieth century, holism [15, 16] commanded a high level of attention in the scientific domain, which focuses on systems thinking and its derivatives, such as chaos theory and system complexity. New or so-called emergent behaviors increasingly focused on the study of complex systems in biology, psychology, or sociology, claiming that neither the behavior nor the properties of a system can be deduced from its constituents alone.

Holistic methodology, rather than the reductionist paradigm, has been widely accepted in studying complex systems and system complexities with unexpected behaviors occurring (so-called emergence of behavior or intelligence). Unexpected behaviors and phenomena surprise people or have unexpected effects on the system whole and collection of parts. Holism holds that a system and its properties cannot be viewed as a collection of parts, but that the system functions as a whole and this functioning cannot be fully understood solely in terms of the component parts. Unlike the argument that the property of a system as a whole can be predicted from a good understanding of all its constituents, a holistic approach such as the so-called complexity theory perceives the need to explore unpredictable features in a system that emerge from individual components.

One key driver of the emergence of surprising behavior or phenomena results from the coupling relationships embedded between system parts that drive the interactions between parts. This is typical in complex systems, in which visible and invisible couplings and interactions connect system members to form an organic whole. When a system also has system members on a large scale, even weak couplings and interactions can evolve into significant unexpected emergent behavior. Typical system complexities including openness, human involvement, hierarchy, nonlinearity, and uncertainty further impact on the couplings and interactions and make a system more unpredictable.

Holism and reductionism are often viewed as opposite philosophies. Holism takes a top-down approach, while reductionism centers on a bottom-up paradigm. Whether a holistic or a reductionist approach is appropriate is very much dependent on the status, characteristics, scale, and other system complexities in a specific system or phenomenon. For open complex systems, especially open complex giant systems, both methodologies have roles to play. However, they are not sufficiently capable, which leads to another view that is systematology.

3.4 Systematology

The respective focuses and advantages of top-down-oriented holism and bottom-up-based reductionism only address certain challenges in complex systems. Both holism and reductionism thus have only limited advantages and tools for understanding and explaining partial system complexities. The challenges faced by holism may lie in its limitations in understanding local functioning and composition mechanisms and its simplified way of grabbing concrete, specific system particles and predictable behaviors. Such needs are widely apparent in simple systems, or subsystems of a complex system.

Compared to holism, reductionism is a very mature methodology, with many effective tools and theories, as well as successful case studies, in scientific and social areas. By focusing on decomposing a complex whole into parts, reductionism has the advantage of reducing system complexity and understanding the functioning of individual parts, which are not the focus of holism. The strengths of reductionism are very necessary for analyzing, designing, and implementing complex systems, as most existing engineering technologies, tools, and systems have been built on top of reductionism.

Holism, however, is good at creating an overall view of the whole rather than its parts, and centers on system-level complexity, providing the means to capture system-level characteristics including emergence and nonlinearity, and underlying drivers including interactions and coupling relationships between parts. These features are not centered in reductionism.

Clearly, it is essential to combine the strengths and advantageous tools of holism and reductionism, both of which are needed to understand a complex problem. This forms a new methodology: systematology [2–9].

Systematology argues that (1) a good understanding of system members not only contributes to forming an overall picture of the whole, but also reduces the complexity of understanding and makes a complex problem understandable or solvable; (2) a complex system goes beyond the sum of its parts as a result of various system complexities, and unexpected and unpredictable system characteristics, special theories, and tools need to be developed to address such complexities and characteristics. While the sum of parts is not equal to the whole, collections of parts do contribute to the whole.

The purpose of systematology is to combine the valuable qualities of holism and reductionism and to form a loop for systematic thinking and exploration by integrating bottom-up and top-down approaches. For a complex problem, it is not possible to obtain complete understanding in a one-off loop. Hence, the loop may appear as an iterative and qualitative-to-quantitative progression:

- Starting from a top-down understanding of major characteristics, often in a qualitative way, to grasp the main challenges and core functions and functioning components.
- Once a basic understanding is achieved (which may be wrong or biased), focus may be moved to (quantitatively) understanding those major (or likely to be major) components (subsystems) by a reductionist approach.
- Understanding the major components helps to deepen the understanding of specific system complexities and to adjust the original "estimation" (a guess which may or may not be correct) toward a more appropriate understanding of the whole and parts and their interactions and coupling relationships.

- Such deepening and adjusting actions may be conducted in several rounds before a confident picture about the whole and parts and their relations are extracted.
- Findings in each step are accumulated and synthesized, with conflicts or diffused cognition mitigated or managed toward building a common consensus.

The above qualitative-to-quantitative cognitive process is called *metasynthesis*. In Chap. 5, we will explain the process, challenge, theories, and techniques required to enable metasynthesis.

3.5 Summary

System methodologies are crucial for effectively understanding, engineering, and computing systems. Typical philosophies for handling systems include reductionism, holism, and systematology. Reductionism takes a bottom-up approach, while holism favors top-down thinking. Systematology represents a way to combine the strengths in reductionism and holism.

Complex systems raise critical challenges to existing system methodologies. Further development of systematology is necessary to address the typical challenges in open complex giant systems, including openness, giant scale, sociality, human involvement, evolution and dynamics, and ubiquitous intelligence. These issues will be addressed in Chap. 5 in terms of the methodology of metasynthesis, which has been proposed to handle open complex giant systems, and Chaps. 6, 7, 8, 9, 10, 11, and 12 in terms of computing and engineering techniques and tools.

References

1. Auyang, S.Y.: Foundations of Complex-system Theories: In Economics, Evolutionary Biology, and Statistical Physics. Cambridge University Press, Cambridge (1999)
2. Cao, L., Dai, R.: Open Complex Intelligent Systems (in Chinese). Post & Telecom, Beijing PRC (2008)
3. Dai, R., Li, Y., Li, Q.: Social Intelligence and Metasynthetic System (in Chinese). Post & Telecom, Beijing PRC (2013)
4. Dai, R.: Qualitative-to-quantitative metasynthetic engineering (in Chinese). Pattern. Recogn. Artif. Intell. **6**(2), 60–65 (1993)
5. Qian, X., Yu, J., Dai, R.: A new scientific field–open complex giant systems and the methodology (in Chinese). Chin. J. Nat. **13**(1), 3–10 (1990)
6. Qian, X.: Revisiting issues on open complex giant systems (in Chinese). Pattern Recogn. Artif. Intell. **4**(1), 5–8 (1991)
7. Qian, X.: Building Systematology (in Chinese). Shanghai Jiaotong University Press, Shanghai PRC (2007)
8. Cao, L., Zhang, C., Zhou, M.: Engineering open complex agent systems: a case study. IEEE Trans. Syst. Man Cybern. C: Appl. Rev. **38**(4), 483–496 (2008)
9. Cao, L., Dai, R., Zhou, M.: Metasynthesis: M-space, M-interaction and M-computing for open complex giant systems. IEEE Trans. Syst. Man Cybern. A **39**(5), 1007–1021 (2009)

10. http://en.wikipedia.org/wiki/Reductionism
11. Brigandt, I., Love, A.: Reductionism in biology. In: Zalta, E. (ed.) The Stanford Encyclopedia of Philosophy (2012). http://plato.stanford.edu/archives/sum2012/entries/reduction-biology/
12. Jones, R.H.: Analysis & the Fullness of Reality: An Introduction to Reductionism & Emergence. Jackson Square Books, New York (2013)
13. Kauffman, S.: Beyond reductionism: reinventing the sacred. Zygon J. Relig. Sci. **42**(4), 903–914 (2007)
14. Carroll, J.W.: Chapter 13: Anti-reductionism. In: Beebee, H., Hitchcock, C., Menzies, P. (eds.) The Oxford Handbook of Causation. Oxford Handbooks Online, p. 292. ISBN 019927973X (2009)
15. http://en.wikipedia.org/wiki/Holism
16. Mastin, L.: Holism, The Basics of Philosophy (2008). http://www.philosophybasics.com/branch_holism.html

Chapter 4
Computing Paradigms

4.1 Introduction

To analyze, design, and implement problem-solving solutions for complex systems, we need effective computing paradigms. The methodologies for engineering complex systems have been evolving toward:

1. Addressing increasingly complex problems and building corresponding systems
2. Providing more user-friendly interfaces
3. Supporting enterprise application integration

Usually, a computing paradigm (methodology) is proposed on top of a core metaphor and concept. With the proposal of core concepts such as objects, components, services, agents, agent service, organization, and cloud, corresponding software engineering methodologies have also been proposed or are the subject of study at the same time. Objects [1], components [2], services [3], and agents [4] are very popular but high-level abstraction conceptions. They have been or are currently used by software academic designers and industrial architects to construct software systems to model the real world. Subsequently, methodologies including object-oriented methodology, component-based methodology, service-oriented methodology, and agent-oriented methodology have been proposed to analyze, design, and implement the complexities in complex systems (usually engineering systems).

In addition, increased attention has been paid to other general concepts in the social science, economics, and cultural domains; for instance, behavior, organization, autonomy, sociality, cloud, and service. These have formed new computing paradigms, including autonomic computing, behavior computing, social computing, and cloud computing. More recently, to address the system complexities in open complex systems, metasynthetic computing has been proposed.

In this chapter, the concepts, basic principles, and strengths and weaknesses of the above core concepts and computing paradigms are discussed. The aim

L. Cao, *Metasynthetic Computing and Engineering of Complex Systems*,
Advanced Information and Knowledge Processing,
DOI 10.1007/978-1-4471-6551-4_4

is to provide basic concepts to understand what computing and engineering methodologies are available and which of these are the most suitable for analyzing, designing, and implementing open complex systems.

4.2 Objects and Object-Oriented Methodology

Objects are defined as computational entities that encapsulate certain states, are able to perform actions or methods on those states, and communicate by message passing. In implementation, a *class* is the design-time component that a developer works with. "Object" is an instance of a class, or in other words a "class" is a template for "objects."

We may define an object as a computational entity which stores data and those operations (methods) that manipulate the data. It is a unit of instantiation of a class and has a unique identity. It may have states and the states can be externally observable. It encapsulates its states and behaviors.

Object-oriented analysis and design [1, 5] model the world in terms of objects that have properties and behaviors, and events that trigger operations to change the state of the objects. Objects interact formally with other objects [5].

In *object-oriented programming*, a program execution is regarded as a physical model, simulating the behavior of either a real or an imaginary part of the world. It can basically be expressed in the following formula:

$$\begin{aligned}\text{Object\_Oriented\_Programming} = {} & \text{Polymorphism} + \text{(Some) Late Binding} \\ & + \text{(Some) Encapsulation} + \text{Inheritance} \qquad (4.1)\end{aligned}$$

4.3 Components and Component-Based Methodology

The term *component* has become overloaded over the last few years. The following definitions can be found in academia and industry.

A component is a self-contained, clearly identifiable, physically realizable, predefined entity that provides a well-understood function and is intended to be assembled, using a well-defined architecture and interfaces, with other components to provide more complex functions [62].

A component is a unit of packaging, distribution, or delivery that provides services within a data integrity or encapsulation boundary [6]. A component is a language neutral, independently implemented package of software services, delivered in an encapsulated and replaceable container, accessed via one or more published interfaces. While a component may have the ability to modify a database, it should not be expected to maintain state information. A component is not platform constrained nor is it application bound [7].

The component-based approach [2] encourages the creation of reusable components and the production of new software systems through the assembly of preexisting components. There are two engineering drivers in the development of a component-based system:

- *Reuse*: The ability to reuse existing components to create a more complex system.
- *Evolution*: By creating a system that is highly componentized, the system is easier to maintain. In a well-designed system, the changes will be localized, and the changes can be made to the system with little or no effect on the remaining components.

In *component-based programming*, the main elements can be expressed in the following formula:

$$\begin{aligned}\text{Component\_Based\_Programming} = {} & \text{Polymorphism} \\ & + \text{(Really) Late Binding} \\ & + \text{(Real, Enforced) Encapsulation} \\ & + \text{Interface Inheritance} \\ & + \text{Binary Reuse}\end{aligned} \tag{4.2}$$

Many component-based developments are object oriented. The possible tension between the two approaches arises because the component-based method aims to encapsulate functions in large components that have loose couplings. Object-oriented development is also about encapsulation, but with objects, and there tends to be highly complex coupling between objects.

4.4 Services and Service-Oriented Methodology

A *service* is a function that is well defined and self-contained and does not depend on the context or state of other services. Software services are discrete units of application logic that expose message-based interfaces suitable for access across a network. Service-based architectures [3] permit very flexible deployment strategies, rather than requiring all data and logic to reside on a single computer. The service model allows applications to leverage networked computational resources. An application architecture that depends on interoperable services can provide high-value business logic and state management.

Although *Web services* [8] are the most popular way to implement services, a service does not have to be a "Web service" – that is, a Web service can exist which is not a service. Many people use Web services as a transport protocol, so "services" and "Web services" should not be treated interchangeably.

In Web services, services are autonomous platform-independent computational elements that can be described, published, discovered, orchestrated, and programmed using XML artifacts for the purpose of developing massively distributed interoperable applications.

Service-oriented methodology [3] is the newly-emerging paradigm for distributed computing and e-business processing. It has evolved from object-oriented and component-based computing to enable agile networks of collaborating business applications, distributed within and across organizational boundaries, to be built.

A *service-oriented architecture* is essentially a collection of services. These services communicate with one another. The communication can take the form of simple data passing, or it can involve two or more services coordinating an activity. Some means of connecting services to each other is needed.

Combined with recent developments in the area of distributed systems, workflow management systems, business protocols, and languages, services can provide the automated support needed for e-business integration, both at the data level and the business logic level. They also provide a sound support framework for developing complex business transaction sequences and business collaboration applications.

The application of service-oriented computing to Web resources is manifested by Web services to provide a loosely coupled model for distributed processing.

4.5 Agents and Agent-Oriented Methodology

An agent is an encapsulated computer system [9] or *cellular automata* situated in a specific environment; it is capable of flexible, autonomous actions in that environment which enable it to meet its design objectives. This definition means that agents are:

- Clearly identifiable problem-solving entities with well-defined boundaries and interfaces.
- Situated (embedded) in a particular environment over which they have partial control and observability—they receive inputs related to the state of their environment through sensors, and they act on the environment through effectors.
- Designed to fulfill a specific role—they have particular objectives to achieve.
- Autonomous—they have control over both their internal states and their own behaviors.
- Capable of exhibiting flexible problem-solving behaviors in pursuit of their design objectives—being both reactive (able to respond in a timely fashion to changes that occur in their environment) and proactive (able to opportunistically adopt goals and take the initiative) [10].

Compared with other computational entities, an agent is a computer system with two important capabilities.

1. It is at least to some extent capable of autonomous action—of deciding for itself what it needs to do to satisfy its design objectives.
2. It is capable of interacting with other agents, not simply by exchanging data but by engaging in the kind of analogous social activity in which we all engage every

day of our lives: cooperation, coordination, negotiation, communication, and the like.

As we know, an agent is a goal-driven entity with autonomy and self-control of its capabilities. From this point of view, *goal-oriented analysis* [11] is compatible with the motivation of agent-oriented methodology. Goal-oriented requirements analysis covers both functional requirements and nonfunctional requirements [12]. Goal-oriented analysis in TROPOS [13] includes the early and late requirements analyses; in this work, however, we adopt it for early requirements analysis only.

Since goal-oriented analysis focuses on the early stage of requirements analysis and on the rationalization of the development process, it can be used for the early requirements analysis of agent-based systems. In the following section, we will therefore discuss goal-oriented analysis of early functional and nonfunctional requirements. Please note that we introduce the principles of goal-oriented analysis here and discuss its modeling techniques in Chap. 7, together with detailed visual modeling case studies. One reason for this strategy is that it has been popularly used for the last several years, for example, in TROPOS and the *i** framework; another is that we adopt this as the default approach and compare it with *organization-oriented analysis*, which will be discussed in detail in Chap. 6.

4.5.1 Goal-Oriented Requirements Analysis

Lamsweerde [12] provided a guided tour of *goal-oriented requirements engineering*. A *goal* is an objective to be achieved by the system under consideration. Goals may be formulated at different levels of abstraction, ranging from high-level, strategic concerns to low-level, technical concerns. Goals also cover different types of concerns: *functional* concerns are associated with the services to be provided, and *nonfunctional* concerns are associated with quality of service—such as safety, security, accuracy, performance, and so forth.

The benefit of *goal modeling* is to support heuristic, qualitative, or formal reasoning schemes during requirements engineering. *Goal-oriented analysis* [13–15] focuses on the description and evaluation of *alternatives and their relationship* to the organizational objectives behind a software development project. Capturing these interdependencies between organizational objectives and the detailed software requirements can facilitate the tracing of the origins of requirements and can help make the requirements process more thorough, complete, and consistent.

Many techniques have been proposed by academia for goal-oriented requirements engineering. Most goal-oriented modeling uses graphical notations; this can be termed *diagrammatic modeling* or *visual modeling*. Diagrammatic modeling can facilitate an intuitive understanding of scenarios. The *i** model [16, 17] develops graphical notations to analyze the early (functional) requirements and late (nonfunctional) requirements. In TROPOS [18], the *i** model is used for early requirements analysis, and the *strategic dependency model* and *strategic rationale model* are used for late requirements analysis.

Diagrammatic models, however, enclose implicit, incomplete, and imprecise information which undermines the understanding of the problem and blocks later design and implementation. To this end, formal specifications may complement these models. RML [19] and KAOS [14] are formal requirements modeling languages. Their formal semantics constitute a solid foundation for building goal-oriented sophisticated analysis tools.

Some goal-oriented approaches link agent and goal together. In KAOS, *responsibility* links are introduced to relate the goal and agent submodels. A goal may be assigned to alternative agents through organizational relationship responsibility links; this allows alternative boundaries to be explored between the software and its environment. In the *i** framework [16, 17], various types of entity dependency links are defined to model situations in which an entity depends on another for a goal to be achieved, a task to be achieved, or a resource to become available. TROPOS, first proposed in Bresciani et al. [20] and refined in Kolp et al. [18], is transformed into an agent-oriented methodology which takes organizational metaphor.

A systematic integration of visual modeling and formal modeling can make the requirements engineering complete and precise. [21] introduced the ATOS (an acronym for AUML to PROMELA/SPIN) approach. ATOS introduces a textual notation of AUML; it can be translated to an extended finite state machine which can be processed by a model checker. It has reportedly been used to perform formal verification of the *AUML sequence diagram* specification of the interaction protocols of multiagent systems. [22] discussed the possibility of intermixing formal and informal specifications in order to guide and support *conceptual modeling*; the work describes a modeling language for lightweight usage of formal verification techniques when performing conceptual modeling in agent-oriented methodology. Gans [23] and Wang [24] proposed a framework which combines *i** and Congolog [25] frameworks to exploit the complementary features of the two frameworks for complete and precise modeling.

In Chap. 9, we discuss *integrative modeling* which combines visual modeling and formal specifications in analyzing and designing agent-based systems. The integrative modeling covers both functional and nonfunctional requirements.

4.5.2 Agent-Oriented Software Engineering

The agent-oriented approach to software engineering requires the decomposition of the problem into multiple, autonomous components that can act and interact in flexible ways to achieve their set objectives. The key abstraction models that define the agent-oriented mind-set are agents, interactions, and organizations. Explicit structures and mechanisms are often used to describe and manage the complex and changing Web of organizational relationships that exist between the agents.

The argument in favor of an agent-oriented approach to software engineering includes:

- Showing that the key abstractions of the agent-oriented mind-set are a natural means of modeling complex systems
- Showing that agent-oriented decompositions are an effective way of partitioning the problem space of a complex system
- Showing that the agent-oriented philosophy for modeling and managing organizational relationships is appropriate for dealing with the dependencies and interactions that exist in complex systems

Research in the area of agent oriented software engineering (AOSE) has expanded significantly in the past few years. Several groups have begun to address the problem of modeling agent systems with appropriate abstractions, and defining the methodologies for MAS development. Shehory and Sturm [26] provide a survey of this topic.

Several methodologies, for instance, MaSE [27], MESSAGE [28], TROPOS [19, 22], and GAIA [29, 30], have been proposed for the analysis and design of MAS. Some of these agent-oriented methodologies are discussed below.

4.5.2.1 MaSE

In the analysis phase, the MaSE methodology [27] utilizes requirements to define use cases, application goals and subgoals, and eventually to identify the roles to be played by agents and their interactions. In the design, agent classes and agent interaction protocols are derived from the outcome of the analysis phase, leading to a complete architecture of the system. MaSE is only suitable for closed agent systems.

4.5.2.2 MESSAGE

The MESSAGE methodology [28] defines an organization in terms of a *structure*, determining the roles to be played by the agents and their topological relations (i.e., the interactions occurring between them). In MESSAGE, an organization is also characterized by a *control entity* and a *workflow structure*, which determine the laws to which agent actions and interactions have to conform during execution.

4.5.2.3 TROPOS

The TROPOS methodology, first proposed in [20] and refined by Kolp [19, 22], recognizes that the organizational structure is a primary dimension for the development of agent systems and that an appropriate choice of structure is needed to meet both functional and nonfunctional requirements. TROPOS methodology defines uniform and coherent guidelines for the activities of both early and late requirements engineering.

4.5.2.4 GAIA

GAIA methodology, proposed by Wooldridge et al. [29] and refined by Zambonelli et al. [30], is founded on the view of a multiagent system as a computational organization consisting of various interacting roles. It provides a clean separation between the analysis and design phases. It defines a fully elaborated role model, together with an accurate description of the protocols in which the roles will be involved, an overall organizational structure, global organizational rules, and the modeling of the environment.

The GAIA methodology is both general, in that it focuses on conceptual modeling so that it is applicable to a wide range of MAS, and comprehensive, in that it deals with both the macrolevel (societal) and the microlevel (agent) aspects of systems. It exploits the organizational abstractions to provide clear guidelines for the analysis and design of complex and open software systems. Table 4.1 presents a comparison of the above four methodologies.

Table 4.1 Comparison of agent-oriented methodologies

	GAIA	MaSE	MESSAGE	TROPOS
Metaphor	Organization		Organization	Organization
Modeling	Late requirements engineering, modeling of the environment	Use cases	UML, AUML	Early and late requirements engineering
Analysis phase	Role model, protocols	Application goals and subgoals, agent roles, interaction	Structure, role, topological relations	Role, organizational structure, functional and nonfunctional requirements, structural dependencies
Design phase	Global organizational rules	Agent classes, agent interaction protocols, system architecture	Control entity, workflow structure	
Agent systems	Open agent systems	Closed agent systems	Open agent systems	
Limitations	Early requirements analysis	Open	Modeling the organizational rules, designing the organizational structure	Global laws

4.5.3 Issues in Agent-Oriented Software Engineering

We classify the base or abstraction foundation of existing AOSE approaches into two categories:

1. Goal-oriented abstraction
2. Organization-oriented abstraction

Goal-oriented abstraction takes the abstraction and representation of agents' and systems' goals as the key in engineering MAS. We believe this is sufficient and powerful for modeling goal-centered requirements. However, it is insufficient and incomplete as a comprehensive software engineering tool, because the goal is only one element or focus of the agent organization.

To handle comprehensive complexities, *organization-oriented abstraction* is an appropriate option. For *organization-oriented abstraction*, *organizational metaphor* is the key to conceptualizing and analyzing MAS.

In this section, we discuss a number of issues in the existing agent-oriented software engineering approaches following goal-oriented abstraction.

Goal-oriented requirements analysis is proposed as a generic methodology for requirements beyond object-oriented analysis. When it is used for agent-oriented methodology, it must be updated and extended with respect to the agent-based paradigm. In the following, we take the *i** framework and TROPOS, which typically follow goal-oriented abstraction, as examples to discuss the issues that arise when they are used to analyze OCAS.

The *i** framework and TROPOS are effective techniques for analyzing goal-oriented requirements. The emergence of the organizational metaphor is helpful for understanding and viewing complex software systems, in particular OCAS. For supporting the abstraction of OCAS, however, it is necessary to add more explicit capabilities to the current goal-oriented *i** model and TROPOS in terms of the following issues and challenges.

- Internal differences exist between the actors of an organization. It is claimed that actors in the *i** framework can take the form of a role, an agent, or a position. However, as will be discussed in more detail in Chaps. 6 and 7, the actors in an organization also present differences; the role of an actor is determined by its scenarios and goals. For instance, if an actor takes the form of an agent, then the actor should satisfy the criteria for being an agent. Agents cannot be introduced into the strategic dependency model without explicit definition when it is brought into the agent-oriented abstraction. If this occurs, the goal and role distribution in an agent system is not embodied; rather, *i** modeling is simply deployed in agent-oriented requirements engineering.
- Explicit definitions of actors are essential for a system viewed as an organization. Some new types of stakeholders need to be declared explicitly in terms of their roles and scenarios in an organization. For instance, service-oriented abstraction needs to be considered in modeling actors of an open system.

The explicit modeling of differentiated actors assists us to capture and present them to understand their constituents and their roles in an organization.

- New patterns are required in modeling the organizational rules and relationships in an organization. In *i** modeling, a hidden basic hypothesis is that a software system can be simplified as a dependency network. Thus, all relationships between actors are expressed as dependencies. In reality, dependency is only one of the possible relationships in a soft organization. Other interdependencies, such as control, peer, benevolence, and ownership [31], and constraints may also coexist in an organization. Therefore, besides dependency, other possible relationships are also needed to completely express the range of organizational relationships.
- More interactive patterns need to be explicitly supported in modeling an organization. Multiple types of interactive activities and links exist which reflect heterogeneous or homogeneous social interactions in an organization. For simplicity of representation of linkages and activity patterns in an organization, it is necessary for us to abstract link specifications to explicitly represent the multiform interactions between actors.

New modeling mechanisms are therefore required to model MAS, and especially OCAS. From a requirements analysis viewpoint, the new modeling candidate should encompass model-building blocks which can represent organizational elements beyond goals and newly emergent actors in an agent computational organization. Probably one of the most appropriate directions is *organization-oriented analysis*. As a phase of software engineering, organization-oriented analysis will form part of the whole science of *organization-oriented software engineering* including the abstraction, analysis, design, and implementation of complex agent systems in terms of the theory of AAMAS. We develop an approach called *OSOAD* (*organization- and service-oriented analysis design*) as one option for implementing organization-oriented software engineering.

4.6 Relations Among Agents, Objects, Components, and Services

Components can essentially be taken as "natural extensions" of objects. An object is a design-time artifact for the developer to work with before its compilation into one or more components. It is what the developer works with and understands. A component is an assembly made up of a number of objects. What makes the collection of objects a component? The fact that they are in the same assembly! In implementation, a component may be an actual DLL or exe instantiated in the system at run time to execute the designed functionality. It is what the machine works with and understands.

An idea maintained by some researchers is that if you are debugging functionality, then you work with objects, and if you are tuning systems or implementing locking and transactions, then you work with components.

A service, however, is a product that cannot be understood as either a component or an object: it concerns activity rather than structure. The service features of being platform neutral and having a self-describing nature, and particularly the ability to enable business collaborations, provide significant competitive advantages.

The main distinctions between the view of a "traditional" object and agent are as follows. A "traditional" object can be thought of as exhibiting autonomy over its state: it has control over it, but it does not exhibit control over its behavior. An agent, however, embodies a stronger notion of autonomy than an object. For an object, the decision lies with the object that invokes the method, while for an agent, the decision lies with the agent that receives the request. This distinction has been summarized as "an object does it for free; an agent does it because it wants to." Second, "traditional" objects cannot deal with flexible behaviors, whereas an agent is capable of integrating reactive, proactive, and social behaviors [32, 33] into one system. The third distinction is that there is only a single thread of control in an object-oriented system, while in a multiagent system, each agent is assumed to have at least one thread of control. The system is therefore inherently multi-threaded.

In contemporary thinking, the distinction between objects and agents is becoming less sharp with time; for instance, some researchers have proposed the component-based design of intelligent agents [34]. Nevertheless, it is clear that agent-based abstractions suit when the attention is shifted from object/components to complex systems [35].

Table 4.2 further summarizes the relationships between OO, CB, and AO.

4.7 Autonomic Computing

Autonomic computing aims to build the self-managing characteristics of distributed computing resources. An autonomic system can adapt to unpredictable changes while hiding its intrinsic complexity from users. The main purpose of designing autonomic systems is to enhance the self-management capability of a system to reduce management complexities. Accordingly, an autonomic system makes decisions on its own based on high-level policies and rules embedded in the system. Components in such a system tend to be autonomic as well, which means they can constantly check its status, optimize its behavior, and automatically adapt to changing environments and conditions.

To enable autonomic computing, several functions should be enabled by the corresponding system frameworks: autonomic components that can interact with one another; self-managing, self-regulating, and self-configuration mechanisms; system policies and rules; adaptivity through sense-effect mechanism and planning;

Table 4.2 Comparison between object-oriented, component-based, and agent-oriented methods

	Object oriented	Component based	Agent oriented
Origin	Semantic net	Semantic net	Symbolic AI and behavior-based AI
Computational entity	Objects	Components	Agents
State parameters for an entity	Unlimited	Unlimited	Intentional stance, like belief, desire of an agent
Activity	Passive	Passive	Proactive with self-control thread, automated
Computational process	Message passing and method response	Message passing and method response	Message passing and method response
Message types	Unlimited, handle messages by method vocation	Unlimited, handle messages by method vocation	Speech act
Language abstraction elements	Objects, classes, modules	Reuse, design patterns, application framework	Agents, class, modules, design patterns, framework, organization, roles, society, goal
Modeling abstraction mechanism	Fine object as an action entity, method invocation used for describing interaction. Static organizational modeling, no semantics	Stronger abstraction mechanism, e.g., component, reuse, design patterns, application frameworks	Agents as coarse and automated computing entity, social ability (organization, roles, etc.), dynamic organizational modeling
Analysis and design	Abstraction in fine granularity; object model, dynamic model, function model	Abstraction in coarser granularity; component library, framework library, object bus	Coarser abstraction; role model, interaction model, agent model, service model, acquaintance model
Encapsulation	State and behavior	State and behavior, application framework	State and behavior, behavior activation
Organizational relationship	Static syntactic inheritance	Static syntactic and structural inheritance	An interactive network with inter- and intra-subsystem and subsystem element interactions, multiple organizational relationships (hierarchy, marketing, etc.)
Interaction	Syntactic interaction, invoking methods or functions, simple message passing	Syntactic interaction, invoking methods or functions, message passing	Interaction on knowledge and social levels

(continued)

Table 4.2 (continued)

	Object oriented	Component based	Agent oriented
System problem-solving	Event/behavior driven; design-time decision; no automated problem-solving; predefined execution	Event driven; design-time decision-making	Goal-driven, automated, and flexible problem-solving; active decision-making at run time, reasoning ability
Complexity in problem-solving	Generic system, with predefined interactive relationships	Not strong enough for modeling complex systems	Building complex distributed systems (data, ability, and control)
System property	Somehow encapsulation, autonomy, passivity, and interaction	Somehow encapsulation, autonomy, passivity, and interaction	Autonomy, reactivity, sociality, proactiveness; loss of control, greater freedom, uncertainty, and indeterminism
Interrelationship		Evolution (specialization) of OO	Evolution (specialization) of OO and CB

interactions between systems and environments; and awareness of environment. A typical autonomic system embeds following basic characteristics:

- Automatic: making its own decisions through self-contained design and self-controlled operation
- Adaptive: adjusting its states, functions, and configuration to changing conditions and environment
- Aware: sensing the context and environment and having an effect on state or context change

IBM defines several basic functional areas for automatic systems:

- Self-configuration: automatic configuration of components
- Self-healing: automatic discovery and correction of faults
- Self-optimization: automatic monitoring and control of resources to ensure optimal functioning with respect to the defined requirements
- Self-protection: proactive identification and protection from arbitrary attacks

Due to the complexity of maintaining a self-managing system for handling complex enterprise applications, multiple levels of system functions may be arranged. Different levels of autonomy can then be arranged, with some levels more dependent on manual control, while others rely on automatic control.

Existing work on this issue is mainly characterized by agent-based systems and agent-based modeling. Accordingly, the system frameworks are either centralized or cluster-based architectures. Such frameworks make it essentially complex, costly, or even impractical to build automatic systems for complex problem-solving.

Due to the focus on self-managing design, the corresponding architects are unable to anticipate, design, and maintain the complexity of interactions, system behaviors, and emergence of intelligence. The problems surrounding existing autonomic computing lie in the reduction of management costs, weak ability in handling complex logics, and providing innovative services. The exclusion of human and domain experts in the system design and problem-solving, and the limitation on openness and sociality support, essentially make such autonomic systems very ineffective in handling open complex systems. In addition, no work has been reported on handling large-scale systems.

4.8 Organizational Computing

Organizational computing utilizes an organizational model for computation, that is, the principles, methods, and practices of human and social organizations are used by computers or computing systems to understand, design, implement, and evaluate complex systems. To achieve this, core concepts such as organization, goals, norms, organizational rules, organizational structures, organizational interactions, policies, constraints, and commitments are key elements in understanding organizational formation, evolution, dynamics, inference, and impact.

Hewitt et al. [36] argued that organizations of restricted generality (ORGs) have been proposed as a foundation for organizational computing. In general:

- ORGs mirror the structure of large-scale human organizations.
- ORGs are a natural extension of Web services, which are the standard for distributed computing and software application interoperability in large-scale organizational computing.
- ORGs are structured by organizational commitment [37–39], which is a special case of physical commitment that is defined as being information pledged.
- In many cases, humans will take part in the operation of an ORG. For example, in a credit card ORG, a particular credit decision may be reviewed by a human before being decided.

It is argued that large-scale organizational computing requires reflection and strong paraconsistency [39, 50] for organizational practices, policies, and norms:

- Norms express how systems can be used and tested in practice.
- Policies express overarching justification for systems and their technologies.
- Practices express implementations of systems.

Reflection is required in order that the practices, policies, and norms can mutually refer to each other and make inferences. To this end, it is argued that [36] direct logic is a powerful inference system for large-scale organizational computing that includes unstratified reflection and strong paraconsistency.

Issues remain for the community about how openness, hierarchy, sociality, evolution, human–machine cooperation, and ubiquitous intelligence can be further explored. The next task is to consider how the principles, methods, practices, and findings identified in organizational study can be transferred to the research on open complex systems.

4.9 Behavior Computing

Existing management information systems and enterprise applications do not support the storage of behaviors very well. The entity of physical or social behavior is usually decomposed to multiple transactions without protecting the semantics and complete behavior journey. Behavior as a very soft buzzword and is widely used without a clear definition and systematic representation. Such "implicit" behavior in transactional data is not consistent with the "explicit" semantic existence in business. Hence, it is crucial to develop computing techniques for the explicit and in-depth quantification and informatics of behaviors.

With the concept of behavior and the introduction of an abstract behavior model, the representation, modeling, data analysis and mining, and learning and decision-making of behaviors are becoming doable and increasingly *useful*, *essential*, yet *challenging* in ubiquitous behavioral applications and problem-solving. They form a new computing opportunity, necessity, and technological innovation which we refer to as *behavior computing* or *behavior informatics* [32, 33, 40–42], for the explicit and in-depth understanding and analysis of genuine behavior-oriented actions, operations, and events associated with many challenging business problems.

Behavior computing, or behavior informatics, consists of methodologies, techniques, and practical tools for representing, modeling, analyzing, learning, discovering, and utilizing human, organismal, organizational, societal, artificial, and virtual behaviors, behavior interactions and relationships, behavioral networks, behavioral patterns, and behavioral impacts, and for forming and decomposing behavior-oriented groups and collective intelligence and the emergence of deep behavioral intelligence in conjunction with their environments.

Behavior computing contributes to the in-depth understanding, discovery, applications, services, and management of behavior intelligence. In more detail, behavior computing addresses the following key aspects, as shown in Fig. 4.1.

- Extracting behavioral data: In preparing behavioral data, behavioral elements hidden or dispersed in transactional data need to be extracted and connected, then converted and mapped into a behavior-oriented feature space, otherwise called a *behavioral feature space*. In the behavioral feature space, behavioral elements are presented in behavioral itemsets; hence it is necessary to map and convert transactional data to behavioral data.
- Representing and modeling behavior: This involves developing behavior-oriented specifications for describing behavioral elements and the relationships between the elements. The specifications reshape the behavioral elements to suit the presentation and construction of behavioral sequences. Behavioral modeling

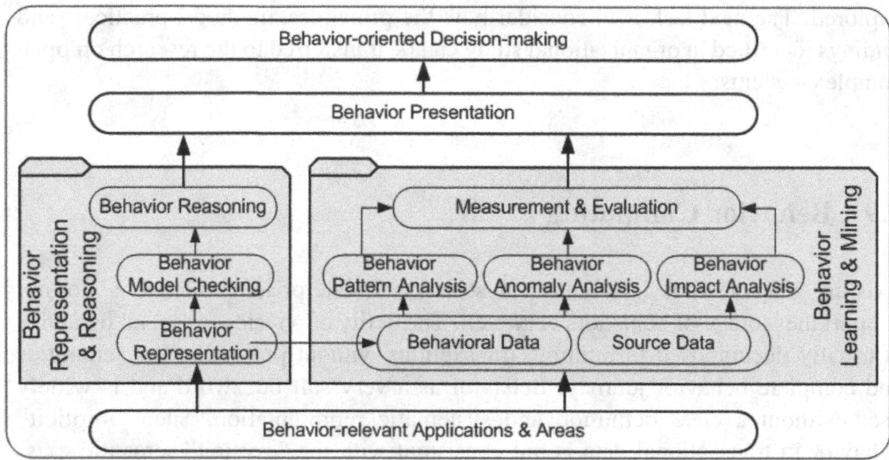

Fig. 4.1 Behavior computing research map

also provides a unified mechanism for describing and presenting behavioral elements, impacts, and patterns.

- Analyzing behavioral impact: In analyzing behavioral data, one may be particularly interested in those behavioral instances that are associated with a high impact on business processes and/or outcomes. Behavioral impact analysis [41] features the modeling of behavioral impact.
- Discovering behavioral patterns: There are in general two methods of behavioral pattern analysis. One is to discover behavioral patterns without consideration of the behavioral impact, and the other is to analyze the relationships between behavior sequences and particular types of impact.
- Emerging behavioral intelligence: To understand behavioral impacts and patterns, it is important to scrutinize behavioral occurrences, evolution, and life cycles, as well as the impact of particular behavioral rules and patterns on behavioral evolution and the emergence of intelligence. An important task in behavioral modeling is to define and model behavioral rules, protocols, and relationships and their impact on behavioral evolution and intelligence emergence.
- Understanding behavioral networking: Multiple sources of behavior may form a type of behavioral network.
- Particular human behavior is normally embedded into such a network to fulfill its roles and affect a particular situation. Behavioral network analysis aims to understand the intrinsic mechanisms inside a network; for instance, behavioral rules, interaction protocols, convergence and divergence of associated behavioral itemsets, as well as their effects such as network topological structures, linkage relationships, and impact dynamics.
- Simulating behaviors: To understand all the above mechanisms that may exist in behavioral data, simulation plays an important role in enabling the observation of dynamics, the impact of rules/protocols/patterns, the emergence of behavioral intelligence, and the formation and dynamics of a social behavioral network.

- Presenting behaviors: From analytical and business intelligence perspectives, behavioral presentation aims to explore the presentation means and tools that effectively describe the motivation and interest of stakeholders in the particular behavioral data. In addition to the traditional presentation of patterns, such as associations, visual behavioral presentation is a major research topic. It is of great interest to analyze behavioral patterns in a visual manner.

The above tasks form a clear field structure and research map of behavior computing. A generic process of computing behaviors will complement classic approaches toward a more comprehensive and in-depth understanding of behavior and problem-solving. A business application first converts entity relationship-oriented transactional data to behavior feature-oriented data through behavioral modeling. Behavioral patterns, exceptions, dynamics, and impacts are then analyzed by the development of corresponding behavior-based analytic and learning methods. The outcomes are then presented as behavioral patterns, rules, or visual diagrams, and/or transformed into decision-support business rules to disclose the interior driving forces and causes of business problems and their impact.

Behavior is becoming an increasingly important asset to be deeply analyzed and understood to access its explicit and implicit business value and semantic gain which cannot be achieved solely by the transactional data that is usually recorded. The deep values and prospects from computing behaviors may be through:

- Fully disclosing and utilizing the *behavior semantics* which are usually destroyed in recorded transactions and overlooked in behavior analysis
- Fully and deeply exploring the *behavior sequences* and *behavior matrix* of an actor or a group along a certain time period, in which behavior properties are involved
- Deeply engaging and learning the explicit and especially hidden *social relationships* governing behavior formation, structuring, networking, evolution, and emergence of behavior intelligence
- Deeply discovering behavioral patterns, exceptions, relational patterns, and changes of individuals, groups, or the global population against behavior formation, evolution, and revolution
- Effectively capturing and quantifying the relationships between behaviors, behavior evolution, and their impacts as well as measuring the *impact and performance* of behaviors and behavior dynamics on business objectives
- Deeply understanding the *belief, desire, and intent* behind behaviors conducted and the impact caused
- Actively detecting, predicting early, and intervening in *unexpected behaviors* of individuals, groups, or cohorts to convert them to the expected direction and impact

To access the above prospects, cross-disciplinary efforts need to be made. In addition to informatics and analytics, the theories, methodologies, and tools available in statistics, mathematics, econometrics, marketing, psychology, social sciences, behavioral sciences, behavioral finance, etc. are very necessary. This requires collaboration between disciplines and cross-domain experts as shown in Fig. 4.2.

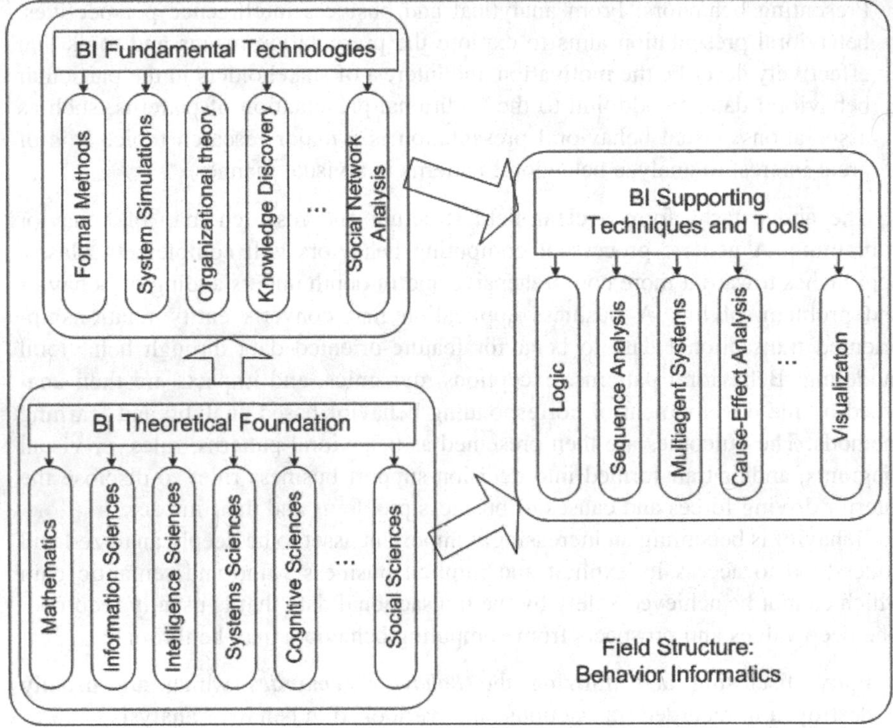

Fig. 4.2 Behavior computing field structure

Unlimited opportunities exist in deep behavior computing in terms of complementing existing behavior analysis, data analysis, event detection, behavior economics, and cognitive study toward data-driven, semantics-oriented, and process-based quantification and formalization of exactly what takes place in the real world. Many application areas [40, 42] from traditional to emergent issues can benefit from it; for instance, exploring the patterns, anomalies, sequencing and intent that drives customer behaviors in retail and online shopping businesses, Web usage and interactions on the Internet, trading behaviors in capital markets, and exceptional activities captured on surveillance systems.

4.10 Social Computing

Social computing has emerged as a recent interest that merges social science with computing science and harnesses the power of computational methods to study social entities, groups, and their behaviors within a social context. Social computing supports the acquisition, representation, processing, analysis, discovery, use, management, and dissemination of information that is distributed across social collectivities such as teams, communities, organizations, and markets.

Naturally, as social entities and collectivities are closely related to behaviors, economy, and culture, social computing is also related to behavior computing, economic computing, and sociocultural computing. *Behavioral, economic, and sociocultural computing* (BESC) aim to develop techniques that study effective methodologies, techniques, and tools for:

- Deeply understanding the cognitive and psychological capacities of human beings, as well as cultural, economic, social, political, legal, online, social media, spatial, environmental, and biological factors, systems, organizations, and institutions
- Effectively representing, modeling, analyzing, simulating, evaluating, and managing behavioral, economic, social, and cultural phenomena, factors, characteristics, dynamics, evolution, consequence, and knowledge
- Developing quantitative and computational methodologies, approaches, algorithms, tools, and systems to gain a deeper understanding of dynamics and evolution, relations and interactions, structures and distributions, rules, patterns and exceptions, effects and consequences, and intelligence and emergence in individual and collective BESC entities, systems, and organizations
- Delivering deep modeling, analysis, and learning of social concepts, including trust, credibility, privacy, and influence, and intrinsic behavior/economic/sociocultural patterns and impacts
- Showcasing the intellectual, value, process, and impact contexts that govern the development and use of science and technology in various BESC applications and domains

Accordingly, BESC focuses on developing interdisciplinary methodologies and approaches that can address system complexities including sociality, hierarchy, interactions, and evolution between subsystems and system components that exist at multiple levels of analysis. More specifically, the focus is on the following areas:

- Computational foundations: including statistical/mathematical/logic/evolutionary/formal methodologies, models, and theories for engineering, representing, visualizing, analyzing, modeling, evaluating, and managing cognitive, behavioral, economic, political, online, social media-based, and sociocultural phenomena, factors, interactions, relations, structures, contexts and situations, systems and organizations and for understanding their formation and evolution, dynamics, self-organization, mobility, learning and adaptation, change and emergence, influence, conventions, trust, norms, protocols, roles, responsibilities, utility, effects, privacy, risk, security, and decision-making and formation processes
- Basic issues: such as group interaction and collaboration, group formation and evolution, group representation and profiling, collective action and governance, cultural patterns and representation, social conventions and social contexts,

influence process and recognition, public opinion representation, viral marketing and information diffusion, and psychocultural situation awareness

- Methodological issues: such as mathematical foundations, verification and validation, sensitivity analysis, matching techniques or methods to research questions, metrics and evaluation, methodological innovation, model federation and integration, evolutionary computing, and network analysis and optimization
- Techniques and approaches: including specific tools, algorithms, models, systems, evaluation metrics for acquisition, extraction, detection, summarization, knowledge representation, visualization, classification, regression, clustering, pattern discovery, factor/event/effect analysis, reasoning, relation/structure/distribution analysis, collaborations/conflicts/negotiation mechanisms and crowding/community formation/grouping and diffusion understanding of BESC-related systems and organizations and for collecting, analyzing, managing, and utilizing BESC-related networking, textural, imaging, sequential, spatiotemporal, multimedia, and graphic and streaming data
- Applications and systems: including innovative use and showcases, simulation and visualization of BESC problems not limited to networking, socialization, globalization, exception and risk management, crisis analysis and management, security, compliance, recommendation, retrieval and search, innovation, and decision support

Typical social computing areas also include social network, media, signal, and service analysis:

- Social networks, media, and services: social networks, semantic Web, mobile social, social media analytics and social media intelligence, service science, quality, architecture, management, tools and case studies, trust, privacy, risk and security in social contexts, social networks/media/service system design, and architectures
- Social signal processing: social intelligence and social cognition, social behavioral modeling, analysis and synthesis, emotional intelligence, cultural dynamics, opinion representation, influence process, reality mining, social signal processing in human–computer interaction, user modeling, interface design, robot cognition and behavior control, mining, learning, and searching social signals
- Applications: collaborative filtering, bookmarking, tagging, and multiagent systems; user-generated contents, blogs, wikis, discussions, etc.; socially adapt interfaces, implicit (human behavior-based) tagging, reality mining systems, and social signal processing system design and architecture

The study of the above issues will hopefully greatly assist in discovering social intelligence, behavior intelligence, organizational intelligence, and relevant social data intelligence for an in-depth understanding of complex systems, especially open complex social systems and problems.

4.11 Cloud/Service Computing

Cloud computing [51–56] forms a distributed computing environment over a network such as the Internet by connecting a large number of computers. An application or computational task is executed on many connected computers at the same time. Cloud computing is further characterized as a number of "services," which are further categorized in terms of software, platform, and infrastructure, and which form major models known as *software as a service*, *platform as a service*, and *infrastructure as a service*. These cloud services may be offered in a public, private, or hybrid network.

Here "cloud" essentially refers to the Internet, and users do not need to know where the computers and applications are on the basis of converged infrastructure and shared services. As the services are essentially run on virtual servers that are transparent to users, such services and servers can be moved around, and scaled up or down on demand or on the fly. Such a process does not affect end-user use, and end users follow a "pay as you go" model to use cloud services.

Cloud computing is built on many existing enabling technologies, such as distributed computing, converged infrastructure, shared services, virtualization, autonomic computing, grid computing, and service-oriented architecture. It provides different resources (software, infrastructure, platform) in terms of measured services and satisfies certain QoS (quality of service) and reliability requirements.

Cloud computing exhibits typical characteristics and built-in functions, including device and location independence, virtualization, multi-tenancy, reliability, scalability and elasticity, security, and application programming interfaces. The National Institute of Standards and Technology (NIST) defines five essential characteristics: on-demand self-service, broad network access, resource pooling, rapid elasticity, and measured service.

Cloud computing systems consist of cloud platform, cloud infrastructure, cloud service, and cloud storage. Cloud computing consumers use predefined configurations, and cloud templates obtain, configure, and deploy cloud services themselves and move applications between clouds through a self-service portal. Networked client devices, such as desktop computers, laptops, tablets, and smartphones, can access services to controlled cloud, private cloud, public cloud, community cloud, or hybrid cloud, based on the permission and security settings.

The common business models of cloud computing include *infrastructure as a service* (IaaS), *platform as a service* (PaaS), and *software as a service* (SaaS). IaaS is the most basic service. PaaS and SaaS rely on IaaS, and SaaS further builds on PaaS. Nowadays, new models have moved toward anything as a service (XaaS), such as strategy as a service, collaboration as a service, business process as a service, database as a service, *network as a service* (NaaS), *communication as a service* (CaaS), and *security as a service* (SECaaS).

Cloud/service computing provides very powerful techniques and tools for supporting the integration of computational resources, connecting to domain experts and necessary information and data anywhere at any time, in a network, essentially the Internet. This setting provides a unique and extremely powerful mechanism for tackling open complex giant systems.

4.12 Metasynthetic Computing

To handle open complex giant systems, opportunities provided by different computing theories and tools need to be carried forward and merged to address respective system complexities. In this book, the theory of metasynthetic engineering proposed by Qian [43–45, 57–61] for solving open complex giant systems is discussed from the computing perspective. It forms the system of metasynthetic computing [60] which will be discussed in Chap. 5.

In Chap. 6, we will introduce an organized framework for abstracting, analyzing, and designing open complex systems by following an organization and agent service-oriented method. Visual and formal modeling methods and tools will then be discussed in Chaps. 7 and 8. To implement the organized framework, Chaps. 10 and 11 will discuss tools for the architectural design and detailed design of complex systems.

References

1. Booch, G.: Object-Oriented Analysis and Design with Applications, 2nd edn. Addison-Wesley, Reading (1995)
2. Norris, M., Davis, R., Pengelly, A.: Component-Based Network Systems Engineering. Artech House, Norwood (2000)
3. Erl, T.: Service-Oriented Architecture: A Field Guide to Integrating XML and Web Services. Pearson Education, Upper Saddle River (2004)
4. Iglesias, C., Garijo, M., Gonzales, J.: A survey of agent-oriented methodologies. In: Muller, J., Singh, M., Rao, M. (eds.) Intelligent Agents IV: Agent Theories, Architectures, and Languages. LNAI-1555, pp. 317–330. Springer, Berlin/Heidelberg/New York (1999)
5. Martin, J., Odell, J.J.: Object-Oriented Analysis and Design. Prentice Hall, Englewood Cliffs (1992)
6. Microsoft Corporation. Definition of the term component; http://msdn.microsoft.com/repository/OIM/resdkdefinitionofthetermcomponent.asp
7. Sparling, M.: Lessons learned through six years of component-based development. Commun. ACM **43**(10), 47–53 (2000)
8. Deitel, H.M., et al.: Web Services: A Technical Introduction. Prentice Hall, Englewood Cliffs (2003)
9. Wooldridge, M.: Reasoning About Rational Agents. MIT Press, Cambridge, MA (2000)
10. Wooldridge, M., Jennings, N.R.: Intelligent agents: theory and practice. Knowl. Eng. Rev. **10**(2), 115–152 (1995)
11. Mylopoulos, J., Chung, L., Yu, E.: From object-oriented to goal-oriented requirements analysis. Commun. ACM **42**(1), 31–37 (1999)
12. van Lamsweerde, A.: Goal-oriented requirements engineering: a guided tour. In: Proceedings of the 5th IEEE International Symposium on Requirements Engineering, RE'01, Toronto, pp. 249–263 (2001)
13. Castro, J., Kolp, M., Mylopoulos, J.: Towards requirements-driven information systems engineering: the Tropos project. Inf. Syst. **27**(6), 365–389 (2002)
14. Dardenne, A., Lamsweerde, V.A., Fickas, S.: Goal-directed requirements acquisition. Sci. Comput. Program. **20**, 3–50 (1993)
15. Mylopoulos, J., Chung, L., Yu, E.: From object-oriented to goal-oriented requirements analysis. Commun. ACM **42**(1), 31–37 (1999)

16. Yu, E.S.K.: Modeling organizations for information systems requirements engineering. In: Proceedings of the 1st IEEE International Symposium on Requirements Engineering (RE'93). San Diego, pp. 34–41 (1993)
17. Yu, E.: Towards modeling and reasoning support for early-phase requirements engineering. In: Proceedings of the 3rd IEEE International Symposium on Requirements Engineering (RE'97). Annapolis, pp. 226–235 (1997)
18. Kolp, M., Giorgini, P., Mylopoulos, J.: A goal-based organizational perspective on multiagent architectures. In: Intelligent Agents VIII: Agent Theories, Architectures, and Languages. LNAI-2333, pp. 128–140. Springer, New York (2002)
19. Greenspan, S., Borgida, A., Mylopoulos, J.: A requirements modeling language and its logic. Inform. Syst. **11**(1), 9–23 (1986)
20. Bresciani, P., Perini, A., Giorgini, P., Giunchiglia, F., Mylopoulos, J.: A knowledge level software engineering methodology for agent oriented programming. In: Proceedings of the 5th International Conference on Autonomous Agents, pp. 648–655. ACM, New York (2001)
21. Koning, J.L., Romero-Hernandez, I.: Generating machine processable representations of textual representations of AUML. In: AOSE 2002, Bologna, Italy (2002)
22. Perini, A., Pistore, M., Roveri, M., Susi, A.: Agent-oriented modeling by interleaving formal and informal specification. In: AOSE 2003, Melbourne, Australia (2003)
23. Gans, G., Jarke, M., Kethers, S., Lakemeyer, G.: Modeling the impact of trust and distrust in agent networks. To appear in Proceedings of the 3rd International Bi-Conference Workshop on Agent-Oriented Information Systems (AOIS-2001), Interlaken (2001)
24. Wang, X.Y., Lespérance, Y.: Agent-Oriented Requirements Engineering Using ConGolog and i. In: AOIS 2001, Interlaken, Switzerland (2001)
25. Giacomo, D., Lespérance, G., Levesque, Y., ConGolog, H.J.: A concurrent programming language based on the situation calculus. Artif. Intell. **121**, 109–169 (2000)
26. Shehory, O., Sturm, A.: Evaluation of modeling techniques for agent-based systems. In: Proceedings of the 5th International Conference on Autonomous Agents, pp. 624–631. ACM, New York (2001)
27. Wood, M., Deloach, S.A., Sparkman, C.: Multiagent systems engineering. Int. J. Softw. Eng. Knowl. Eng. **11**(3), 231–258 (2001)
28. Caire, G., Coulier, W., Garijo, F., Gomez, J., Pavon, J., Leal, F., Chainho, P., Kearney, P., Stark, J., Evans, R., Massonet, P.: Agent-oriented analysis using message/uml. In: Proceedings of the 2nd International Workshop on Agent-Oriented Software Engineering. LNCS-2222, pp. 119–135. Springer, New York (2002)
29. Wooldridge, M., Jennings, N.R., Kinny, D.: The GAIA methodology for agent-oriented analysis and design. J. Autonom. Agents Multi-Agent Syst **3**(3), 285–312 (2000)
30. Zambonelli, F., Jennings, N.R., Wooldridge, M.: Developing multiagent systems: the GAIA methodology. ACM Trans. Softw. Eng. Methodol. **12**(3), 317–370 (2003)
31. Zambonelli, F., Jennings, N.R., Wooldridge, M.: Organisational abstractions for the analysis and design of multi-agent systems. In: Proceedings of the 1st International Workshop on Agent-Oriented Software Engineering, Limerick, pp. 127–141 (2000)
32. Cao, L.: In-depth behavior understanding and use: the behavior informatics approach. Inf. Sci. **180**(17), 3067–3085 (2010)
33. Cao, L., Zhao, Y., Zhang, C.: Mining impact-targeted activity patterns in imbalanced data. IEEE Trans. Knowl. Data Eng. **20**(8), 1053–1066 (2008)
34. Brazier, F.M.T., Jonker, C.M., Treur, J.: Principles of component-based design of intelligent agents. Data Knowl. Eng. **41**(1), 1–27 (2002)
35. Wooldridge, M.: An Introduction to Multiagent Systems. Wiley, Chichester (2002)
36. Hewitt, C.: Large-scale organizational computing requires unstratified reflection and strong paraconsistency. In: Sichman, J.S. et al. (eds.) Coordination, Organizations, Institutions, and Norms in Agent Systems III. LNCS-4870, pp. 110–124, Springer, Berlin/Heidelberg (2008)
37. Jennings, N.: Commitments and conventions: the foundation of coordination in multi-agent systems. Knowl. Eng. Rev. **8**(3), 223–250 (1993)
38. Noriega, P.: Agent mediated auctions: the fishmarket metaphor. Ph.D., Universitat Autonoma de Barcelona (1997)

39. Singh, M., Huhns, M.: Service-Oriented Computing: Semantics, Processes Agents. Wiley, Chichester (2005)
40. Cao, L., Motoda, H., Srivastava, J., Lim, E., King, I., Yu, P.S., Nejdl, W., Xu, G., Li, G., Zhang, Y. (eds.): Behavior and Social Computing. LNCS-8178. Springer International, Switzerland (2013)
41. Cao, L., Ou, Y., Yu, P.S.: Coupled behavior analysis with applications. IEEE Trans. Knowl. Data Eng. **24**(8), 1378–1392 (2012)
42. Cao, L., Yu, P.S. (eds.): Behavior Computing: Modeling, Analysis, Mining and Decision. Springer, London (2012)
43. Qian, X., Yu, J., Dai, R.: A new scientific field–open complex giant systems and the methodology (in Chinese). Chin. J. Nat. **13**(1), 3–10 (1990)
44. Qian, X.: Revisiting issues on open complex giant systems (in Chinese). Pattern Recogn. Artif. Intell. **4**(1), 5–8 (1991)
45. Qian, X.: Building Systematology (in Chinese). Shanghai Jiaotong University Press, Shanghai PRC (2007)

Chapter 5
Metasynthesis

5.1 Introduction

The problems and systems we tackle in our daily business are becoming increasingly complex, and as a consequence, existing methodologies and tools are similarly confronted by heightened challenges.

5.2 Open Complex Giant Systems

Complex systems have been recognized as one of the greatest challenges in science and technology, both now and in the future [1–5]. They seriously affect the future of systems, men, and cybernetics [6]. As a result, a new scientific field, namely, the *science of complexity* [7, 8], has emerged which focuses on the studies of complex systems. This is also evidenced by many emergent research centers for complex systems. A very special part of the complex system family, *open complex giant systems* (OCGS) [2, 3, 9–11] were proposed as a new field. Their typical instance is the Internet [12]. The Internet demonstrates system complexities that are *open* through interactions [13] with the environment; *giant,* consisting of billions of hyperlinks, transactions, and surfers from every corner of the world; *dynamic,* with a speed of evolution beyond our imagination; *adaptive* [14] to problem-solving and consensus building; *uncertain* about the current state and next step; and *societal* [15], involving humans, communities, and organizations [16] with a variety of cultures, traditions, religions, politics, laws, policies, and social norms.

Problem-solving within OCGS is very challenging due to the intrinsic system complexities of such systems. In fact, many of them are unrecognized or unperceived, for instance, how collective intelligence can emerge from the interaction between a large variety of system components. Furthermore, from the perspective of OCGS

L. Cao, *Metasynthetic Computing and Engineering of Complex Systems*,
Advanced Information and Knowledge Processing,
DOI 10.1007/978-1-4471-6551-4_5

problem-solving philosophy, we need to consider the cooperation between human beings and systems and study what roles human beings can play to better handle OCGS.

In general, the history of human social activity, literature aggregation, and exploration in less complex problems has provided us with effective methodologies, philosophies, and technologies which guide us toward understanding an unrecognized and unperceived problem step by step. An empirical methodological conclusion from such efforts is the establishment of a new field of science: *open complex giant systems* and its methodology, *qualitative-to-quantitative metasynthesis*, proposed by a group of distinguished Chinese scientists in the 1990s [2, 3, 9, 11].

The proposition of qualitative-to-quantitative metasynthesis results from a number of critical challenges in dealing with OCGS, real-life giant engineering experiences, multidisciplinary complementation, and creative thinking and cognition in building a modern scientific and technological system. We interpret them below:

1. The system complexities of OCGS challenge the traditional problem-solving methodologies, which have been designed mainly for less complex systems. One therefore realizes the importance of involving human intelligence and the interaction between humans and machines in the problem-solving.
2. The establishment of metasynthesis benefits from the long-term and real-life experiences of the theory founders who have been engaged in leading and conducting many giant technical and social system engineering projects, such as designing the a-bomb and h-bomb. Such experiences deliver invaluable lessons about integrating relevant data, information, human groups, and machinery systems into the problem-solving process.
3. The formation of metasynthesis profits from the interaction and complementation among experts from multiple disciplines. As a result, the problem-solving methodology has to involve intrinsic knowledge and experiences from many domain experts in relevant areas.
4. The initiators of metasynthesis have made great advances in building modern scientific and technological systems [17] and, in particular, creating breakthroughs and foundations for the cognitive sciences and systems science. Through these efforts, they have realized that the problem-solving methodology for a new and challenging field involves three layers of a scientific field, namely, basic research, technological research, and engineering technologies.

The principal idea of *qualitative-to-quantitative metasynthesis* (herein, *metasynthesis*) is shown in Fig. 5.1. Due to system complexities that are beyond an individual's capability to handle, domain experts are often invited to discuss possible solutions and the directions to take for any given OCGS problem.

Experts collect knowledge and experiences, and initiate a preliminary understanding of, and solution to, the problem. This understanding is likely to be empirical, incomplete, imprecise, and even biased. Such *qualitative* understanding is the starting state of our cognition of the problem. Following the theory of metasynthesis, relevant experts are invited to join in an online interaction space (called *metasynthesis-space*, or *M-space* for short, see more in Sect. 5.5) to discuss

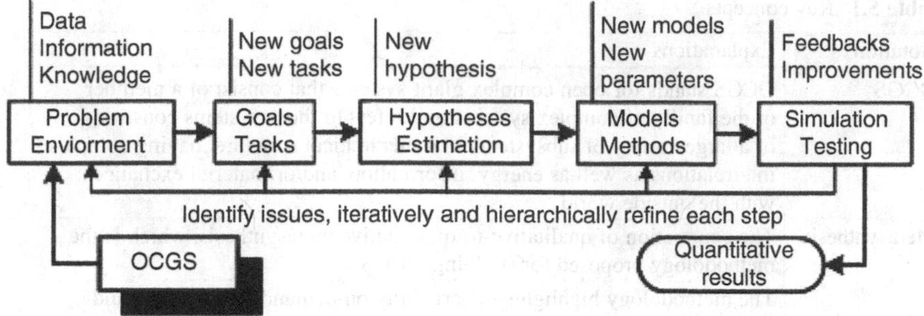

Fig. 5.1 An OCGS problem-solving methodology: qualitative-to-quantitative metasynthesis

problem-solving solutions. Experts utilize their knowledge, experience, tools, and resources collected for this M-space to further deepen their understanding.

Significantly, an expert in one domain can exchange discussion, argument, or negotiation with one or more experts in other domains by knowledge exchange, integration, and fusion.

The interaction is likely to be iterative. During each iteration, interaction between domain experts may trigger new understanding and possibly creative cognition of the underlying problem. For each such iteration, new objectives, approaches and methods, data and resources may emerge, as well as progress in developing periodically consequent models, methods, and knowledge about the system and problem-solving. Up to a certain stage, such periodical models, methods, and knowledge can be either partially or completely represented in quantitative definitions, theorems, formulae, equations, and variables. If these can be further coded into computerized languages and systems, we are ready to convert the original qualitative understanding to semiquantitative understanding and finally up to full-quantitative knowledge and systems about the underlying problems and problem-solving methods. The understanding of an OCGS is therefore an intelligence emergence in a qualitative-to-quantitative process based on social cognitive interaction.

The theory of metasynthesis actually reflects the working mechanism of social cognitive interaction between many relevant domain experts. It thereafter advocates the development of corresponding intelligent information processing and systems to support such working mechanisms. From the perspective of social cognitive interaction, we briefly discuss the principles of metasynthesis-based problem-solving for dealing with OCGS in this work. We summarize the framework and process of social cognitive interaction-based metasynthesis and intelligence emergence for problem-solving.

The principles we discuss are critical in handling the system complexities of OCGS, which are among the greatest challenges in current systems sciences and cognitive sciences [18].

Table 5.1 lists key concepts and their abbreviations as used in this chapter. Figure 5.2 shows their relationships. The three key components, M-space,

Table 5.1 Key concepts

Notations	Explanations
OCGS	OCGS stands for open complex giant systems that consist of a member of the family of complex systems and refers to those systems consisting of a large variety of subsystems in a hierarchical structure, having complex interrelations as well as energy, information, and/or material exchange with the outside world
Metasynthesis	The contraction of qualitative-to-quantitative metasynthesis, which is the methodology proposed for studying OCGS
	The methodology highlights the crucially on-demand involvement and seamless synergy of the relevant expert group, data, information, knowledge, and computer systems, as well as the scientific theory of various disciplines, and human experience and knowledge. This creates a system in itself. The methodology was originally called a *metasynthetic engineering* method because of the technical perspective. As a result of the involvement and significant role of social intelligence in OCGS problem-solving, it can be considered as *metasynthetic social intelligence engineering*
M-interaction	M-interaction is the short form of *metasynthesis interaction*, which is the problem-solving mechanism of metasynthesis based problem-solving. It describes the activities of human–computer interaction, human–human interaction, and computer–computer communications in M-space following the theory of metasynthesis
M-space	M-space is the short form of *metasynthesis-space*, which is a problem-solving system for handling OCGS in terms of the metasynthesis methodology. Such a system looks like a workshop hall for metasynthesis social intelligence engineering (its original translated English term is *hall for workshop of metasynthetic engineering* (HWME)). With the rapid development of Internet technologies, HWME can be built as a *Cyberspace for Workshop of Metasynthetic Social Intelligence Engineering*; nowadays, an effective option for a practical system is to combine both physical halls and cyberspaces into a *M-space*
M-computing	M-computing is the short form of *metasynthetic computing*. It is an approach to the analysis, design, and implementation of OCGS which facilitates the study and development of M-spaces following qualitative-to-quantitative metasynthesis

M-interaction, and M-computing consist of a systematic framework for instantiating the theory of metasynthesis in handling the OCGS problem-solving. An M-space is the OCGS problem-solving system following the theory of qualitative-to-quantitative metasynthesis. The working mechanism of M-space is M-interaction. M-interaction involves interactions such as human–machine interaction, machine–machine interaction, and human–human interaction as needed during the problem-solving.

The above interactions may involve social cognitive interaction, during which social cognitive intelligence emerges. Such social cognitive intelligence plays a significant role in OCGS problem-solving. Through M-interaction, system components in M-space interact with each other for OCGS problem-solving.

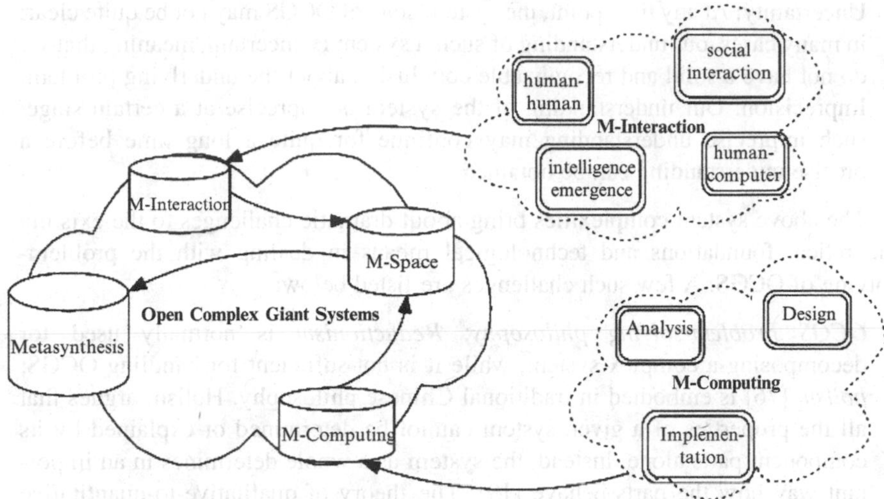

Fig. 5.2 A concept map

The implementation of M-space and M-interaction relies on the techniques in M-computing. M-computing consists of engineering approaches to the analysis, design, and implementation of M-space and M-interaction. We will explain the above concepts further in subsequent sections.

5.3 OCGS System Complexities

The system complexities of OCGS consist of *openness*, *giant scale*, *hierarchy*, *human involvement*, *societal characteristics*, *dynamic characteristics*, *uncertainty*, and *imprecision*. They are introduced as follows:

- Openness: An OCGS exchanges energy, information, and materials with its external environment.
- Giant scale: An OCGS is composed of hundreds or even millions of system constituents and components.
- Hierarchy: There are usually many levels in an OCGS. In some cases, the number of levels is unknown. It may consist of many sub-OCGS, which may further include sub-sub-OCGS.
- Human involvement: Relevant human beings are an essential constituent of an OCGS.
- Societal characteristics: Many social factors such as laws, politics, organizational factors, and business processes are embedded in an OCGS.
- Dynamic characteristics: OCGS is dynamic in the sense that it may change its states, working mechanism, constituents, and internal and external interaction mechanism at any time, often beyond one's imagination.

- Uncertainty: At any time point, the system state of OCGS may not be quite clear; in many cases, our understanding of such a system is uncertain, meaning that we do not have a solid and recognizable conclusion about the underlying problem.
- Imprecision: Our understanding of the system is imprecise at a certain stage; such imprecise understanding may continue for quite a long time before a precise understanding can be obtained.

The above system complexities bring about dramatic challenges to the existing theoretical foundations and technological means in dealing with the problem-solving of OCGS. A few such challenges are listed below:

- *OCGS problem-solving philosophy*: *Reductionism* is normally used for decomposing a complex system, while it is not sufficient for handling OCGS; *holism* [76] is embodied in traditional Chinese philosophy. Holism argues that all the properties of a given system cannot be determined or explained by its component parts alone. Instead, the system as a whole determines in an important way how the parts behave [19]. The theory of qualitative-to-quantitative metasynthesis advocates the combination of reductionism with holism and builds up so-called systematism [20] as the methodological philosophy.
- *Human–machine relationship*: Traditionally, we tend to build machine-centered systems such as automated systems. With the increase of system complexities, one realizes the importance of human–machine interaction, even though this is insufficient for handling OCGS. Due to these intrinsic complexities, an OCGS problem-solving system, namely, an M-space, consists of both human beings (a group of domain experts) and machine components. Both parties help and collaborate with each other, but human beings are in control of the problem-solving. We call this "a human–machine-cooperative human-centered mode" [1].
- *The power of ubiquitous intelligence*: In handling OCGS, ubiquitous intelligence such as human qualitative intelligence, machine quantitative intelligence, social intelligence, domain intelligence, and network intelligence are all involved and play different but essential roles [21, 22] in the problem-solving. The problem-solving is a process of multiple types of intelligence interactions and emergences.
- *Collective intelligence and social cognitive interaction*: In the problem-solving of OCGS, a collection of experienced domain experts and their effective interactions are essential; this involves the working mechanisms for social cognitive interactions, group expert-based problem-solving, and the development and emergence mechanisms of social intelligence systems.
- *OCGS problem-solving methodologies*: How can a problem-solving system for OCGS problems be built? The answer is to build an M-space. Then, what is an M-space?
- *Dynamic system theories*: As a problem, OCGS is dynamic.
- As a problem-solving system, an M-space is also dynamic. We need to study new dynamic system theories for such systems and describe the dynamics of system goals, organizational relationships, interaction modes, and system states.

From the engineering perspective, we also face many challenges, for example:

- *Large scale of system simulation and modeling*: It is essential to develop simulation tools, languages, and evaluation systems to simulate the working mechanism for a large scale of social intelligence emergence and group expert interaction-based problem-solving.
- *Large scale of system analysis and design methods*: There is a need for a large scale of system analysis and design methodologies, tools, and evaluation systems.
- *Human-centered computing*: How can domain experts be supported to take the leading problem-solving role in an M-space? How can the dynamic human–machine task allocation and cooperation be supported?
- *Integration of computing paradigms*: To analyze, design, and implement OCGS, it is necessary to combine many types of computing paradigms. How should they be integrated? An appropriate guideline is to follow the metasynthesis theory, but how can this be implemented? A solution for engineering OCGS is to develop M-computing techniques that integrate relevant computing paradigms, but how?
- *Knowledge science, engineering, and management*: How can domain knowledge, ad hoc knowledge, meta-knowledge, and knowledge from data and information be captured, represented, transformed, discovered, and used?
- *Online M-space infrastructure*: Distributed M-space is necessary because of the wide involvement of problem-solving experts, resources, and tools. How, then, should such an online M-space be built?

In the following sections, we try to report the lessons and understanding learned in the development of an M-space for OCGS problem-solving from the perspective of social cognitive interactions.

5.4 Knowledge and Intelligence Emergence

A typical system feature in open complex systems is the emergence of knowledge and intelligence [77, 78]. Knowledge and intelligence emergence is regarded as the built-in working mechanism behind and resultant effects of interactions between diverse system complexities, as discussed in the section above. It differentiates complex systems from simple intelligent systems. Below, the examples and discussions in Cui (2006) [77, 78] are used to explain the above argument by elaborating the working process in a hall for workshop of metasynthetic engineering (HWME), which is claimed to be one of the most effective problem-solving systems for handling open complex giant systems [9, 17, 18, 20, 24, 31], to address typical questions including:

- Where does the knowledge come from in a HWME?
- What is the process of production, evolution, and transferral in a HWME?

In an HWME, many domain experts from different fields and domains are involved in the entire problem-solving and decision-making process. They follow certain seminar rules and norms [31–33] and interact, cooperate, negotiate, debate, and compromise with each other. The problem-solving and decision formation process involves diverse intelligence, including data and resources, domain knowledge, organizational and social factors, network/Web infrastructure and resources, and human qualitative intelligence. An ecosystem is formed within and outside an HWME. Such an ecosystem works like an artificial society, having the goal to understand and solve the underlying problem, through human–machine cooperation and metasynthetic engineering in which humans may take the leading roles, by following certain social constraints including social norms and behavioral rules.

Knowledge emerges from the interactions with information and data resources, and from the thinking exchanged within the expert group. These interactions may take place:

- Between different information providers and sources
- Between different thoughts shared in the group
- Between subsystems and system elements
- Between the system and its environment
- Between human thinking and machine reasoning and simulation

In addition, knowledge emerges from the interactions between different types of knowledge in the HWME and the underlying problem:

- Knowledge at different levels
- Knowledge in different domains and on different topics
- Implicit and explicit knowledge
- Facts and connections between facts
- Certain and uncertain knowledge
- Qualitative and quantitative knowledge
- Low-level and high-level knowledge
- Local and global knowledge

Knowledge is the product of thinking, discovery, learning, and reasoning. Different types of thinking and knowledge production processes or mechanisms will generate a diverse quality and quantity of knowledge at different levels, with respective characteristics. As a major driver, thinking plays an important role which takes different forms, owns various features, and generates respective effects:

- Systematism: individual thinking, group thinking, and social thinking
- Cognitive aspect: logical thinking, creative thinking, imaginary thinking

The thinking process acquires, processes, and verifies existing knowledge and generates and invents new knowledge. This process involves the interactions discussed above (collaboration and/or negotiation) between individuals and groups and between human decision-makers and the environment.

Cui (2006) developed a formal language to describe the process of thinking-driven knowledge production in a HWME. It involves both syntactic and semantic grammars, which are believed to be complementary. Here is the approach. Assume Ω represents the complete state space in the physical world holding the underlying problem and $\sum$ is the fact set. An expert i in the HWME has his/her knowledge function k_i in the space Ω, $k_i(\omega)$ represents the knowledge owned by expert i when the true world state is ω. Let us define an information function I in the world state Ω:

$$I(\omega)::=\left\{\omega^{'}\in\Omega:k(\omega)=k(\omega)\right\} \tag{5.1}$$

This indicates that the information associated with the true world state ω is the state set in which the state ω is closely associated with the decision-making participant i and $I(\omega)$ contains all possible states ω' that are relevant to the true state ω. It means that although the true state ω is there, a participant can usually only understand one or certain possible states ω'. Sometimes, there may be a big gap between ω and ω'.

This is the typical case in big data understanding. It is similar to the story of a group of blind people recognizing an elephant. Each blind person can only "see" a local picture, which seems explicit to them, although it could be very biased and incomplete. The goal of HWME is to achieve a near complete or complete understanding of the true state space, namely, to narrow the gaps between ω and ω'.

$I(\omega)$ represents the information set about state ω in the space Ω from participant i. Assume the union of all facts in $I(\omega)$ forms set K and define a knowledge function $K:\sum\rightarrow\sum$ as follows: for any event E, KE refers to the union of all events contained in event E in space to satisfy Functions (5.2), (5.3), (5.4), and (5.5):

$$KE\subset E \tag{5.2}$$

$$E\subset F\Rightarrow KE\subset KE \tag{5.3}$$

$$\sim KE\subset K\sim KE \tag{5.4}$$

$$K(I_aE_a)=I_aK(E_a) \tag{5.5}$$

Formula (5.2) means that a participant can only know a certain part of an event; (5.4) shows that if the participant knows nothing, then he/she does not know E. Only when everything about the event known to the participant takes place does the participant obtain a full understanding of the event. (5.5) shows that often we know something but do not get the full picture; thus, there are things unknown to us. When E is contained in another event F, and participant i knows E, then i can reason about F and can thus know F too, as shown in Formula (5.3).

There is hierarchy in knowledge, the same as in an open complex system. The extent of knowledge hierarchy may increase as a result of the process of discussions within a group to reach a deeper understanding of a problem. The following section discusses this situation between two participants only, i and j. Figure 5.3 shows the initial status of knowledge in a seminar. Participant i knows the knowledge space A, which is also known to other people. B represents the blind knowledge which is unknown to i but known to other people, i knows the knowledge space C which is hidden from others, and D is fully unknown to both i and others.

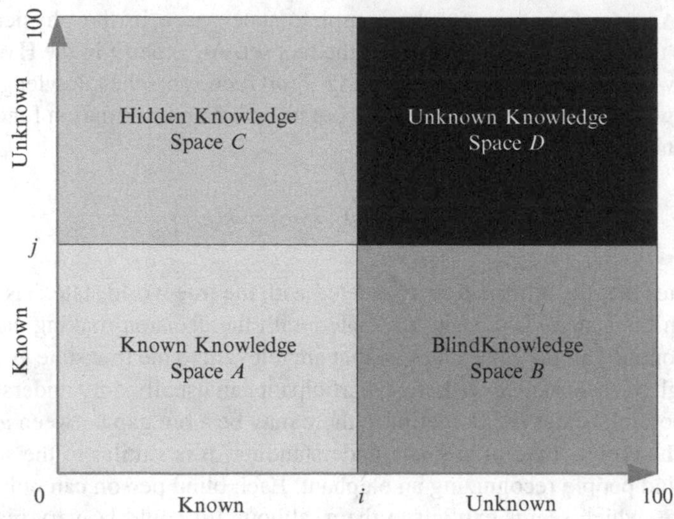

Fig. 5.3 Knowledge map: initial interaction status

In the seminar process, knowledge hierarchy h_i for participant i is built on a limited number of knowledge layers $I^1 \times I^2 \times \ldots$, which are associated with the information set $I(\omega)$:

- h_i^{m-1} is the mapping of h_i^m to the state space I^{m-1}, indicating that i's information and knowledge in every layer is consistent with that in the previous layer.
- i knows what j knows, namely, knowledge space A.
- i knows j's knowledge hierarchy h_j.

Assuming j's knowledge hierarchy set is $\left(I, h_j^1, h_j^2, \ldots, h_j^{m-1}\right)$, if h_i^m contains $\left(I, h_j^1, K, h_j^{m-1}\right)$ during the seminar process, then h_j^{m-1} also contains $\left(I, h_i^1, K, h_i^{m-2}\right)$, and the interactions between i and j become feasible.

If i and j can interact, then the interactions between i and j will be deepened and widened with the progress of discussions. Through the interactions, i expands its knowledge space at a certain knowledge layer from known A to part of the blind knowledge space C, which was originally unknown to I, through discussing and learning from j who knows C. Similarly, j develops the understanding of unknown knowledge in hidden space B through understanding and learning from i. i and j may also develop the understanding of unknown knowledge in unknown space D due to the mutual inspiration between them and the production of new knowledge. Building on known knowledge in A, and the expanded understanding of originally unknown knowledge in B and C, the interactions between i and j produce higher levels of knowledge h_i^{m+1} and h_j^{m+1} in region D (Fig. 5.4).

Figure 5.5 further illustrates the process of generating knowledge through information passing and group interactions. Each cell (atomic knowledge base for an individual) represents a participant who sits in a certain domain, takes certain roles and responsibilities, and owns knowledge at a certain level that can be shared

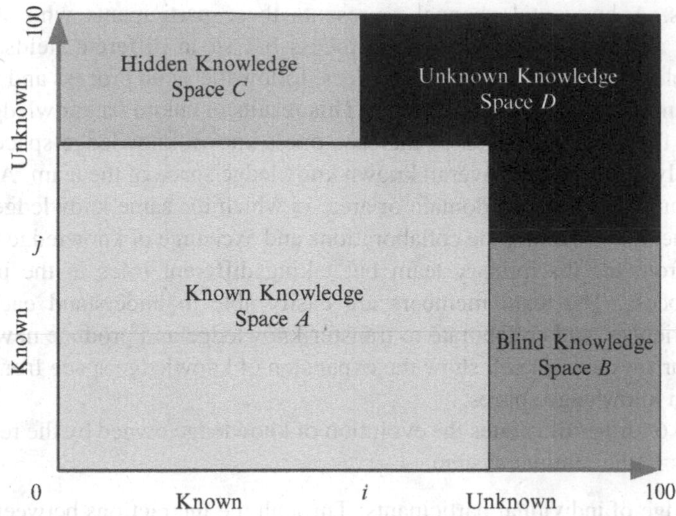

Fig. 5.4 Knowledge map: group interaction status

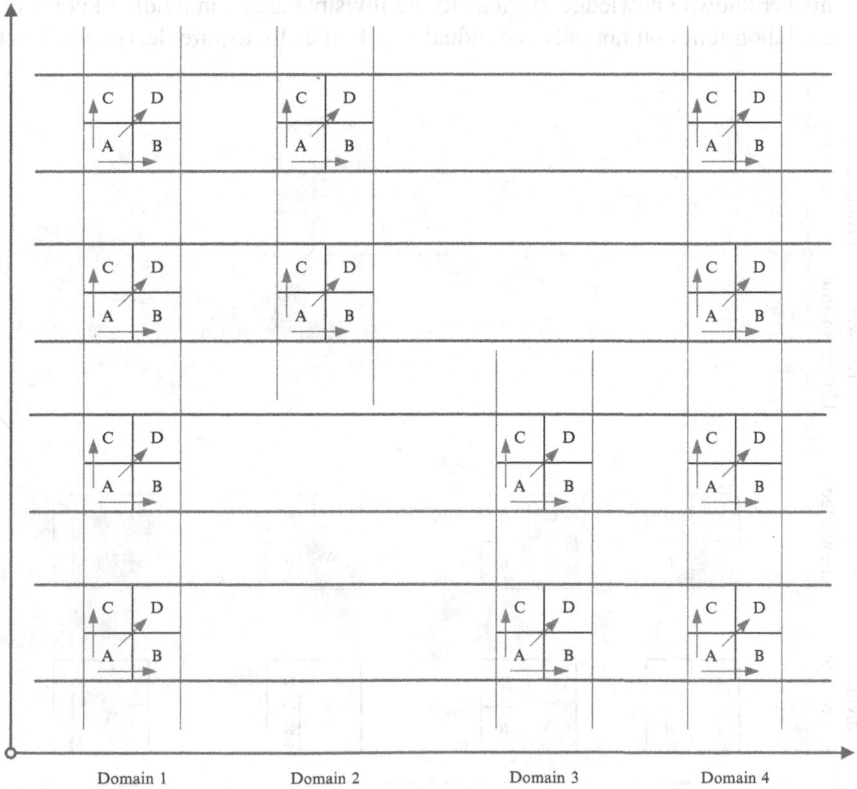

Fig. 5.5 Knowledge network in group interactions

with others. A horizontal channel represents those participants who share the same roles and responsibilities in the process but sit in different fields. Taking the same role, they share similar experiences, follow the same process and policies, and share knowledge, opinion, and tools. This results in unknown knowledge being transferred from one to another to increase one's known knowledge space A, and consequently increasing the overall known knowledge space of the team. A vertical channel represents the same domain or area, in which the same knowledge base is shared in the field, enabling the collaborations and exchange of knowledge between members from the disciplinary team but taking different roles in the problem-solving process. The team members are easily able to understand each other, share experiences, and collaborate to transfer knowledge and produce new knowledge. The arrows in each cell show the expansion of knowledge space from known to unknown knowledge spaces.

Figure 5.6 further illustrates the evolution of knowledge owned by the respective participants in the seminar system.

- Knowledge of individual participants: Through the interactions between participants in the same role or in the same field, each participant obtains increasing knowledge from others about blind, hidden, and unknown information, and his/her known knowledge expands to the invisible areas; individual knowledge evolution relies on not only individual capabilities to acquire, learn, and invent

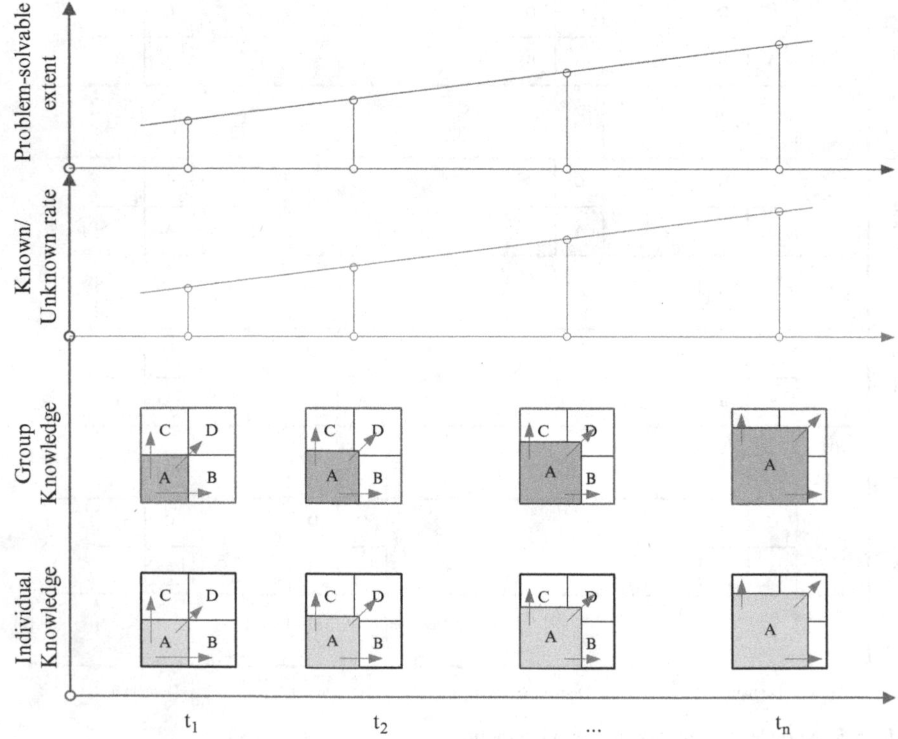

Fig. 5.6 Knowledge production through metasynthetic engineering

knowledge but also on sharing and learning from others; the capacity to be inspired and collaborate may put some individuals in a more advantageous position in the team.
- Knowledge of the expert group: The knowledge of the team of participants develops to cover areas originally unknown to the team. Note that the evolution of group knowledge depends on the availability of individual knowledge, the capability of individuals to expand their unknown knowledge, the interactions and sharing of knowledge between members, and the creative thinking within the group to invent new knowledge and tools. At a certain stage, the group enjoys the emergence of group knowledge about invisible and unknown areas.
- Known/unknown knowledge rate: Also called visible/invisible knowledge rate, this rate increases during the group interaction process as a result of the teamwork; the proportion of unknown and invisible knowledge reduces progressively.
- Problem-solvable extent: Consequently, the solvable extent of the underlying problems increases during the process; the extent of the increase depends on the contribution and knowledge expansion speed of both individual and group participants.

5.5 Theoretical Framework of Metasynthesis

To deal with the system complexities of OCGS, Qian et al. proposed the problem-solving methodology of *qualitative-to-quantitative metasynthesis* [2, 3, 9, 23, 24]. They also proposed that a feasible and technical solution for OCGS problem-solving was to build an M-space. An M-space is a human–machine-cooperative human-centered intelligent problem-solving workspace and artificial computational organization which consists of human beings, computers, and relevant computing tools. In an M-space, human beings help computers, for instance, by supervising the modeling process, and computers support human beings. The key working process of an M-space is social cognitive interaction-based problem-solving in which all relevant domain experts interact, collaborate, communicate, and negotiate with one another like multiple agents toward the problem-solving.

We interpret the theoretical framework of an M-space from system and cognition perspectives. The system framework of social cognitive interaction-based metasynthesis consists of the following key points:

System Framework: M-Space

- An M-space consists of human beings and computers, in which humans and computers are system constituents.
- The capability of an M-space results from the metasynthesis of all system constituents.
- In an M-space, many collaborative groups may emerge that are formed based on the requirements of problem-solving.
- The members of each group may change with the dynamics of the system and its problem-solving process.

- There is hierarchy in an M-space; some layers are stable, for instance, responsibilities, roles, and permissions, while others are dynamic.
- An M-space is open in the sense that both itself and its problem-solving process are dynamic.
- To support the human–machine-cooperative working mechanism of an M-space, it is necessary to have efficient and detailed indexing and searching capabilities. The resources for indexing are dynamic; some are existing, while others may be instantly added by system constituents in a problem-solving process.
- An M-space is capable of receiving messages from its environment.
- There are effective communications between an M-space and its environment and among the system members in an M-space.
- The problem-solving mechanism of an M-space is achieved through the information exchange among system constituents and between an M-space and its environment.
- An M-space needs to provide capabilities such as information storage, access, representation, search, analysis, discovery, inference, transfer, use, and management of resources, data, information, meta-knowledge, and empirical data that may be in qualitative and quantitative, and structured and ill-structured forms.
- An M-space supports distributed cooperation and processing, situated perception, effect and inference, run-time internal and external interactions, and dynamic adaptation or control.

In addition, an M-space involves the following key cognitive characteristics from the perspective of social cognitive interaction:

Cognitive Framework: Social Cognitive Interaction-Based M-Space

- An M-space has goals; goals present characteristics such as hierarchy, relative certainty, and dynamic evolution.
- System constituents of an M-space have cognitive capability and social requirements such as beliefs, desires, intentions, reputation, credit, thinking (convergent and divergent), inference, judgment, self-learning, and learning from others.
- In an M-space, solving a problem is achieved through effective interaction, collaboration, and cooperation between the human being-based subsystem and the computerized subsystems.
- In some cases, the problem-solving is human centered, while for other problems, automated computer systems play major roles.
- Each system constituent has specific cognition, experiences, and beliefs about the world; constituents may share their cognition, but may reach no consensus.
- A constituent has a desire to learn from others, while others can also think independently.
- There are certain rules, norms, and policies that must be respected by all system members in the hierarchical job allocation and cooperation.
- There may be domain-specific organizational rules and relationships that must be followed in the problem-solving process of an M-space.

- There is cognitive evolution, restriction, and integration during the cognitive interaction, which assists consensus building or conflict resolution.
- An M-space is capable of effectively and systematically importing, stimulating, emerging, and integrating intelligence, as well as aggregating, summarizing, and exporting goals and outputs.

5.6 Problem-Solving Process in M-Space

The fundamental process of metasynthesis-based OCGS problem-solving is shown in Fig. 5.7. Many relevant experts are invited to login to the M-space server. They choose their respective topics and sessions of interest or join in the M-space-based discussions, or *M-interactions*, as requested. The interaction needs to follow certain interaction templates, scripts, and protocols. There are two sorts of action to be taken in the interaction. One is to exchange ideas with other experts online through brainstorming, negotiation, or even debate. The other is to call relevant models, methods, and computing tools to simulate and test their ideas and hypotheses. As a result, a member generates initial individual results or decisions based on the above discussion and computing results. These are merged with other members' conclusions through social cognitive interaction and consensus building to form the final problem-solving results.

The above basic process of metasynthesis-based OCGS problem-solving is explained as follows:

Basic Process: OCGS Problem-Solving in M-Space

1. Define or understand what the problem is and what the objectives are.
2. Call relevant domain experts to engage in online M-interactions through an M-space system.

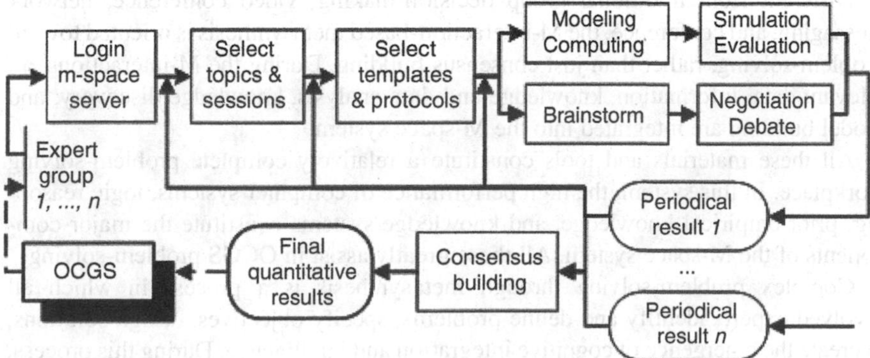

Fig. 5.7 Principle of M-space as an OCGS problem-solving system

3. Obtain a preliminary understanding of the problem through the M-interactions and by involving experts' empirical knowledge and intuition.
4. Propose methods for analyzing complex problem structures.
5. This is subject to the expertise and imaginary thinking of all attendees and the involvement of knowledge from more experienced domain experts.
6. Based on the structural characteristics, quantify the problem analysis by involving domain and prior empirical knowledge progressively, step by step.
7. Build quantitative and semiquantified local or global models for the problem processing; the models come from the involved experts' intelligence and experience and attest the rules existing in the problem and relevant data.
8. Aggregate or integrate the local/global models into system models if the expert group agrees with the local/global ones.
9. Simulate the system models and evaluate the models' reliability using the involved expert group; if the expert group is not satisfied with the models, go back to step (3) or whichever step is necessary to generate more suitable models; the modeling process stops when the expert group agrees with the performance of the model.

Except for step (1), all steps are conducted in an M-space. The key characteristic and function of M-interactions are the social cognitive interaction, collaboration, and negotiation between all the involved experts. We call this form of discussion-based problem-solving *M-interactions*. M-interactions are based on social cognitive interaction among the involved domain experts as well as between human beings and computer systems. They involve the whole process of problem-solving. Through the M-interaction process, the individual intelligence is upgraded and aggregated into collective intelligence. In fact, M-interaction-based problem-solving is learning-based, in particular, group-learning based. During the process, all the involved experts communicate, argue, and negotiate with each other. This not only improves an individual's understanding of the problem, but triggers the emergence of collective intelligence from the expert group, which is likely to outperform the problem-solving capability of individual experts. The collective intelligence plays a critical role in solving the problem.

Different from traditional group decision-making, video conference, network messaging, and conference, the M-interaction-based metasynthesis is oriented toward problem-solving, rather than just consensus building. During the M-interactions, all relevant data, information, knowledge and data analysis, knowledge discovery, and model building are integrated into the M-space system.

All these materials and tools constitute a relatively complete problem-solving workplace. In this system, the high performance of computer systems, logic reasoning, prior empirical knowledge, and knowledge systems constitute the major components of the M-space system. All these greatly assist in OCGS problem-solving.

Complex problem-solving through metasynthesis is a process in which all involved experts identify and define problems, specify objectives, design solutions, to create the emergence of cognitive integration and intelligence. During this process, it is important to utilize qualitative knowledge, intelligence, domain knowledge, expertise, quantitative computing, network intelligence and computing, and social intelligence and computing. Another important aspect is an interaction environment that supports fair, free, and open communication, negotiation, coordination, integration, and consensus building of cross-domain and hierarchical group thinking and collective intelligence. Effective norms and policies for considering the evolution of effective prevention, deviation rectification, barrier avoidance, stimulus, infighting, and argument are also necessary.

From the cognitive evolution and intelligence emergence perspective, the following process summarizes the group cognitive evolution for complex problem-solving:

Cognitive Process: M-Interaction-Based Cognitive Evolution and Intelligence Emergence

1. Open M-interactions.
2. Issue discussion topics.
3. FOR each topic.
4. Open broad-based discussions of the problem using brainstorming.
5. Model initial and target problem status alternately through brainstorming and nomination in the expert group.
6. Obtain initial approaches and solutions for the problem-solving based on brainstorming and *nominal group techniques* [25] within the expert group.
7. Obtain qualitative understanding of the problem through the use of deep discussions and arguments in the expert group with the involvement of the above learned initial approaches, domain knowledge and expert intelligence.
8. Form separate M-space discussion sessions focusing on specific topics by dividing the expert group; for each session, try to build semiquantified and quantitative understanding of the issues on the basis of the above qualitative understanding by using deep discussions and negotiation.
9. Fuse the output from all sessions into a higher level of preliminary quantitative understanding of the problem through deep discussions and negotiation between the expert groups.
10. Reorganize the expert group into separate sessions to structure specific issues identified in the above steps.
11. Repeat steps (8) and (9) to progressively structure and quantify the problems.
12. ENDFOR.
13. Aggregate the output to obtain the main solutions using methods such as nominal group technique.
14. Conduct computer simulation and evaluation of the main solutions by deep discussions and negotiation in the expert group.
15. Review and rank the resulting solutions based on the satisfaction of technical and business expectations, and go back to any step from (2) to (12) if necessary to refine the solutions.
16. Obtain the decision-support solutions based on negotiation and nomination in the expert group.

17. Summarize the qualitative structural principle of the problem-solving using methods such as brainstorming and nominal group technique.
18. Output the finally agreed findings.
19. Close the M-interactions.

5.7 Social Cognitive Intelligence Emergence in M-Space

5.7.1 *Individual Cognitive Model*

Individual cognitive capability is defined by three key factors. One is *individual cognitive degree*, which means the extent to which one understands and grasps the whole picture of an object or issue. The whole picture may include aspects such as scope, internal dynamics, and external environment. The second factor is *object openness degree*, which indicates to what extent an object has been completely understood by human beings. The third is the *individual cognitive methods*, namely, how one perceives an object.

Let α be one's individual cognitive degree with "1" reflecting 100 % understanding of an object, while "0" represents nothing. Any value of β in [0, 1] indicates the degree to which one understands the problem. In practice, we often set up several levels based on a qualitative estimation of our understanding of the problem, for instance, "complete," "partial," and "unknown."

Let β be the degree of an object openness with "1" indicating a fully solved status, while "0" represents an unknown status. Any value of β in [0, 1] reflects the degree to which a problem has been solved. Similarly, we often use words like "well solved," "partially solved," and "unsolved" to reflect the openness of a problem. In general, the openness of an object is determined by social cognitive power rather than a particular individual.

Let ν_i $(i=1,..N, N$ is the number of members in an M-space) be an individual cognitive method used by a member i. A major individual cognitive method is *learning* with γ indicating the significance and contribution of method ν_i to the understanding of an underlying problem. In general, learning can take one of three forms: (1) self-learning, (2) ex-learning, and (3) creative learning.

Self-learning refers to the learning process that is basically conducted by individuals. *Ex-learning* is driven by the support, supervision, and coaching of other senior members. *Creative learning* is a process and action that triggers new ideas based on the existing knowledge learned. The relationship between the above three key factors is as follows:

One understands a problem based on the cognitive methods ν_i to reach a certain cognitive degree α. The individual cognitive degrees of a collection of people determine the level of understanding and problem-solving of the problem. The overall problem understanding is determined by the level of collective cognition in relation to the thinking of the most prestigious group experts specialized in the problem, or by the genuine knowledge extended to the whole society.

If a problem has the status of only one person knowing it, then the problem-solving degree Φ (viz., the problem openness β) is determined by that person's initial cognitive degree α_0, degree of object openness β_0, and contribution (γ_i) of the cognitive method v_i used.

$$\Phi = p\left(\alpha_0, \beta_0, \gamma_i \middle| v_i\right) \tag{5.6}$$

The goal of individual cognitive modeling is to find an appropriate *p-function* $(p(\cdot))$ to reflect the principle of a person's cognition in understanding a problem.

In the following sections, we will further discuss social cognitive interaction and cognitive intelligence emergence.

5.7.2 Social Cognitive Interaction Model

The cognitive methods of a group of experts consist not only of individual learning but also of *social interaction*. Through the social interaction of involved experts, the experts themselves improve their individual cognitive degrees and, at the same time, enhance the collective cognitive intelligence as a group about the underlying problem.

Social cognitive interaction consists of two key components: social cognitive interaction methods and group cognitive interaction protocols. Examples of *social cognitive interaction methods* are "heuristic discussion," "brainstorming," and "debate." They may be organized through seminars, workshops, video conferences, or online workshops or seminars.

Heuristic discussion is a form of guided discussion in which a mentor guides the process and determines the key milestones of each session to steer the discussion in possibly the right directions. The mentor is usually an experienced authority in the field who may either partially know the answer to the problem or be aware of the right direction to pursue. His/her role in the discussion is to guarantee that the discussion is productive, efficient, and deliverable, and to avoid obvious misunderstanding, unnecessary debate, or going astray.

Brainstorming encourages the free communications of ideas and is helpful for fostering an environment that encourages the emergence of new and perhaps conflicting notions. In a *debate*, parties may take an antagonistic position against reaching a possible solution. In contrast, a more rational debate may build consensus among members to finally reach an agreement. There are several key elements in a social cognitive interaction: the number of participants, norms and policies, and the interaction protocols. There are a few types of discussion, for instance, peer to peer and multiparty.

Members may also be grouped based on their specialities. In some cases, hierarchical groups may be organized to reflect the difference between knowledge and experience and corresponding role differences in M-interaction-based problem-solving.

Certain interaction norms, rules and policies are essential for productive problem-solving. We can nominate a number of rules and policies for interaction sessions. They vary according to background, structure, culture, organizational constraints of the attendees and the problem openness. For instance, in some situations, the following rules and policies are recommended to members and groups involved in problem-solving:

Member norm: sample norms, rules, and policies for social cognitive interaction

- Do not automatically accept anything from an authority; all points need to be checked and evidenced
- Always think of the conditions of using a concept, conclusion, method, result, etc.
- Do not use hypotheses without evidence
- Always believe the limitation and potential of cognition, and think of critical and creative ideas

Interaction protocol: sample codes of conduct for social cognitive interaction

- The lower level of subgroups must respect the decisions of a higher level
- Chairpersons have the ultimate authority to decide the process and policies to be used
- Senior members carry more weight in determining a solution than junior members
- Creative thinking, if recognized by the majority of people, has more weight in determining a solution

As a problem-solving system, certain interaction protocols are needed for social cognitive interaction in an M-space. For example, the following basic protocols may be formed to guide the interaction:

Interaction protocol: sample codes of conduct for social cognitive interaction

- The lower level of subgroups must respect the decisions of a higher level
- Chairpersons have the ultimate authority to decide the process and policies to be used
- Senior members carry more weight in determining a solution than junior members
- Creative thinking, if recognized by the majority of people, has more weight in determining a solution

To support the formalization of social cognitive interaction, we further define a kind of interaction ontology based on descriptive logic. An interaction operator represents a type of interaction mode $\widetilde{m}$ used by relevant members. Examples of interaction operators are as follows:

Interaction operator: sample operators representing interaction modes

- Disjoint ($\widetilde{d}$): two sorts of cognition disjoint from each other
- Overlap ($\widetilde{o}$): two sorts of cognition at least partially overlap each other
- Include ($\widetilde{i}$) one cognition is a class or part of another

Interactions following a disjoint mode are likely to lead to disagreement, while modes of overlap and inclusion more likely to converge ideas into a consolidated form.

5.7.3 Cognitive Intelligence Emergence

The above defined mechanisms provide a foundation for us to describe cognitive intelligence emergence. Suppose α_{i0} represents the initial cognitive state of member i on the target problem, member i has authority weight μ_i, and he/she interacts with other n members by an interaction mode $\widetilde{m}_j$. In a discussion session, they follow the interaction protocols φ_j and norm η_j. Furthermore, for the interaction trend, let $\wedge$ seek the common points while reserving differences, $\vee$ indicates conflict debate. Then, we can define a kind of algebra given in the Backus–Naur form to describe the social cognitive interaction that may happen in an M-space:

$$\phi ::= 0|\phi|\phi_1 \wedge \phi_2|\phi_1 \vee \phi_2 \tag{5.7}$$

0 stands for the unknown status of the problem, ϕ is the cognitive degree determined by one member only, $\phi_1 \wedge \phi_2$ indicates two members interacting and seeking common points (through *convergent thinking*), $\phi_1 \vee \phi_2$ reflects the two members who cannot reach an agreement but go in opposite directions because of conflicted understanding (*divergent thinking*).

In this way, we build up the following model to describe a social cognitive interaction process and corresponding intelligence emergence from the initial cognitive states of individual members to a resulting state of the problem through social cognitive interaction. As a result of metasynthesis in the M-space, the problem-solving degree Φ (namely, the problem openness β) is described as follows:

$$\Phi = \beta = g\left(\alpha_{i0}, \beta_0, \gamma_{i,j}, \mu_i \middle| \widetilde{m}_i, \varphi_j, \eta_j, v_{i,j}\right) \tag{5.8}$$

i refers to the member i, and j is the number of current interaction sessions (an interaction session refers to the time during which a group of attendees conducts M-interactions on a specific topic or problem).

The above model indicates that the current problem-solving status Φ is an emergent effect of collective cognition accumulated from the interaction between N members. Φ is determined not only by a member's cognitive capabilities as indicated by the initial cognitive openness degree of the underlying problem β_0 but also by a particular member i's initial cognitive degree α_{i0}, individual cognitive method v_{ij}, and weight γ_i of an individual learning method v_i. It is also affected by the group interaction mode $\widetilde{m}_i$, interaction protocols φ_j, and social impact factors such as a member's cognitive authority weight μ_i in the interaction.

We therefore say that social cognitive problem-solving capability is determined not only by initial individual problem-solving degrees, individual interaction modes, and authority but also by social cognitive interaction and creative capability.

The emergence of cognitive intelligence in an M-space thus follows a certain cognitive working mechanism formulated by *g-function*. The function reflects the impact of cognitive interaction trend Ø, interaction protocol ρ, norm η, and mode $\widetilde{m}_i$. A key challenge of social cognitive interaction-based metasynthesis is to find such a good function that reflects the rules of collective intelligence aggregation and emergence in M-space.

5.8 Thinking Pitfalls in M-Interactions

There are a few thinking pitfalls that may affect the effectiveness of M-interaction-based metasynthetic social intelligence engineering and its resulting performance. They may consist of the following: clanthink [26, 27], dependent thinking, divergent thinking [28], groupthink [29, 30], linkthink [26, 27], rigid thinking, spreadthink [26, 27], and cynicism and negativity. We briefly introduce here the symptoms and strategies for dealing with *dependent thinking*, *groupthink*, *divergent thinking*, and *rigid thinking* during M-interactions in an M-space.

Dependent thinking is a mode of thought that relies on the ideas of a "more important" or "more senior" person, e.g., a mentor or authority, while self-dependence and self-discovery are lacking. A person with dependent thinking preferences tends to avoid viewpoints outside the comfort zone created by a leader or authority. Even though the person may have their own critical and creative ideas, they try to hide them and cater instead to the thinking of the other "more important" person. To avoid dependent thinking in an M-space, the following strategies can be used:

Strategy: avoiding dependent thinking in M-interactions

- Encourage separated or anonymous expression of ideas, and ask those people likely to have dependent thinking to express their ideas at an early stage of the discussion process
- Conceal or withhold the ideas of the authority until the end of a session

(continued)

- Distribute information and techniques to dependent thinkers in a uniform way with others, and avoid the circulation of messages that might carry an impression of exercising leadership
- Encourage them for being positively involved in discussions and freely expressing their different thoughts
- Take actions to build their confidence and reputation and to encourage different ideas and creative thinking

Divergent thinking is a thinking process or method which is usually applied with the goal of generating new ideas. Divergent thinking is often used for creative and problem-solving purposes in conjunction with convergent thinking. There are different methods in divergent thinking. The strategies for encouraging divergent thinking include brainstorming, keeping a journal, free writing, and subject mapping.

In an M-space, divergent thinking plays a critical role in fostering creative insights into the various aspects of a topic, e.g., problem definition, goals, approaches, and techniques. It typically occurs in a spontaneous, free-flowing manner, such that ideas are generated in a random and unorganized fashion. As a result, it may also lead to negative impact on problem-solving because of over-divergence without convergence building. The prevention of over-divergent thinking is to guide the thinking toward convergence and organize ideas and information using convergent thinking, i.e., putting the various ideas back together in an organized and structured way. In M-interactions, the following techniques may be helpful for not only maintaining divergent thinking but also balancing it with the final convergence of thoughts:

Strategy: avoiding over-divergent thinking in M-interactions

- Maintain the extent of local divergence and global convergence
- Develop a universal or transparent communication platform to maintain transparency between different domain languages, terminologies, and results to build mutual and consistent understanding
- Encourage the identification of individuals with similar thinking to establish certain foundations, and expand group thinking to include the importation of other members' ideas
- Encourage the negotiation and debate of various thoughts to benefit from the clarification and influence of other members
- Develop information integration and fusion that crosses heterogeneous domains and systems for the transparent exchange of ideas,with mutually recognizable evaluation systems to measure the significance of results

Groupthink [29, 30] is a type of thought exhibited by group members who try to minimize conflict and reach consensus without critically testing, analyzing, and evaluating ideas. During groupthink, members of the group avoid promoting viewpoints outside the comfort zone of consensus thinking. A variety of motives for this may exist, such as a desire to avoid being seen as foolish or embarrassing, or to avoid angering other members of the group. Groupthink may cause groups to make hasty and irrational decisions, where individual doubts are set aside for fear of upsetting the group's balance. In an M-space, group thinking may have a negative impact by limiting divergent and creative thinking. The prevention of groupthink can be achieved by adopting the following strategies:

Strategy: avoiding group thinking in M-interactions

- At least one group member should be assigned the role of chair to lead the discussion. This role should be assigned to a different person for each meeting
- Leaders should assign each member the role of "critical evaluator." This allows each member to freely air objections and doubts
- Superiors should not express an opinion when assigning a task to a group
- The organization should set up several independent groups, working on the same problem
- All effective alternatives should be examined
- Each member should discuss the group's ideas with trusted people outside the group
- The group should invite outside experts into meetings. Group members should be allowed to discuss with and question the outside experts

Rigid thinking is a kind of thought that tends to follow patterns, predefined or authorized rules, and existing manners. It is not sensitive to critical and creative ideas. People with rigid thinking tend to stick to existing ideas, empirical cognition, and patterns and do not consider a variety of conditions and situations. As a result, such people do not want to accept new ideas, suggestions, and knowledge. The following strategies can be used in M-interactions to avoid rigid thinking:

Strategy: avoiding rigid thinking in M-interactions

- Request and remind the conditions of using a model, method, and case
- List and ask for prerequisites for drawing a conclusion
- Encourage more experienced people to discuss with each other openly, in order to disclose the intrinsic limitations and constraints of each argument, model, method, and case used, to reach a more complete and in-depth understanding of the problem
- Train thinking capabilities such as divergent thinking, critical thinking, and creative thinking
- Build up the feedback from the initial conclusion to consequent pitfalls, and analyze the causes and effects by involving other experts
- Encourage comparison with other options and perform evaluations by considering conditions and constraints

5.9 M-Computing: Engineering OCGS

M-computing consists of the engineering methodologies and techniques [70–75] for reifying the theory of qualitative-to-quantitative metasynthesis, carrying out M-interaction, and constructing the M-space to tackle OCGS. From the computing perspective, M-computing presents a new computing paradigm for OCGS-oriented problem-solving, which involves but is not limited to the following objectives and tasks:

- Objectives and Tasks: M-Computing
- Studying methodologies for guiding the system analysis, design, and implementation of M-space
- Studying the modeling, representation, communication, mediation, and negotiation of M-interaction
- Studying the mechanisms for human group/community formation and evolution, human group/community interaction and intelligence emergence, and social network formation and evolution
- Studying the interaction mechanisms and impact of human group/community, context/environment and interaction of inter-/intra-group/community, between systems and human groups, and between systems and the environment on problem complexities and problem-solving
- Studying the mechanisms for representing, simulating, and supporting social cognitive networks in OCGS and its problem-solving
- Studying the cognitive elements such as beliefs and intentions and their evolution, their interaction mechanisms, relations and processes, and cognitive convergence and divergence in M-interaction
- Studying the mechanisms and tools for consensus building and social cognitive intelligence emergence for the problem-solving
- Studying the mechanisms and tools for supporting M-interaction-driven problem-solving and resolving possible conflicts
- Studying human role allocation, group management, and process/resource dispatching for the best possible stimulation and utilization of both individual and group intelligence during the problem understanding and solving
- Developing software engineering techniques and tools for engineering the M-space, which may involve social and organizational factors, as well as existing computing paradigms
- Developing mechanisms, tools, protocols, policies, and norms for supporting M-interaction-driven problem-solving and M-space
- Developing mechanisms and tools for project management, risk control, and performance evaluation during M-space construction
- Developing M-computing paradigms and considering the use and integration of existing paradigms in building the M-space

With deeper studies into OCGS and the above tasks, many open issues await further development, for instance:

- How should an OCGS be analyzed and modeled? What formal methods are necessary for the analysis and design of OCGS? How should the architecture and M-interactions be designed?
- To engineer an M-space, what existing techniques can be used? What else needs to be developed?
- In studying the system infrastructure and architecture of an M-space, what are the roles of existing techniques, such as agent-based technique [31–33], three-layer client/server structure [34], and system supports for human–computer interaction, business logics, applications, knowledge/data resources, and service management?
- What methodologies and approaches are suitable for system analysis and design mechanisms for engineering an M-space? For instance, are some existing approaches suitable or adaptable, such as the OSOAD methodology for engineering open complex agent systems, which integrates organization, agent service-oriented analysis, and design approaches [1, 31–33]?
- An OCGS is initially likely to be a grey box or even a black box. How can an artificial system [35] be built that can accurately simulate a real system? If such an artificial system can be obtained, it is then possible to understand OCGS with limited risk and costs, and one day to work directly on the real system.
- How can organizational and societal factors and issues be handled in M-interactions? In building an artificial M-space system, what new techniques are needed to support organizational computing and social computing [36]?

5.10 Discussions

The theory of metasynthesis involves many disciplines, for instance, cognitive science, systems science, artificial intelligence, information systems, human–machine interaction, and machine learning. In the past twenty years, great efforts have been made in the sphere of both theoretical foundations and engineering techniques by multidisciplinary researchers from areas such as information technology, systems science, and cognitive science. Such efforts have promoted the development of both theories and applications of metasynthesis, leading to an increasingly mature scientific field.

Researchers from different domains have contributed to the development of metasynthesis. For example, as an expert in quantitative economics and key contributor in the field of OCGS and metasynthesis, Yu summarized the related progress of approaches, theories, techniques, engineering [37], and case studies [38]. As an expert in systems science, Gu studied system modeling [39], the relations between metasynthesis and systems science [40], and the relationship between metasynthesis and the knowledge sciences [41], and summarized his view of theories and applications on metasynthesis [42].

Our recent effort developed software engineering methodologies and techniques for engineering open complex intelligent systems such as M-spaces [1, 31–33]. We have also proposed the concept of *metasynthetic computing* [43] as an engineering approach to implementing metasynthesis.

In addition, comprehensive research work has been conducted from both theoretical and technological aspects to investigate critical issues in OCGS and metasynthesis, for instance:

- Building a theoretical framework of metasynthesis and metasynthetic social intelligence engineering as an OCGS problem-solving methodology [44], a conceptual system [45] and model design [46], and knowledge reconstruction [47]
- Building metasynthetic intelligent systems, open complex intelligent systems [1, 44, 48], and the science of social intelligence [49]
- Studying cognitive working mechanisms [18] and supporting tools for, in particular, divergence [50] and convergence of domain expert group thinking, social cognitive interaction and its problem-solving mechanisms, as well as strategies for avoiding the pitfalls of social cognition within a group of experts [34, 51]
- Studying the emergence of metasynthetic wisdom in cyberspace [52] and the relationship between systematology and the creative development of Chinese medicine [53]

As a new scientific field, and also as a result of the challenges of OCGS, the research on OCGS and metasynthesis is at an early stage. Although there are many issues to be studied with regard to three components, M-interaction, M-computing, and M-space, we list a few here which we believe are fundamentally important:

- Appropriate representation mechanisms and tools are necessary for modeling, constraining and presenting M-interactions, in particular, social interactions such as group interaction and group-computer network interaction.
- With regard to individual cognition and social cognitive interaction models, what are the *p-function* and *g-function*?
- To support M-interaction-driven problem-solving, how can a balance be achieved between freedom of self-organization during the discussion process and the attainment of a final agreement with problem-solving solutions?
- Many issues in relation to M-computing are listed in Sect. 5.9, for instance, techniques and tools for analyzing, designing, and implementing M-space.
- In building M-space-based problem-solving systems for OCGS, how can domain knowledge, human intelligence, and organizational intelligence be involved and represented in M-space?

In recent years, more and more areas and problems have been studied by utilizing the theory of metasynthesis. Many successful case studies have been reported so far. A typical area is military decision support and military systems [54]. For instance, the theory of metasynthesis and its problem-solving system M-spaces has reportedly been used in the Chinese manned space program [55], military strategic decision support [56], military supply chain management [57], complex space military decision systems [58, 59], and equipment support [60] and demonstration [61].

Other applications of metasynthesis include Chinese character recognition [62], business intelligent systems [63], intelligent building systems [64], agile supply chain management [65], regional sustainability [66], population, the analysis and evaluation of public opinion on the Internet [67], resources and environmental economics [68], and digital urban planning [69].

References

1. Cao, L.B., Luo, D., Luo, C., Zhang, C.: Systematic engineering in designing architecture of telecommunications business intelligence system. In: Abraham, A., Koppen, M., Franke, K. (eds.) Design and Application of Hybrid Intelligent Systems, pp. 1084–1093. IOS Press, The Netherlands (2003)
2. Qian, X.S., Dai, R.W.: Emergence of Metasynthetic Wisdom in Cyberspace. Shanghai Jiaotong University Press, Shanghai (2007)
3. Qian, X.S., Yu, J.Y., Dai, R.W.: A new scientific field—open complex giant systems and the methodology. Chin. J. Nat. **13**(1), 3–10 (1990)
4. Simon, H.A.: The Sciences of the Artificial, 3rd edn. MIT Press, Cambridge, MA (1996)
5. Wegner, P.: Why interaction is more powerful than algorithms. Commun. ACM **40**(5), 80–91 (1997)
6. Guo, X.Z., Ge, J.L.: The determination and treatment of divergence experts in expert group thought. Complex Syst. Complexity Sci. **4**(1), 1–5 (2007)
7. Tianfield, H.: Formalized analysis of structural characteristics of large complex systems. IEEE Trans. Syst. Man Cybern. A Syst. Hum. **31**(6), 559–572 (2001)
8. Wang, S.Y., Yu, J.Y., Dai, R.W., Wang, C.W., Qian, X.M., Tou, Y.J.: Open Complex Giant Systems. Zhejiang Science & Technology, Hangzhou (1996)
9. Miao, D.S.: Epistemological foundations for metasynthesis. J. Syst. Dialectics **11**(1), 37–42 (2003)
10. Wang, B., Chang, X.Q., Li, Y.Z.: Conceptual HWME for space military systems. J. Acad. Equipment Command Technol. **14**(1), 46–49 (2003)
11. Ye, H.M., Dai, G.Z.: Analysis and evaluation of public opinion on Internet based on metasynthesis. Comput. Eng. Appl. **41**(16), 216–217 (2005)
12. Dai, R.W., et al.: Metasynthetic Spaces for Macroeconomic Decision-Support Utilizing Qualitative-to-Quantitative Metasynthesis, 1999–2004. China NSF large grant.
13. Warfield, J.N., Teigen, C.: Groupthink, Clanthink, Spreadthink and Linkthink. Institute for Advanced Study in the Integrative Sciences, George Mason University, Fairfax (1993)
14. Hipel, K.W., Jamshidi, M.M., Tien, J.M., White, C.C.: The future of systems, man, and cybernetics: application domains and research methods. IEEE Trans. Syst. Man Cybern. C Appl. Rev. **37**(5), 726–743 (2007)
15. Gao, H.X.: Knowledge reconstruction methods in hall for workshop of metasynthesis engineering. Ph.D. dissertation, Chinese Academy of Science, Beijing (2003)
16. Qian, Z.Y., Yang, G.Y., Wei, D.S., Cheng, L.Z.: The practice of meta-synthesis—the research method summarization of 'Developing Strategies of Chinese Manned Space Program'. Eng. Sci. **8**(12), 10–15 (2006)
17. Qian, X.S.: On Cognitive Sciences. Shanghai Science & Technology, Shanghai (1996)
18. Qian, X.S., Tsien, H.S.: Revisiting issues on open complex giant systems. Pattern Recogn. Artif. Intell. **4**(1), 5–8 (1991)
19. Zhu, H., Zhou, M.C.: Role-based multi-agent systems. In: Gonzalez, R.A., Chen, N., Dahanayake, A. (eds.) Personalized Information Retrieval and Access: Concepts, Methods, and Practices, pp. 254–285. Information Science Reference, Hershey (2008)
20. Qian, X.S.: On Somatology and Modern Science and Technology. Shanghai Jiaotong University Press, Shanghai (1998)
21. Zhao, X.Z., Guo, R.: Metasynthesis for military systems. Syst. Eng.-Theory Pract. **24**(10), 127–130 (2004)
22. Zhu, H., Zhou, M.C.: Roles in information systems: a survey. IEEE Trans. Syst. Man Cybern. C Appl. Rev. **38**(3), 377–396 (2008)
23. Cui, X.: WWW-based model design in hall for workshop of metasynthesis engineering. Ph.D. dissertation, Chinese Academy of Science, Beijing (2004)
24. Dai, R.W., Cao, L.B.: Internet: an open complex giant system. Sci. China Ser. E **33**(4), 289–296 (2003)

25. Dai, R.W., Li, Y.D.: Hall for workshop of metasynthesis engineering and system complexity. Complex Syst. Complexity Sci. **1**(4), 1–24 (2004)
26. Warfield, J.N.: Twenty laws of complexity: science applicable in organization. Syst. Res. Behav. Sci. **16**(1), 3–40 (1999)
27. Warfield, J.N.: An Introduction to Systems Sciences. World Scientific, Singapore (2006)
28. Baer, J.: Creativity and Divergent Thinking: A Task-Specific Approach. Lawrence Erlbaum, Hillsdale (1993)
29. Holland, J.H.: Hidden Order: How Adaptation Builds Complexity. Perseus Books, Washington, DC (1995)
30. Janis, I.L.: Victims of Groupthink. Houghton Mifflin, Boston (1972)
31. Cao, L.B., Dai, R.W.: Open Complex Intelligent Systems. Posts Telecom, Beijing (2008)
32. Cao, L.B., Zhang, C.Q., Dai, R.W.: Organization-oriented analysis of open complex agent systems. Int. J. Intell. Control Syst. **10**(2), 114–122 (2005)
33. Cao, L.B., Zhang, C.Q., Dai, R.W.: The OSOAD methodology for open complex agent systems. Int. J. Intell. Control Syst. **10**(4), 277–285 (2005)
34. Lei, Y.J., Li, X.Y.: Metasynthesis for military supply chain management based on knowledge management. Commer. Res. **17**, 146–149 (2006)
35. Robbins, S.P.: Organization Behavior: Concepts, Controversies and Applications, 7th edn. Prentice-Hall, Englewood Cliffs (1996)
36. Wang, D.L.: Group thinking in hall for workshop of metasynthesis engineering. Technical Report, Chinese Academy of Science, Beijing, (2001)
37. Yu, J.Y., Tu, Y.J.: Case studies: qualitative-to-quantitative metasynthesis. Syst. Eng.-Theory Pract. **22**(5), 1–7 (2002)
38. Yu, J.Y.: Qian's studies on open complex giant systems. Syst. Eng.-Theory Pract. **12**(5), 8–12 (1992)
39. Gu, J.F., Tang, X.J.: Metasynthesis and knowledge sciences. Syst. Eng.-Theory Pract. **22**(10), 2–7 (2002)
40. Gu, J.F., Tang, X.J.: Theories and applications on metasynthesis. J. Syst. Dialectics **13**(4), 1–7 (2005)
41. Giddens, A., Duneier, M., Appelbaum, R.M.: Essentials of Sociology. Norton, New York (2006)
42. Gu, J.F., Tang, X.J.: Modeling of metasynthetic systems. Complex Syst. Complexity Sci. **1**(2), 32–42 (2004)
43. Cao, L.B.: Some critical issues in agent-based open giant intelligent systems. Ph.D. dissertation, Chinese Academy of Science, Beijing (2002)
44. Dai, R.W.: Systematology and Creative Development of Chinese Medicine. Science Press, Marrickville (2008)
45. Li, Y.D.: Paradigms of meta-synthetic computing. In: Cao, L.B., Dai, R.W., Gorodetsky, V. (eds.) Proceedings of 1st IEEE International Workshop on EOCS-MCP/32nd Annual International Computer Software and Applications Conference, COMPSAC 2008, pp. 862–867. IEEE Computer Society Press, New York (2008)
46. Cheng, X., Zhang, N.: The use of open complex giant systems for studying population, resources and environmental economics. Shandong Soc. Sci. **10**, 106–107 (2006)
47. Du, C.Z.: The application of metasynthesis in intelligent building systems. Intell. Build. **6**, 22–26 (2003)
48. Boumen, R., de Jong, I.S.M., Mestrom, J.M.G., van de Mortel-Fronczak, J.M., Rooda, J.E.: Integration and test sequencing for complex systems. IEEE Trans. Syst. Man Cybern. A Syst. Hum. **39**(1), 177–187 (2009)
49. Dai, R.W.: Qualitative-to-quantitative metasynthetic engineering. Pattern Recogn. Artif. Intell. **6**(2), 60–65 (1993)
50. Gu, J.F., Wang, H.C., Tang, X.J.: Metasynthesis Method System and Systems Sciences. Science Press, Marrickville (2007)
51. Waldrop, M.: Complexity. Simon and Schuster, New York (1992)
52. Qian, X.S.: Building Systematism. ShanXi Science & Technology, Taiyuan (2001)

53. Dai, R.W.: Science of Social Intelligence. Shanghai Jiaotong University Press, Shanghai (2007)
54. Zhang, J.T., Wang, D.L., Wang, H.A., Dai, G.Z.: HWME for agile supply chain management. J. Syst. Eng. **18**(6), 515–520 (2003)
55. Qian, X.S., Yu, J.Y., Dai, R.W.: A new discipline of science—the study of open complex giant system and its methodology. Chin. J. Syst. Eng. Electron. **4**(2), 2–12 (1993)
56. Delbecq, A.L., Van de Ven, A.H., Gustafson, D.H.: Group Techniques for Program Planning: A Guide to Nominal Group and Delphi Processes. Scott, Foresman, Glenview (1975)
57. Jin, G.J., Qu, G.H., Yang, Z.X.: Hall for workshop of digital urban planning based on meta-synthesis. J. Jilin Univ. Eng. Technol. Ed. **37**(4), 981–984 (2007)
58. Jiang, Z.J., Wu, S.H., Li, D.H.: Hall for workshop of meta-synthetic engineering for navy equipment demonstration. Mil. Oper. Res. Sys. Eng. **21**(2), 3–6 (2007)
59. Wang, F.Y., Carley, K.M., Zeng, D., Mao, W.J.: Social computing: from social informatics to social intelligence. IEEE Intell. Syst. **22**(2), 79–83 (2007)
60. Yu, J.Y.: Approaches, theories, techniques and engineering about metasynthesis. Trans. Syst. Eng. Inf. **5**(1), 3–10 (2005)
61. Janis, I.L.: Groupthink, Psychological Studies of Policy Decisions. Houghton Mifflin, Boston (1982)
62. Weaver, W.: Science and complexity. Am. Sci. **36**, 536 (1948)
63. Cao, L.B., Dai, R.W.: Software architecture of the hall for workshop of metasynthetic engineering. J. Softw. **13**(8), 1430–1435 (2002)
64. Dong, H.: The theory of metasynthesis and its applications in military practice. J. Nat. Def. Univ. **8**, 27–39 (2005)
65. Yu, T.G.: Preliminary research on HWME for equipment support systems. Ordnance Ind. Autom. **24**(6), 65 (2005)
66. Cao, L.B., Dai, R.W., Gorodetsky, V. (eds.). Proceedings of 1st IEEE International Workshop on EOCS-MCP/32nd Annual International Computer Software and Applications Conference, COMPSAC 2008, Turku, Finland (2008)
67. Xiao, B.H., Wang, C.H., Dai, R.W.: Metasynthetic approach for handwritten Chinese character recognition. Int. J. Inf. Technol. Decis. Making **1**(4), 621–634 (2002)
68. Chen, P.: Metasynthesis and strategic development and planning of regional sustainability. China Soft Sci. **10**, 106–111 (2005)
69. Jin, X., Li, Y.Z., Ma, H.G.: Metasynthesis for complex space military decision systems. Complex Syst. Complexity Sci. **2**(2), 87–92 (2005)
70. Cao, L.B.: Metasynthetic computing for solving open complex problems. In: Cao, L.B., Dai, R. W., Gorodetsky, V. (eds.) Proceedings of 1st International Workshop on IEEE EOCS-MCP/ 32nd Annual International Computer Software and Applications Conference, COMPSAC 2008, pp. 896–901. IEEE Computer Society Press, New York (2008)
71. Cao, L.B., Zhang, C.Q., Zhou, M.C.: Engineering open complex agent systems: a case study. IEEE Trans. Syst. Man Cybern. C Appl. Rev. **38**(4), 483–496 (2008)
72. Dai, R.W., Wang, J., Tian, J.: Metasynthesis of Intelligent Systems. Zhejiang Science Technology, Hangzhou (1995)
73. Li, Y.D.: Design and implementation of Hall for workshop of metasynthesis engineering. Ph. D. dissertation, Chinese Academy of Science, Beijing (2003)
74. Li, X., Dai, R.W.: Conceptual system structure and metasynthesis. Technical Report, Chinese Academy of Science, Beijing (1998)
75. Li, Y.M.: Forecasting in complex economic systems and the roles of metasynthetic engineering. J. Lanzhou Univ. (Social Sci.) **32**(5), 131–133 (2004)
76. Holism. [Online]. Available: http://en.wikipedia.org/wiki/Holism
77. Cui, X., Dai, R.: Human-centered hall for workshop of metasynthesis – artificial society (part II). Complex Syst Complex Scie **2**, 13–24 (2006) (in Chinese)
78. Cui, X., Dai, R.: Human-centered hall for workshop of metasynthesis – artificial society (part I). Complex Syst Complex Sci **2**, 5–12 (2006) (in Chinese)

Chapter 6
OSOAD Methodology

6.1 Introduction

Computing and engineering complex systems requires effective methodologies that can cater for the specific system complexities within those systems. In particular, such a methodology should address the characteristics of complex systems including openness to the environment, interactions, relationships and rules, and sociality and support both analysis and design of complex problem-solving systems.

In this chapter, organizational metaphor is taken to abstract and map a complex system to a complex organization. Accordingly, the main constituents and system complexities in a complex system are represented in terms of a so-called ORGANISED framework [1, 2]. The ORGANISED framework captures major system characteristics and complexities including organizational rules, goals, actors, norms, interactions, structures, environment, and dynamics.

Organization-oriented analysis and agent service-based design are also introduced for the analysis of complex problems and designing the corresponding problem-solving system. These form a software engineering methodology, *organization- and service-oriented analysis and design* (OSOAD) [1–8], for analyzing and designing complex systems. This chapter presents the details of OSOAD.

6.2 Organizational Abstraction

An *organization* is a structure that describes how the members of the organization relate to one another and interact to enable the cooperative achievement of a common goal. As the analogy of a distributed computer system as a distributed human organization has indicated, the most appropriate metaphor [9] for the analysis, understanding, and design of agent-based software systems is that of a human organization based on organizational theory [10].

L. Cao, *Metasynthetic Computing and Engineering of Complex Systems*,
Advanced Information and Knowledge Processing,
DOI 10.1007/978-1-4471-6551-4_6

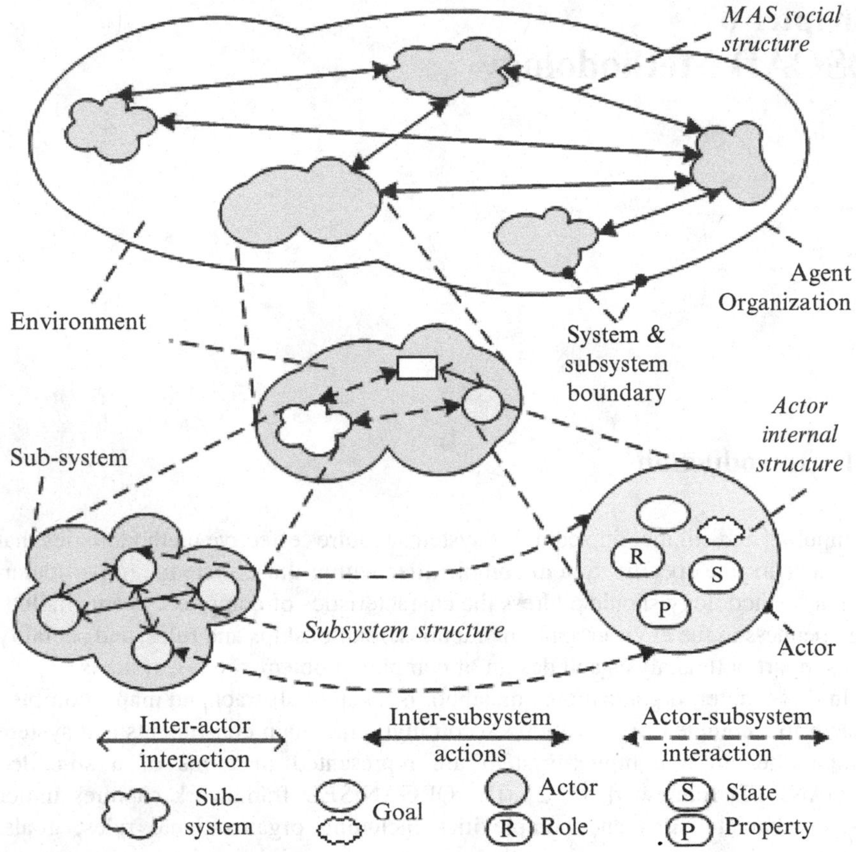

Fig. 6.1 Agent organization

With the *organizational metaphor*, the problem-solving system for handling an open complex system is an open complex agent-based software system which can be viewed as an artificial organization. Such agent systems are abstracted and mapped as artificial organizations in terms of the following generic elements: organizational rules, goals, actors, norms, interactions, structures, environment, and dynamics (ORGANISED) [4, 11] from the perspective of agent methodology. Other elements such as beliefs, intentions, policies, laws, self-organization, emergence, planning, and the like may also be considered in different agent-based scenarios. Figure 6.1 gives an organizational view of agent systems.

6.2.1 Actors

Actors are active or passive stakeholders who play different roles in the organization and their environment. An actor can be all or part of a subsystem or a subsystem component located at different granularities within the organization. As shown in Fig. 7.1, an actor can be a subsystem at the medium level, a subsystem component at the lowest level, or an object in the environment. An actor can also be a human being or an abstract concept such as a problem space.

Unlike the conception of actor in the *i** framework and TROPOS [34], in which all actors take similar positions, our actor is much more comprehensive and comprises the conceptions of actor, entity, and role in other methodologies. This design is mainly for simplicity and the later consideration of system design. Our actors are also much more specific. Actors in a system are decomposed and specified as different types in terms of roles and functions. In Chap. 7, we will present a comprehensive but reasonable classification of the various actors in an open agent organization.

6.2.2 Environment

The *environment* of an agent organization comprises actors, principles, processes, and forces [12]. In general, all actors are located in, and interact with, their environment. Environment is a relative concept in a system. It can be the globally external world of an agent system, but it can also encompass all the entities and interactions inside or outside a subsystem. Environment can be another artificial system which has relations with the target system; it can also be a social system such as that of human beings. In terms of its characteristics, the agent environment may be characterized as comprising accessibility, determinism, uncertainty, diversity, controllability, volatility, continuity, locality, temporality, or spatiality. Environment can play a significant role in determining whether an agent organization is open, semi-open, semi-closed, or closed.

Environment is what the system is immersed in. Modeling the environment involves the determination of all the entities and resources that the organization can exploit, control, or consume when it is working toward the achievement of its organizational goals.

6.2.3 Interaction

Interaction is a kind of social activity which connects two or more actors through an interaction protocol by following certain rules. Interactions can occur at various levels or across multiple levels in an agent organization, and can be defined as inter-role, interagent, interservice, intra-subsystem, inter-subsystem, and intersystem interactions.

Interactions are closely related to organizational rules and relationships. An interaction is an activity embodiment of a certain organizational relationship between specific actors with organizational rules. Such activity can take the form of coordination, cooperation, communication, negotiation, mediation, matchmaking, teamwork, coalition, resource access, or conflict resolution.

Furthermore, interactions usually follow social patterns and protocols. A social pattern defines certain manifest interactive behaviors for a specific purpose at a level of granularity within the organization. A protocol captures the constraints governing the interaction between actors. Patterns and protocols can be extracted from aspects of behavior, information, service, business, and the like.

6.2.4 Organizational Rules

Just like human society, every agent organization also has a system of rules for managing its actors, their activities, and the evolution of the organization. *Organizational rules* may take social patterns as norms [13], policies [14], laws [15], and constraints. They are distributed over and between components, subsystems, and interaction activities globally and locally. For instance, norms can be divided into perceptual, denotative, evaluative, cognitive, and behavioral categories [13]. Constraints may take such forms as preconditions, invariant, cardinalities, fulfillment conditions, postconditions, and exceptions.

Precondition defines the conditions at the moment at which an actor begins to execute an activity or desire a goal. Invariant constraints define conditions that should hold throughout the lifetime of all instances. Cardinalities specify the number of instances of a certain component existing in a system. Fulfillment constraints define the conditions of the moment at which the activity is executed or the goal is achieved. A postcondition is a fact that must always be true immediately after the execution of an activity or the fulfillment of a goal. An exception describes a fact which will have currency if a rule is not respected.

From the perspective of social laws [15] and policies, the liveness and organizational safety rules [9] are defined. Here, we specify another two sets of rules: (1) structural rules, specifying the organizational relations between actors and roles, and (2) problem-solving rules, specifying how an organization solves its problems.

6.2.5 Organizational Structure

Certain organizational relationships and structures will be generated as a result of the rules within an organization. *Organizational structure* explicitly presents high-level emergent patterns of an organizational architecture and the internal relationships between members. It encompasses the topology of interaction patterns and relationships and the control regime of the organization's activities. In an organization, multiple subsystems are linked by inter-subsystem interactions. A subsystem may comprise sub-subsystems linked by intra-subsystem interactions. A sub-subsystem may further consist of lower-level subsystems. The lowest level of subsystems is likely to consist of multiple subsystem components which cooperate in intra-component interactions. The subsystem components are the atom entity in a system.

Organizational structures may take different forms according to different classification methods. Gasser [16] defines four types of organizational structure: centralized, market-like, pluralistic community and community with rules of behavior. Taking consideration of the topology and couplings in an organization, Ferber [17] classifies structure into groups such as fixed hierarchical predefined structure, variable peer-to-peer evolutionary structure, variable peer-to-peer predefined structure, and evolutionary structure. In general, the organizational structures of agent organizations can be grouped into three main categories: hierarchical, egalitarian or peer, and hybrid structures.

Similarly, *organizational relationships* can be categorized from different viewpoints. In [18], organizational relationships consist of communication, authority, and proximity. In the eyes of Ferber [17], organizational relationships include acquaintance, communicational, subordinate, operative, informational, confidential, and competitive. However, Zambonelli [19] classified organizational relationships as control, peer, benevolence, dependency, and ownership.

6.2.6 Organizational Goal

An *organizational goal* determines the overall common motivation and objective of an agent organization. Global organizational goals are hierarchically decomposed and fulfilled directly or indirectly by individual agents. Individual goals do not apply uniformly to all actors but reflect the perspectives and interests of each stakeholder. In an organization, all actors and their environment interact in terms of certain rules to achieve their individual goals and ultimately fulfill the collective objectives. The goals in an organization can be classified as functional of nonfunctional. Nonfunctional goals refer to qualities of service, development objectives, architectural constraints, and so on.

6.2.7 Organizational Dynamics

Dynamics is the collective emergence of all actors interacting in the light of rules and goals. In an agent organization dynamics determine the overall behavioral patterns, intelligence emergence, and problem-solving capabilities of the organization. From the perspective of systems, an agent organization may take the form of a static system, a discrete-event dynamic system, or a continuous-time dynamic system. Dynamic systems may be driven by predefined laws and policies; they may also act as self-organizing, center-controlled, or stochastic bodies.

6.3 Organization-Oriented Analysis

Before we proceed with *organization-oriented analysis*, it is important to take a look at current organization-related software engineering and discuss the challenges it faces. We will then be able to understand what pure organization-oriented analysis is and why organization-oriented analysis is required.

6.3.1 Challenges for Current Organization-Related Software Engineering

Agent-oriented software engineering has been a continuously popular research topic in the agent-related research area. As a result of emphasizing organizational metaphor, several agent-oriented methodologies, for instance, GAIA, MaSE, MESSAGE, and TROPOS, have taken organizational metaphor into account and are making the transformation to organization-oriented abstraction. However, none of the existing methodologies have evolved to a stage that can be recognized as, or clai, to be, a generic tool for the analysis, design, and implementation of colorful agent-based systems. This situation applies even less when middle- and large-scale open agent systems are taken into consideration.

Recently, there has been some retrospection and discussion [20] in relation to issues in current agent-oriented methodologies. We believe these issues are closely related to the metaphor applied to building up the methodology. As part of the work to figure out these and other potential issues, some AOSE researchers [9, 21] have referred to organizational theory [10] and artificial social systems [22] for the purpose of seeking original and effective support in rethinking AOSE. This trend is probably triggered by the philosophy that an agent system—especially a middle- or large-scale open system—looks more like a human organization. This trend has resulted in the adoption of organizational metaphor in the research on *organization-oriented analysis, design, and implementation* (OADI) of agent systems. Holders of GAIA, MaSE, MESSAGE, and TROPOS are adopting ideas from human organizations and organizational theory and trying to apply them to OADI.

Nevertheless, the above work is still preliminary and it seems there is far to go to reach the final stage. This supposition is based on the following observations of current OADI:

- The mechanisms in current methodologies are not developed in terms of OADI. Almost all of them are created on the basis of traditional agent theory or non-AOSE requirements.
- The organizational abstraction is not yet complete. Certain important organizational semantics such as the environment, the organizational rules, and the like have not been covered in most of the methodologies.

- Not every phase of OADI has yet been investigated clearly. For instance, most methodologies do not support both architectural and detailed designs.
- None of the existing methodologies provide full support and smooth transformation for phases in OADI. For instance, implementation and model checking are not covered in most methodologies.
- Most methodologies only develop graphical notations for modeling. Formal specifications and integrative modeling that embed both visual and formal specifications are essential for complete and precise analysis.

Building an OADI system will be neither an easy nor a short-term project. In this work, we tentatively rethink the organizational abstraction of agent systems, especially open complex systems. We try to develop a an OADI system in terms of purely organizational abstraction.

6.3.2 *What is Organization-Oriented Analysis?*

The analysis phase involves defining in great detail what the whole information system needs to accomplish to provide the organization with the desired benefits. Across all system development methodologies, the activities which are included in the analysis phase are essentially the same. These essential activities, as shown in Fig. 6.2, include [23] the following: (1) gathering information, (2) defining system requirements, (3) prioritizing feasibility and discovery, (4) generating and evaluating alternatives, and (5) reviewing recommendations with management. These activities may be carried out throughout the analysis phase, rather than only at the beginning, or sequentially.

In contrast, the concrete content and fulfillment requirements for performing the above essential activities normally differ from one methodology to another. Using organization-oriented abstraction for agent system analysis, we summarize the key questions to be answered for the above activities in Table 6.1. The list of activities is taken from [23].

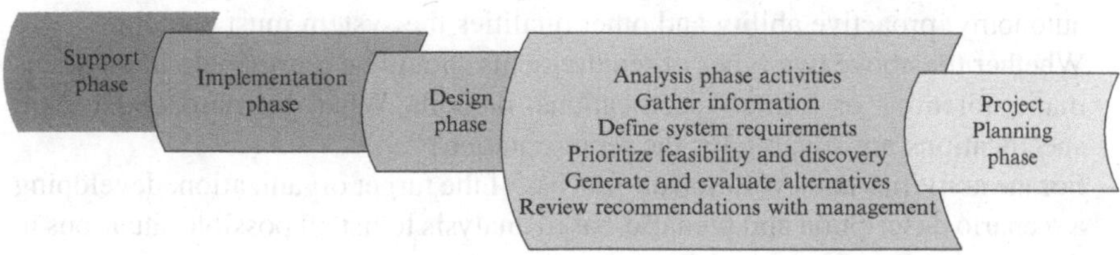

Fig. 6.2 The activities in the analysis phase

Table 6.1 Analysis phase activities and key questions

Analysis phase activities	Key questions
Gather information	Do we have all the information, goals, beliefs, intention, insight, desire, and environment we need to define what the system must do?
Define system requirements	What functional (stakeholders, goals, rules, policies, constraints, etc.) and nonfunctional (global qualities of the system) details do we need the system to include?
Prioritize requirements	What are the most important goals and interaction activities of the system?
Prototype for feasibility and discovery	Can the proposed organization-oriented model-building technology deal with what we think we need to do in the system? Have we built prototypes to ensure that users fully understand the potential of the technology?
Generate and evaluate alternatives	What is the best way for multiagent methodology to create the agent-based system?
Review recommendations with management	Should we continue with the design and implementation of the system we have proposed?

In object-oriented analysis, in general, the following modeling techniques are used to perform requirements analysis: use case and scenario description and use case diagrams, interaction diagrams, state charts, class diagrams, and so forth.

The following aspects must specifically be addressed in organization-oriented analysis:

- Organizational abstraction of agent organization in terms of organizational metaphor.
- Organizational decomposition in terms of organizational abstraction, goals, functions, or self-defined criteria.
- For functional requirements, building model-building blocks for organizational actors, interaction, organizational rules, organizational structure, environment, and goals; if required, beliefs, intention, desire, norms, policies, constraints and so forth may also need to be supported.
- For nonfunctional requirements, extracting what and how to support flexibility, autonomy, proactive ability and other qualities the system must contain.
- Whether the above two types of requirements should be represented in diagrammatic notations or formal specifications, or both. What diagrams and formal specifications are suitable for the representation?
- For the activities and evolutionary process of the target organization, developing a scenario description and scenario-based analysis to list all possible situations in the organization.
- For the state transformation process of stakeholders in the organization, listing the state chain sequence of each stakeholder to grasp the state transition in the organization.

6.4 Agent Service-Oriented Design

From the analysis and comparison of computational entities—objects, components, services, and agents—we get to know that services and agents are more suitable for modeling high-value business logic, state management, and automated and flexible functional elements in a complex software system. In this section, we define the concept *agent service*, and *agent service-oriented design* in our work. Lastly, we will give our reasons for advocating agent service-oriented design.

6.4.1 *Agent Service, Services of Agent, and Services of Service*

The strengths and benefits of agents and services with respect to "objects, components, services, and agents" are different. Services are discrete units of application logic that expose message-based interfaces suitable for access across a network, while agents are suited to model stakeholders with autonomous actions and cooperative ability to archive their design objectives. Therefore, both agent and service should be useful for abstracting the different OCAS stakeholders. More importantly, we believe in and advocate the combination of agent and service. This combination of agent and service, which we call *agent service*, demonstrates much greater power for modeling complex agent systems.

There are two meanings in our definition of *agent service*. On one hand, it is a loosely coupled entity; it represents one integrated abstract computational concept, which combines the functionalities of both agents and services. On the other hand, the aggregated entity is composed of two stakeholders—agent and service; it encompasses two computational units, and in some cases, one computational unit is relatively independent of the other. This is our conception of *agent service*.

When we say agent service, we need to clarify it with respect to a similar conception such as services of agent. For this consideration, we need to clarify another aspect of the concept "service". Apart from the meanings of service discussed above, service can also represent the dynamic functions and activities an agent can perform. In this case, a set of services belongs to an agent. Thus, we need to differentiate services which belong to an agent from agent service, so we call the former *services of agent*.

As a stand-alone computational unit, service also has services to perform. We call this *services of service*, which refers to the dynamic functions and activities a service will undertake.

6.4.2 *Why Agent Service-Oriented Design?*

As discussed earlier, current agent-oriented software engineering has met with many critical challenges resulting from the complexities in OCAS. OCAS is not a simple complex system; it could well be a very large open agent-based system. Here, *very large* has the following meanings: (1) There are probably dozens if not hundreds of types of agents and services. (2) The population of subsystems and subsystem components in the system could be very big, for instance, in the hundreds or thousands of agents and services. In addition, multiple levels exist within the system, and the number of business organizations and users involved on the Internet could be extremely big.

Importantly, the following specific challenges exist in current agent-oriented system design approaches in modeling OCAS:

- Services not explicitly split and supported.
- No explicit distinction between architectural design and detailed design in many approaches; most focus on detailed design.
- Most approaches focus on designing agents, but the interagent and the intra-agent mechanisms are not explicitly distinct.
- Management of agents and agent organization is not covered or is weakly supported.
- Most existing agent-based approaches cannot handle the development of the specific requirements of F-TRADE very well. For instance, one problem is how to support the Internet-based plug and play of agents and services. The main challenges come from the following software complexities:
- It is an open system; however, most of the current agent-based technologies under investigation are actually only suitable for closed or semi-open systems.
- It must be interoperably oriented toward enterprise applications. This brings new open problems for "traditional" agent-based architecture; for instance, the issue of how to locate, transport, and interoperate with those distributed heterogeneous enterprise applications.
- It is lodged in an Internet-based environment. Most agent systems are standalone systems, or at most local network-based. The F-TRADE is designed to be Internet-oriented. In this environment, new technology is required to locate, transport, and interoperate services and applications on the Internet.
- Computational entities with interoperable, flexible, and automated capabilities are needed in a network-based computing environment.

The conception of agent service, agent service architectural design, and detailed design is proposed to deal with software complexities in building open enterprise-automated systems.

6.4.3 What is Agent Service-Oriented Design?

In general, the activities which will be performed in the system design consist of the following: (1) design and integrate the network, (2) design the application architecture, (3) design the user interfaces, (4) design the system interfaces, (5) design

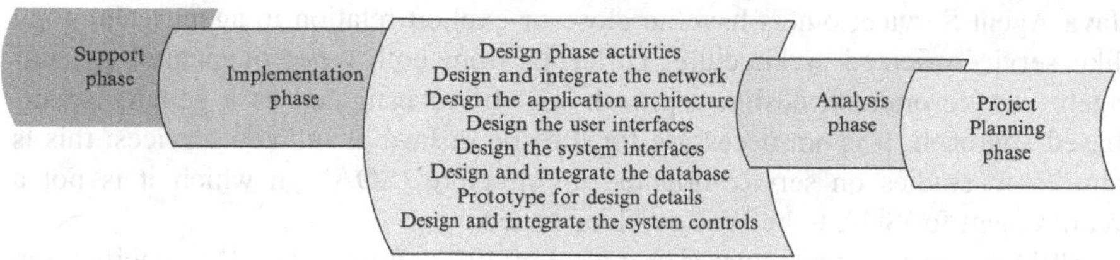

Fig. 6.3 The activities in the design phase

and integrate the database, (6) create a prototype of the design details, and (7) design and integrate the system controls. Figure 6.3 illustrates these activities.

Using the above definition of agent service, services of agent, and services of service, we advocate a system design method in our work which we call *agent service-oriented design*. *Agent service-oriented design* follows a top-down principle and is composed of two levels of system design mechanism. One is *agent service-oriented architectural design*, which focuses on high-level or global system design. Then, we go into detail and perform *agent service-oriented detailed design*, which investigates how to implement details in terms of computational unit agent service and other building blocks.

The current issues in developing open agent systems have shown that at this stage, there is no standard or commonly recognized development approach for solving specific problems. In reality, it is the work of every research group in the multiagent field to develop suitable agent-based approaches in terms of its own particular requirements and the generic agent-oriented methodology. In this process, it would be smart to absorb and aggregate any useful contributions from other approaches and extend them to achieve the final requirements for the specific problem.

In our observation of organization-oriented software engineering of agent systems, services as well as agents will be investigated for modeling business logics, organizing and managing interoperation in enterprise computing, wrapping legacy applications, and the like. Thus, it can be seen that agent services are two fundamental computational units. In addition, an agent system and its environment will be explicitly abstracted and modeled in the metaphor of organization and society [24]. Activities of agents and services, roles, organizational hierarchy and structure, local and global organizational rules, and interactions inside the organization and with external humans and applications are the main tasks in abstraction design. The above work of system design forms the approach of agent service-oriented design. We believe it encapsulates intrinsic strengths for dealing with such open enterprise complex software systems as F-TRADE.

The agent service-oriented design approach is a new research direction. We have listed some insights of this approach in the above architectural design and detailed design. In addition, relevant work can be accessed from FIPA Abstract Architecture [25], Java Agent Services [26], Web services [27], and service-oriented architecture [28]. Some of these technologies focus on specific services like Web services and

Java Agent Service; others have no close or explicit relation to agent technology like service-oriented architecture. Differing from both types of technology, our agent service-oriented design approach will be investigated as a generic agent-based approach. It is not necessary for it to be in Java or in Web services; this is similar to studies on service-oriented architecture (SOA), in which it is not a requirement for SOA to be Web services-based.

FIPA Abstract Architecture is proposed by FIPA. It provides (1) a specification that defines architectural elements and their relationships [25]; (2) guidelines for the specification of agent systems in terms of particular software and communications technologies; (3) specifications governing the interoperability and conformance of agents and agent systems. In the specification of FIPA Abstract Architecture, both the abstract agent model and service model are described. Our work for abstract agent and service models will be based on the above FIPA abstraction models, maintaining compatibility with other FIPA specifications related to agents and services. However, as a complete and operable agent-based approach, we will also investigate agent service architectural design and detailed design, which has not been done by FIPA, so that we can build agent service-oriented system design techniques hierarchically and systematically. To this end, we may need (1) to consider more systematic functional supports, and (2) to extend or complement FIPA specifications on demand. To support (1), we will investigate other relevant approaches, adopt effective elements from them, and integrate them into our approach on demand. From this perspective, our agent service-oriented design approach will be an instantiation and extension of FIPA Abstract Architecture specifications and other related agent specifications.

To build the specific modeling techniques for MAS, we will adopt intuition from GAIA and TROPOS. GAIA is good at conceptual modeling in terms of organizational metaphor, while TROPOS provides concrete and complete model-building blocks. Moreover, we will not be limited to GAIA, TROPOS, or FIPA Abstract Architecture since all of them are still under development in supporting open agent systems. Strengths and functionalities from other related organizational abstractions will also be considered in building our system. For instance, Agent UML [29] can be used for modeling interaction protocols; goal-oriented analysis [30, 31] can be conducted in early requirements engineering; and service-oriented architecture and service abstraction can be used for architectural design.

Agents are suited to model entities with autonomous actions and cooperative ability to archive their design objectives, while services are discrete units of application logic related to entities which expose message-based interfaces suitable for being positioned, accessed, transported, and discovered across a network. In the detailed design phase, both an agent model and a service model will be built.

Besides the issues addressed in the above discussion on architectural and detailed design, the following aspects will be investigated in detail to build an agent service-oriented design approach: (1) agent service ontology and representation, (2) naming and registry of agent services, (3) modeling of agent services,

(4) agent service directory, (5) agent service transport, (6) agent service mediation and management, and (7) agent service discovery.

6.4.4 Agent Service-Oriented Architectural Design

System design at the highest level normally takes the form of architectural design, general design, and conceptual design [23]. Architectural design determines the overall structure and design solutions. It specifies in detail how all the system-level activities will actually be carried out, the main system components that exist or are required, and how the structure and interrelationship of these components will be designed and organized. Specification, definition, and organization at this level will follow a methodology but should be independent of the use of specific technology.

In object-oriented architectural design, for instance, the following models—physical data flow diagrams, structure charts, object-interaction diagrams, and other physical models—are usually built to define and organize the application architecture. However, they are independent of specific object-oriented technology, such as whether they will be instantiated in Java-based object-oriented technology or C++-based technology.

For agent service-oriented architectural design, we specify the following tasks:

- Agent service abstract model: This is to define the architecture of agent and service, their types, and the capabilities of the main agent services.
- Agent service functional groups: These are self-contained (possibly active) coarse-grained agent suborganizations that perform certain closely related and aggregated subgoals, functions, and activities.
- Organizational interaction: Defines and describes the interrelationships in agent service functional groups at the subsystem level.
- Organizational management: Simulates the organization and management mechanisms [32, 33] to sketch the organization and management of an agent service in an agent organization, which supports the organic linkage and dynamic evolution of agent service functional groups. It also develops high-level mechanisms for inter-subsystem directory, communication, transport, mediation, transformation, and discovery of agent services.
- Organizational structure: Defines the organizational topological structure and interactive linkage at the subsystem level in order to link all agent service functional groups.
- Organizational architectural frameworks: Specify reusable organization-oriented patterns at the system framework level.
- System data structure: Describes how to organize data sources and the data structures of agents and services at the level of organization or suborganization.

6.4.5 Agent Service-Oriented Detailed Design

The detailed design looks after all the model-building blocks which will eventually be concrete and implementable in terms of the architectural design. Specific design techniques are usually utilized in detailed design to crystallize model elements in the architectural model into the finest grain that is operable for direct implementation.

In object-oriented detailed design, the following models are some of the models usually developed to represent the concrete objects, states, and interactive relationships: package diagram, design class diagram, object database schema. These models are also widely used for agent-based detailed design; however, they are inadequate for representing agent service-based systems.

Agent service-oriented detailed design focuses on developing the internal structure of each agent and service, and how it will achieve its tasks within the system. The focus is on defining capabilities (modules within the agent), in terms of internal events, plans, and detailed data structures.

We define the following modeling tasks for the agent service-oriented detailed design by tracing the models in architectural design:

- Agent service design model: This is to build an instantiated agent model and service model by specifying instantiated and detailed types, roles, activities, architecture, goals, beliefs, intentions, capabilities, and input and output interfaces and event handling, exception handling, and running thread control in terms of the specific implementation language.
- Agent service interaction model: This is to define interagent and intra-agent interaction protocols, patterns, message-passing mechanisms, message content, and communication language.
- Agent service management model: This is to detail and instantiate the interagent service directory, communication, transport, mediation, transformation, and discovery of agent service.
- Agent service data structure model: This is to specify the detailed data structure of each agent or service and interaction of an agent or a service with its data sources.

6.5 Building Organization and Service-Oriented Software Engineering

Agent-based systems for dealing with complex problems are becoming increasingly complicated. We call middle-sized and large-scale agent systems "open complex agent systems". To engineer this type of agent system, we have established that there are many issues with the current agent-oriented software engineering approaches.

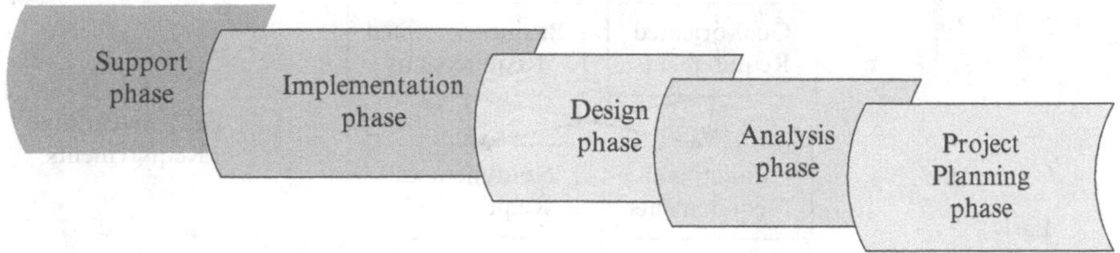

Fig. 6.4 Software development life-cycle phases

We have proposed and discussed the relevant solutions for analyzing and designing OCAS in terms of organizational metaphor. They are *organization-oriented analysis* for system analysis, *agent service-oriented design* for system design, and *Java Agent Service-oriented implementation.* These consist of complete technical solutions for software engineering covering three main life-cycle phases (as shown in Fig. 6.4), namely, the system analysis, design, and implementation of OCAS. We call the integrated solution of the above three phases *organization and service-oriented analysis and design* (OSOAD). Figure 6.5 shows the process of the OSOAD approach.

As shown in Fig. 6.5, the OSOAD approach is composed of the following five phases from the top down: (1) early requirements analysis, (2) late requirements analysis, (3) architectural design, (4) detailed design, and (5) implementation. (1) and (2) form the system analysis, and (3) and (4) consist of the system design:

- Early requirements is concerned with understanding a problem by studying an organizational setting; the output is an organizational model which includes relevant actors, their goals, and dependencies in terms of certain organizational abstraction and organization decomposition rules.
- Late requirements determine how the system is to be described within its operational environment and model-building blocks, along with relevant functions and qualities.
- Architectural design defines the system's global architecture in terms of subsystems, interconnectedness through data, control, and dependencies.
- Detailed design defines the behaviors of each architectural component in further details.
- Implementation, specific development, and deployment in terms of the detailed design.

In the early requirements analysis phase, the integrative requirements [7] consist of goal-oriented requirements and business-oriented requirements. Both contribute to the functional and nonfunctional requirements. After obtaining the basic early requirements of the problem domain, an organizational abstraction of the domain problem is conducted and organizational decomposition is performed in terms of goals or other system functions. An organizational framework called ORGANISED can be used for organizational abstraction and organizational decomposition. ORGANISED is the acronym of the following key organizational members: organizational metaphor, rules, goals, actors, norms, interaction, structure, environment, and dynamics.

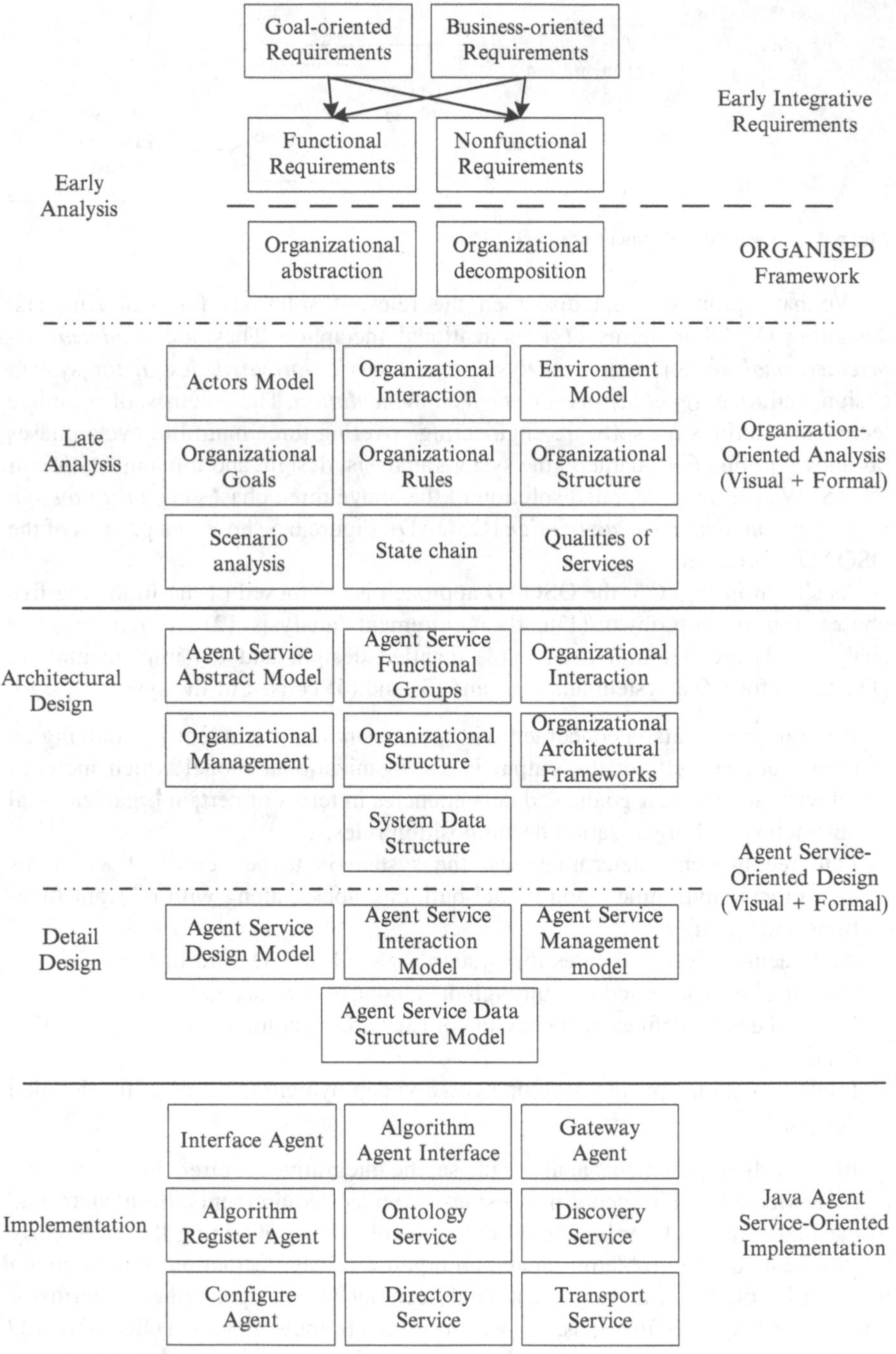

Fig. 6.5 OSOAD approach

The late requirements analysis is fulfilled through organization-oriented analysis, using the organizational metaphor. The organization-oriented analysis approach tries to model a system in terms of the idea that an agent system is an artificial organization. The organizational structures, rules, relationships, actors, interactions, and environments will be embodied in the modeling.

Next is architectural design. The architectural design will be achieved through modeling (1) the agent service abstract model, (2) agent service functional groups, (3) organizational interaction, (4) organizational management, (5) organizational structure, (6) organizational architectural frameworks, and (7) system data structure.

Models such as (1) agent service design model, (2) agent service interaction model, (3) agent service management model, and (4) agent service data structure model will be developed as tasks in the detailed design phase to detail the architectural design.

We integrate the architectural design and detailed design phases into the agent service-oriented design; both follow the same organizational abstraction and the significant modeling technique agent service.

It is recommended that the implementation of the detailed design is performed in Java Agent Service. Taking F-TRADE as an instance, agent classes such as interface agent, configure agent, gateway agent, algorithm register agent, and algorithm interface agent are instantiated; services such as ontology service, directory service, discovery service, transport service, and so forth are programmed in Java Agent Service.

6.6 Summary

Taking an integrative approach that combines organizational computing and agent service-based computing, this chapter has presented the techniques for abstracting open complex systems in terms of organizational metaphor, for analyzing complex systems in terms of organizational computing, and for designing a system using agent services. The techniques and tools presented are intended to create an organization- and agent service-oriented software engineering method for engineering complex systems.

In the following chapters, visual modeling and formal modeling are introduced in Chaps. 7 and 8, followed by integrative modeling, which combines formal modeling and visual modeling. In Chaps. 10 and 11, agent service-based architectural design and detailed design will be introduced.

References

1. Cao, L., Zhang, C., Dai, R.: Organization-oriented analysis of open complex agent systems. Int. J. Intell. Control Syst. **10**(2), 114–122 (2005)
2. Cao, L., Luo, C., Li, C., Zhang, C., Dai, R.: Open giant intelligent information systems and its agent-oriented abstraction mechanism. In: Proceedings of the Fifteenth International

Conference on Software Engineering and Knowledge Engineering (SEKE 2003), San Francisco, 1–3 July 2003, pp. 85–89. ISBN: 1-891706-12-8
3. Cao, L.: Integrating agents, services, organizational and social computing. Int. J. Softw. Eng. Knowl. Eng. **18**(5), 573–596 (2008)
4. Cao, L., Zhang, C., Zhou, M.: Engineering open complex agent systems: a case study. IEEE Trans. Syst Man, Cybern. C. Appl. Rev. (2008), **38**(4), 483–496
5. Cao, L., Dai, R.: Agent-oriented metasynthetic engineering for decision making. Int. J. Inf. Technol. Decis. Making **2**(2), 197–215 (2003), World Scientific Publishing
6. Cao, L., Zhang, C., Ni, J.: Agent services-oriented architectural design of open complex agent systems. In: The 2005 IEEE/WIC/ACM International Conference on Intelligent Agent Technology (IAT'05). Compiegne, France, 19–22 September 2005. doi:10.1109/IAT.2005.28
7. Cao, L., Zhang, C., Luo, D., Chen, W., Zamani, N.: Integrative early requirements analysis for agent-based systems. In: The 4th International Conference on Hybrid Intelligent Systems, IEEE Computer Society, Kitakyushu, Japan, 5–8 December 2004. doi.ieeecomputersociety.org/10.1109/ICHIS.2004.63
8. Cao, L., Li, C., Zhang, C., Dai, R.: Open giant intelligent information systems and its agent-oriented analysis and design. In: Proceedings of the 2003 International Conference on Software Engineering Research and Practice (SERP'03), vol. 2, pp. 816–822, Las Vegas, 23–26 June 2003. CSREA Press. ISBN: 1-932415-20-3
9. Zambonelli, F., Jennings, N.R., Wooldridge, M.: Developing multiagent systems: the GAIA methodology. ACM Trans. Softw. Eng. Methodol. **12**(3), 317–370 (2003)
10. Carley, K.M.: Computational and mathematical organization theory: perspective and directions. Comput. Math. Organ. Theor. **1**(1), 39–56 (1995)
11. Cao, L., Zhang, C., Dai, R.: The OSOAD methodology for open complex agent systems. Int. J. Intell. Control Syst. **10**(4), 277–285 (2005)
12. Odell, J.J., Parunak, H.V.D., Fleischer, M., Brueckner, S.: Modeling Agents and Their Environment. In: AOSE (2002)
13. Stamper, R.: Social norms in requirements analysis—an outline of MEASUR. In: Requirements Engineering: Technical and Social Aspects. Academic, New York (1993)
14. Greaves, M., Holmback, M., Bradshaw, J.: What is a conversation policy? In: Autonomous Agents'99 Special Workshop on Conversation Policies, pp. 118–131. Springer, Berlin/Heidelberg (1999)
15. Shoham, Y., Tennenholtz, M.: On social laws for artificial agent societies: off-line design. Artif. Intell. **73**(1–2), 231–252 (1995)
16. Gasser, L.: Social conceptions of knowledge and action: DAI foundations and open system semantics. Artif. Intell. **47**, 107–138 (1991)
17. Ferber, J.: Multi-agent Systems: An Introduction to Distributed Artificial Intelligence. Addison-Wesley Longman, Harlow (1998)
18. Pattison, H.E., Corkill, D., Lesser, V.: Instantiating descriptions of organizational structures. In: Huhns, M. (ed.) Distributed Artificial Intelligence, pp. 59–96. Pitman, London (1987)
19. Zambonelli, F., Jennings, N.R., Wooldridge, M.: Organisational abstractions for the analysis and design of multi-agent systems. In: Proceedings of 1st International Workshop on Agent-Oriented Software Engineering, Limerick, pp. 127–141. Ireland (2000)
20. Dastani, M., et al.: Issues in multiagent system development. In: Proceedings of the 3rd International Joint Conference on Autonomous Agents and Multi-agent Systems, AAMAS2004, pp. 922–929. ACM Press, New York
21. Giorgini, P., Kolp, M., Mylopoulos, J.: Multi-agents architectures as organizational structures. Int. J. Autonom. Agents Multi-Agent Syst. **13**(1), 3–25 (2006)
22. Carley, K.M.: Artificial social agents. In: Smelser, N., Baltes, P. (eds.) International Encycopedia of the Social & Behavioral Sciences, pp. 811–816. Elsevier, Amsterdam (2001)
23. Satzinger, J.W., Jackson, R.B., Burd, S.D.: Systems Analysis and Design in a Changing World, 2nd edn. Course Technology (Thomson Learning), Boston (2002)
24. Pitt, J. (ed.): Open Agent Societies: Normative Specifications in Multi-agent Systems. Wiley, Chichester (2003)
25. FIPA SC00001L. FIPA Abstract Architecture specification (2002)

26. JSR 87: Java Agent Services. http://www.jcp.org/aboutJava/communityprocess/review/jsr087/
27. Barry, D.K.: Web Services and Service-Oriented Architectures: The Savvy Manager's Guide. Morgan Kaufmann, New York (2003)
28. Erl, T.: Service-Oriented Architecture: A Field Guide to Integrating XML and Web Services. Pearson Education, Upper Saddle River (2004)
29. http://www.auml.org
30. Mylopoulos, J., Chung, L., Yu, E.: From object-oriented to goal-oriented requirements analysis. Commun. ACM **42**(1), 31–37 (1999)
31. Kolp, M., Giorgini, P., Mylopoulos, J.: A goal-based organizational perspective on multiagent architectures. In: Intelligent Agents VIII: Agent Theories, Architectures, and Languages. LNAI-2333, pp. 128–140. Springer, Berlin/Heidelberg (2002)
32. Handy, C.: Understanding Organizations. Penguin, London (1993)
33. Mintzberg, H.: The Structuring of Organizations: A Synthesis of the Research. Prentice-Hall, Englewood Cliffs (1979)
34. Castro, J., Kolp, M., Mylopoulos, J.: Towards requirements-driven information systems engineering: the Tropos project. Inf. Syst. **27**(6), 365–389 (2002)

Chapter 7
Visual Modeling

7.1 Introduction

Visual modeling means to model a system and its components, relationships and interactions, structures and patterns, etc. by setting up visual building blocks. Visual modeling adopts reductionism methodology and focuses on modeling each part, and the interaction between parts, to achieve an overall picture of a complex system.

A number of visual modeling methods have been proposed in software engineering. Typically, it is necessary to apply a çertain metaphor in order to map system complexities in a system to that space. For instance, organization-oriented modeling maps a system to an organization, while service-oriented computing captures services in a system. Agent-based modeling usually maps system components to agents and agent interactions.

In this chapter, taking an organizational computing approach, building blocks are constructed to describe components in a complex system. A complex system is represented in terms of key modeling units, namely: the actor model and actor's role model, environment model, organizational rules, organizational structure, organizational dynamics, interaction, patterns, and system–environment interaction. These components constitute the essential content of this chapter.

7.2 Actor Model

7.2.1 Actor Classification

We explicitly differentiate between actors and classify them in terms of the main categories of entities, functions, and roles in an open agent organization. The actor set consists of the following actors: *human actor*, *workspace actor*, *autonomous actor*, and *service actor*.

L. Cao, *Metasynthetic Computing and Engineering of Complex Systems*,
Advanced Information and Knowledge Processing,
DOI 10.1007/978-1-4471-6551-4_7

- Human (♀): Human beings related to the system.
- Workspace actor (○): Refers universally to the proposed system or problem space.
- Autonomous actor (Ⓣ): System components which fulfill some responsibilities or decision control by themselves.
- Service actor (Ⓣ): Performs activities or functions associated with certain autonomous actors.
- Entity actor (○): May take the form of component, belief, desire, intention, interaction, or self-defined entity.
- Resource (▭): Various databases, knowledge bases, configuration files, etc.

The autonomous actor usually takes its form as agent in an agent-based system. With the introduction of autonomous actors (agents), some functions are implemented autonomously on behalf of human beings. Service actors will be more and more commonly seen in large-scale software systems with the emergence of service-oriented architecture [1]. Table 7.1 lists all the above actors and examples.

Table 7.1 Actor model

Actor name	Symbol	Informal definition	Instances
Human	♀	Human actors are those social actors—human beings— who could be customers, users, administrators and the like who will interact with the system and the granting of permissions in various ways.	Customer, administrator
Workspace	○	Workspace actor is the place where all actors, interactions, and activities will be lodged and undertaken; a common workspace is the proposed system	F-TRADE
Autonomous (agent)	Ⓣ	Artificial actors are system components or agents which fulfill some responsibilities or decision control by themselves; these are usually agents or components	Interface agent, optimizer agent, gateway agent
Service	Ⓣ	Service actors are actors who can perform activities or functions which are associated with certain autonomous actors; these could be services associated with agents	Ontology service, directory service, transport service, discovery service
Entity	○	Entity actors consist of system or subsystem components and abstract entities such as belief, desire, intention, interaction, position, and user self-defined entities	Subsystem components, belief, interaction, user-defined entities
Resource	▭	Various databases, knowledge bases, configuration files, etc.	Databases, knowledge bases, etc.

Every actor has associated attributes (Ⓐ), and creation and inner properties (Ⓟ). Actor agents and services may have intentional (subjectively planned) attributes such as belief, desire, intention, etc. Some actors, such as agents and services, take roles (Ⓡ); an actor may take one or many roles. The following section introduces how to model a role held by an actor.

7.2.2 Role Model

A role is related to basic social commitments or obligations carried out by an agent or a service within the agent society. At this stage, we define only that either an agent or a service can play a role. When an agent or service joins a group or society, it holds one or more roles, thereby acquiring certain commitments that go with those roles. The commitments or obligations of a role on an agent define the restrictions on how the agent acts and communicates with other roles. In general, these commitments are embedded and embodied in some form of interaction protocols and patterns.

A *role model* of agent service in an agency explicitly and precisely describes all relevant tasks, functionalities, activities, and responsibilities [2] relevant to a role taken by the agent service. It also involves protocols and patterns of interaction with other roles in a society of agents.

The informal modeling of roles has been undertaken in GAIA [2]. The basic specifications of a role schema in GAIA are defined in description language in Fig. 7.1. A role schema specifies functionalities, activities, responsibilities, and interaction protocols, which are key elements representing the schema of a role. A formal representation of the GAIA role schema will be presented in Sect. 8.5.

Based on the GAIA role schema, informal role models can be built for each of roles in F-TRADE [47, 51]. For instance, there is a role called PLUGINPERSON which is in charge of the soft plug and play[1] function of algorithms and agents. The role model can be built for the PLUGINPERSON as shown in Fig. 7.2. This specifies the attributes of permissions, responsibilities, protocols, and activities of the role PLUGINPERSON in F-TRADE.

The objective of this role is to plug in an algorithm to F-TRADE which is nonexistent in the current algorithm base. To fulfill this goal, a number of commitments must be performed by the role. For instance, the agent playing this role will execute the protocol ReadRegistration, followed by the activities *ApplyRegistration* and *FillinAttributeItems*, and will finally go through the protocol SubmitAlgoPluginRequest. The role has rights to read the algorithm from a non-plug-in directory and can change the application content for the registration and attributes of the algorithm. As defined in the preconditions, the agent is required to ensure that two safety responsibility constraints are satisfied.

[1] Supports the immediate playing of software plug-ins after being plugged in and registered to the system.

- Role Schema *name of role*

- Description *short description of the role*

- Protocols *activities that an agent performs require interaction with others*

- Activities *corresponds to a unit of action that an agent performs and that does not involve interaction with any other agent*

- Permissions *identifying what can and/or cannot be spent while carrying out the role*

- Responsibilities *the expected behaviors of a role*
- Liveness *"something good happens," that is, describes those states of affairs that an agent must bring about, given certain conditions*

- Safety *"nothing bad happens," that is, an acceptable state of affairs (invariants) is maintained*

Fig. 7.1 GAIA role schema template

Role Schema: **PLUGINPERSON**

Description:

This role involves applying registration of a nonexistent algorithm, typing in attribute items of the algorithm, and submitting plug in request to F-TRADE.

Protocols:

ApplyRegistration, CheckAlgorithmValidity, SubmitAlgoPluginRequest

Activities:

FillinAlgoRegisterOntology

Permissions:

calls	*PluginInterfaces*	*// a plugin interface will be called*
reads	*AlgorithmID*	*// ID of an algorithm is accessed*
fills	*AlgoRegisterOntologies*	*// all ontological items for pluggin an algorithm*

Responsibilities

Liveness:

PLUGINPERSON = (ApplyRegistration).(CheckAlgorithmValidity).
(FillinAttributeItems)+.(SubmitAlgoPluginRequest)

Safety:

- The algorithm agent has been programmed by implementing AlgoInterface agent and ResourceInterface agent, and is available for plug in.
- This algorithm hasn't been plugged into the algorithm base.

Fig. 7.2 Schema for role PLUGINPERSON

7.3 Environment Model

The modeling environment involves the specification of all actors, resources, principles, processes, forces, interactions, and system (or subsystem) boundaries in an agent environment and between an agent and its environment. The environment encompasses resources that the organization can exploit, control, or consume as it moves toward the achievement of its goals. These elements are embodied in the environment model and organizational interaction and structure models.

7.3.1 Characteristics of Agent Environment

The world of an agent environment can vary greatly in different agent systems. To summarize the main features of the environment in variant agent systems, an abstract but comprehensive agent–environment model can be built. We can collect the main features from most of the agent environment and list all of them together on the abstract agent–environment model. Basic characteristics [3, 4] for the agent environment model include accessibility, observability, determinism, uncertainty, diversity, controllability, volatility, temporality (continuity), locality, and spatiality. We summarize these characteristics and list them in Table 7.2.

Table 7.2 Characteristics of agent environment

Feature	Description
Accessibility	To what extent is the environment known and available to the agent? An environment is effectively accessible if the agent can access the environmental state relevant to the agent's choice of action. Another consideration is whether the available resources are ample or restricted
Observability	Whether the uncertain outcomes of actions to the environment are not observable, partially observable, or fully observable
Determinism	To what extent can the agent predict events in the environment? The environment is deterministic when the next state of the environment can be determined by the current state and the actions selected by the agents
Uncertainty	The next state of the environment is unknown, or it is uncertain what will happen to the next action and state
Diversity	How homogeneous or heterogeneous are the entities in the environment?
Controllability	To what extent can the agent modify its environment?
Volatility	How much can the environment change while the agent is deliberating?
Temporality	Is time divided in a clearly defined manner? For example, do actions occur continuously or in discrete time steps or episodes?
Locality	Does the agent have a distinct location in the environment which may or may not be the same as the location of other agents sharing the same environment, or are all agents virtually collocated? Also, how is a particular locality expressed (e.g., coordinate system, distance metrics, relative positioning)?
Spatiality	What designated area is there for agents lodging in the environment? For instance, the region designated to agents could be a flat plane shape or a torus space

Viewed from any one of the above aspects, an agent environment can be:

- Fully, partially, or not observable or accessible
- Static or dynamic
- Controllable or uncontrollable

or

- The next state of the environment could be deterministic, stochastic, or adaptive.
- The task for the environment agent could be episodic or sequential.
- The environment, perceptions, and actions could be discrete or continuous.
- The environment may include single or multiple agents or services.

In the real world of multiagent systems, the agent environment usually takes the form of a synthetic agency with a collection of potential features such as inaccessibility, stochasticity, sequentiality, dynamism, and continuity. For instance, the artificial ant colony exists in an accessible, discrete grid space with continuously changing hormone and stochastic state transferral.

7.3.2 Classification of Agent Environment

Environment provides the surroundings and conditions under which an agent or an agent organization can exist and function. The environment can be classified into types, such as the physical environment or electronic environment, in terms of the niche in which the agents remain, whether in a real world or in a virtual space.

7.3.2.1 Physical Environment

This is a common kind of agent environment which biological (physical or artificial) agents, for instance, ant colony and robots, require for survival. It can be either a physical or ecological niche, and is normally in the real world.

The physical environment provides those principles and processes that govern and support a population of agents [4]. For agents, the principles of the physical environment encompass the laws, rules, constraints, and policies that govern and support the physical existence of agents. The processes of the agent environment refer to autonomous execution units in a computational system, where agents sense the environmental state and perform an action on it to evolve into a new state.

To support the above function and execution of the physical environment, a number of primary components are required. They include (1) a nerve center for the execution, monitoring, security, and management of agents, ontologies, and operations; (2) communication and transportation for interaction, coordination, cooperation, and planning internally and/or externally; and (3) interaction interfaces for human–agent, agent–agent, and agent–environment interaction (AEI), which incorporate interaction channels such as network and media, and signals and interfaces such as sensors, actuators, and controllers.

7.3.2.2 Electronic Environment

This is normally an electronic or virtual agent environment in which a relatively large number of artificial agents exist; for instance, FIPA-compatible agents, or agents made only of code. The ecological niche of this kind of agent environment is normally an electronic, virtual world.

Despite the physical inaccessibility, the electronic environment is similarly under the control of artificial principles and processes, but unlike the physical environment, the principles and processes of the electronic environment require more electronic knowledge and are service-intensive. They are usually encoded and decoded into the electronic computational system, or "artificial agent society", and they also regulate and deregulate the system. Another distinction is that any possible form of principles and processes can be designed, implemented, and tested in an artificial agent environment, either through simulating real counterparts in the real world or worked out without foundation.

In the system analysis of MAS, AEI studies are attracting more and more attention from an artificial social point of view. For an open agent system, whether it is physical or electronic, the interaction between agents and the agent environment is societal. In this situation, an agent environment can also be considered as the social settings of an agent or an agent organization.

7.3.2.3 Social Environment

This is a socio-ecological niche in which social groups of agents coexist, cooperate, and coordinate in a self-organized, self-adaptive, and evolutionary sociocultural body.

In this society, principles and processes are more or less social in that they monitor and maintain the interaction between agents and the agent environment in a bilaterally or multilaterally shared and cooperative context. In this context, communication, coordination, cooperation, negotiation, planning, competition, and the management of agents and the agent environment are implicitly or explicitly self-organized social activities in which multiple agents become involved. All the above social activities form the *social interaction*.

The following services and supports are involved in the social interaction: interaction languages, interaction protocols, interaction patterns, social constraints, rules and policies, and so forth. In other words, social interaction is like a dialogue between agents and the agent environment which follows social orders, policies, principles, and the like.

On the other hand, based on analysis of the characteristics of the agent environment and the behavioral and mechanistic complexity of the environment, other types of agent environment can also be specified—for instance, accessible or inaccessible environments—but we do not list the detail here.

Relevantly, certain types of AEI could also take various forms in terms of the differentiated characteristics of the agent environment. For instance, in rational, reactive, and situated agent systems such as softbot and robots, the sense–actuate (percept–action) mechanism is widely seen in the interaction between agents and

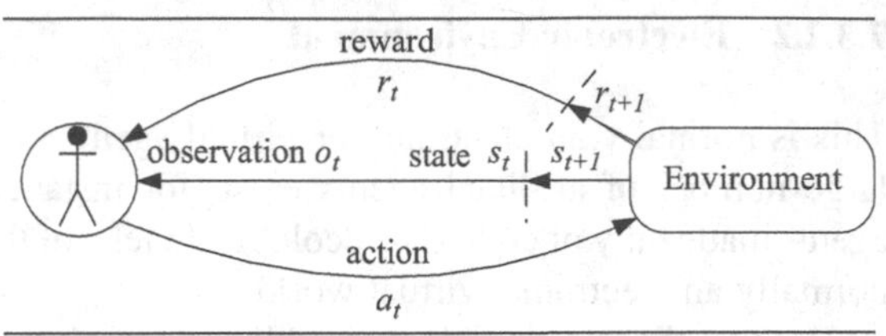

Fig. 7.3 Agent–environment interaction

the agent environment. In these systems, the agent environment is somehow passive compared to agents. However, more active dialogue between agents and environment through message passing is more visible in information agent systems. In such systems, both agents and environment (agents) are active in interaction.

7.3.3 POMDP$_{AEI}$ Model

The modeling environment must also cover the AEI. By introducing the environment observations, an AEI can be modeled as a partially observable Markov decision process (POMDP) [5, 55], which we called the POMDP$_{\mathrm{AEI}}$ model. The POMDP$_{\mathrm{AEI}}$ model is a six-element tuple

$$\mathrm{POMDP_{AEI}} = < \mathrm{S, A, T, R, O, D} >, \tag{7.1}$$

where $\mathrm{O}(s_t, r_t)$ is the finite set of observations related to state s_t and reward r_t and D (a_t, s_{t+1}, d_{t+1}) is the probability of making observation d_{t+1} from state s_{t+1} after having taken action a_t. Figure 7.3 shows the POMDP$_{\mathrm{AEI}}$ model. We will illustrate the AEI dynamics again in Sect. 7.6.

7.4 Modeling Organizational Rules

From the perspective of system design, we define two types of organizational rules: structural rules and problem-solving rules. In the following section, we will discuss them individually.

7.4.1 Structural Rules

Listed below are the four main *structural rules* in agent organizations and their modeling notations. In the notations, symbols A and B represent actors, roles, or goals:

- Control (A→B): B is controlled by A, or B can be achieved by means of A.
- Peer (A↔B): A and B share a peer-to-peer relation.

Table 7.3 Dependency relationships

Dependency	Symbol	Informal definition
Single directional dependency	A→B	A depends solely on B
Bidirectional dependencies	A↔B	A depends on B for some situations; for other situations B depends on A
And/or composition dependency	A, B, C → D	A, B, and C depend on D for one and/or multiple dependencies
And/or decomposition dependency	B, C, D → A	A depends on B, C, and D for one and/or multiple dependencies

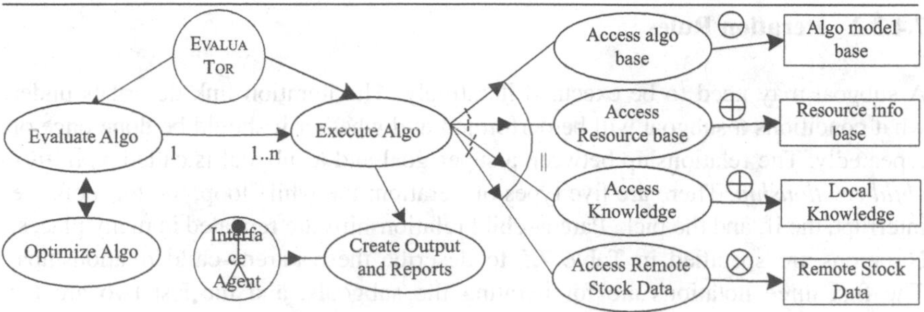

Fig. 7.4 Combination of organization rules

- Ownership (A→B): A owns B.
- Dependency (A→B): A depends on B.

Some of the above relationships normally coexist in certain combinations in a complex system. For instance, Table 7.3 shows four types of combinations of dependency relationships. In Sect. 7.4.3, we will discuss the combination of rules, and Fig. 7.4 will illustrate a number of rule combinations.

In a real agent organization, the above four basic structural rules and their ramifications are usually combined to represent the more complex linkage that exists between multiple actors. In a subsystem, for instance, multiple actors may coexist as follows: actors A and B depend on actor D to conduct a goal, while D depends on C to perform another goal.

7.4.2 Problem-Solving Rules

Some of the main *problem-solving rules* consist of rules for the analysis of means–ends [6], contribution [6], goal decomposition, iteration, and cardinality. The means–ends link represents an alternative way to achieve a goal by elements connected to the goal. In the following description, we only discuss the latter four types of rules individually.

7.4.2.1 Goal Decomposition Rules

Goal decomposition (—+) defines a refinement of a goal temporally or spatially. In the process of decomposition, a goal will be divided into multiple subgoals. In a real dynamic domain, there could be additional constraints, for instance, temporal order, on the subgoals of a goal. We call the relationships between subgoals "sister relationships". *Sister relationship* links are applied to groups of decomposition connections in building goal and structure models. There are four types of combination among subgoals: sequence, alternation, concurrency, and hybrid. They are shown in Table 7.4.

7.4.2.2 Iteration Rules

A subgoal may need to be executed iteratively. The iteration link describes under what conditions a subgoal will be performed and whether it should be done once or repeatedly. The relationship between a super goal and a subgoal is called a *parent–child relationship*. There are five types of iteration: the while loop, the for loop, the interrupt, the if, and the pick. Parent–child relationships are repeated in many places. Notations are specified in Table 7.5 to describe these parent–child relationships. The first three notations are for iterating the subgoals, and the last two are for

Table 7.4 Goal composition rules

Composition	Symbol	Informal definition
Sequence		Subgoals are performed from left to right sequentially to complete the goal; we take sequence to be the default and generally leave it out
Alternation		A goal can be fulfilled by either subgoal A or B
Concurrency		A goal can only be fulfilled by all decomposed subgoals in parallel
Hybrid		A goal is fulfilled by performing multiple subgoals in an order that combines some of the above three relationships

Table 7.5 Iterative rules

Iteration	Symbol	Annotation	Informal definition
While loop	⊕	**while(condition)*	Reiterate the subgoal while the "condition" is satisfied
For loop	⊗	**for(variable, listOfValues)*	List of values for the variable in the subgoal will be held iteratively
Interrupt	⊘	**whenever (variableList, condition)*	Values for the variables in the subgoal will be held whenever the condition is satisfied
If	⊖	*!if(condition)*	The subgoal will be operationalized if the condition is satisfied
Pick	⊖	*!pick(variableList, condition)*	Values for the variables in the subgoal that satisfy the condition will be picked nondeterministically

nondeterministically picking values for the variables in the subgoal that satisfy the condition. Iteration relationships can be notated by a generic formula, so that they can be labeled on diagrams. The generic formula for presenting iteration links is as follows:

$$\text{[Notation] Predicate (Parameter List)} \tag{7.2}$$

7.4.2.3 Contribution Rules

As in the *i** framework, a contribution link describes the impact that a functional goal or subgoal has on other nonfunctional goals. The impact can be negative or positive, and its extension can be partial or sufficient. The notations used for labeling contributions are $\xrightarrow{+,-}$. On the arrow, one or two "+" represent a partial or sufficient positive contribution; while one or two "–" represent a partial or sufficient negative contribution.

7.4.2.4 Cardinality Rules

Fulfilling a goal α may be dependent on achieving one to many goals β. This refers to cardinality constraints on goals in an organization. There are four types of cardinality: mandatory one ($\alpha \rightarrow \beta$), mandatory many ($\alpha \xrightarrow{1 \quad 1...n} \beta$), optional one ($\alpha \xrightarrow{1 \quad [0,1]} \beta$), and optional many ($\alpha \xrightarrow{1 \quad [1...n]} \beta$). We define them in Table 7.6.

7.4.3 Rule Combinations

In reality, it is very common to combine structural and problem-solving rules in an agent system. In building the goal, structure, and later interaction models, most kinds of organizational rules will be combined to describe the dynamics in a real domain. For instance, notations for sister and parent–child relationships and cardinality constraints may be added to the goal composition notations, so that not only the activity of decomposition and composition can be described, but organizational

Table 7.6 Cardinality rules

Cardinality	Symbol	Informal definition
Mandatory one	$A \rightarrow B$	One A must be related to one B
Mandatory many	$A \xrightarrow{1 \quad 1...n} B$	One to many B must be depended on to fulfill A
Optional one	$A \xrightarrow{1 \quad [0,1]} B$	One or none of B will be depended on to fulfill A
Optional many	$A \xrightarrow{1 \quad [1...n]} B$	Optionally, one to many B may be depended on to fulfill A

relationships and constraints among the decomposed elements can also be shown. A relatively complete abstraction of the goal, subgoals, and relevant actors and resources in the target agent organization can then be represented.

Figure 7.4 illustrates the combination of several related organizational rules for representing the relationships for executing, evaluating, and optimizing an algorithm in F-TRADE. To execute an algorithm, the following four subgoals—access algorithm model base, access resource base, access knowledge, and access remote stock data—must be fulfilled in parallel. The requested algorithm will be selected from the algorithm base if it is the right option. The resource information base and local knowledge base will be accessed iteratively while the conditions are satisfied. All stock transactions in the target list will be repeatedly queried from the remote stock data warehouse.

7.5 Modeling Organizational Structure

The normal task for modeling the organizational structure is to build an organizational structure diagram using all the modeling components discussed above and in Chap. 10 on demand. However, since open agent systems are highly complicated, it would be very difficult to construct and pack all the components into one comprehensive organizational structure diagram. An alternative method, where appropriate, is to develop a high-level organizational framework plus multiple low-level subsystem structure diagrams. If the system is sufficiently complicated, multiple hierarchical subsystems (as shown in Fig. 6.1) need to be decomposed and analyzed. The low-level diagram for any target subsystem can then be constructed; this looks like Fig. 7.4 but has more detail. In Chap. 13 of the case study, we will exemplify some of the main functions in F-TRADE through organizational relationship analysis and organizational structure analysis.

In this section, we mainly discuss the GAIRE model. We propose this model to capture the principal organizational members, namely, goal, actor, interaction, rule, and environment, in terms of our proposed ORGANISED framework.

7.5.1 GAIRE Model

GAIRE is acronym of goals, actors, interactions, rules, and environment, which are the main elements in our organizational abstraction framework ORGANISED. We believe these to be the elements that constitute and operate a dynamic organization's formation and evolution. The basic theory for the GAIRE model is as follows. We suppose that an agent system can be physically decomposed into multiple levels. We also assume that the GAIRE elements on every layer can be captured. In building the GAIRE model, our main concerns are the GAIRE elements on all layers, thus we try to capture all the GAIRE items on each layer and build a GAIRE model for that layer. In the GAIRE approach, the work of building the organizational structure is transformed into building a GAIRE model for all the layers.

The algorithm for building multiple layers of the GAIRE model is as follows:

1. Decompose an agent organization into subsystems according to feasibility. If possible and required (if the system includes multiple layers), multiple lower layers of sub-subsystems may be split.
2. Build GAIRE models for every subsystem on the lowest layer. Then, rather than directly building a global organizational structure, we first analyze and capture all the GAIRE items that exist in a component (or subsystem, sub-subsystem) on the lowest layer. We build GAIRE models for each of these components (or subsystems, sub-subsystems).
3. Build the GAIRE model for the lowest layer. The layer GAIRE model is an aggregation of the GAIRE of all components or subsystems.
4. Build GAIRE models for all layers. For instance, we can build a GAIRE model for the second lowest layer, where its components are those subsystems on the lowest layer.
5. Obtain the global GAIRE model. We repeat this procedure from steps 1–4 until we reach the highest level of the GAIRE model. Then, we try to link the multiple layers of GAIRE models to obtain the global GAIRE model.

For instance, assuming there are three layers in total—Layer 0, Layer 1, and Layer 2 from top down—in an agent organization. In Layer 1, there are a total of X (from 0 to $x-1$) subsystems. Items on Layer 2 are components of Layer 1. We can build a GAIRE model for each of these components. For the subsystem S_i in Layer 1, for example, its detailed itemsets of goals, actors, interactions, rules, and environment are G_{1i}, A_{1i}, I_{1i}, R_{1i}, and E_{1i}, respectively. They are as follows:

$$G_{1i} = \{g_{i1}, g_{i2}, \ldots, g_{ij}\} \tag{7.3}$$

$$A_{1i} = \{a_{i1}, a_{i2}, \ldots, a_{ik}\} \tag{7.4}$$

$$I_{1i} = \{i_{i1}, i_{i2}, \ldots, i_{il}\} \tag{7.5}$$

$$R_{1i} = \{r_{i1}, r_{i2}, \ldots, r_{im}\} \tag{7.6}$$

$$E_{1i} = \{e_{i1}, e_{i2}, \ldots, e_{in}\} \tag{7.7}$$

where $g_{i1}, \ldots, g_{ij}$; $a_{i1}, \ldots, a_{ik}$; $i_{i1}, \ldots, i_{il}$; $r_{i1}, \ldots, r_{im}$; and $e_{i1}, \ldots, e_{in}$ are items of goals, actors, interactions, rules, and environments on Layer 2 mapped to subsystem S_i. We then build a GAIRE model for Layer 1. For the X subsystems on Layer 1, we assume that every subsystem has one aggregated goal, actor, interaction, rule, and environment. For subsystem S_i, they are G_{1i}, A_{1i}, I_{1i}, R_{1i}, and E_{1i}, respectively. The GAIRE model for Layer 1 is as follows:

$$G_1 = \left\{G_{10}, G_{11}, \ldots, G_{1(x-1)}\right\} \tag{7.8}$$

$$A_1 = \left\{A_{10}, A_{11}, \ldots, A_{1(x-1)}\right\} \tag{7.9}$$

$$I_1 = \{I_{10}, I_{11}, \ldots, I_{1(x-1)}\} \tag{7.10}$$

$$R_1 = \{R_{10}, R_{11}, \ldots, R_{1(x-1)}\} \tag{7.11}$$

$$E_1 = \{E_{10}, E_{11}, \ldots, E_{1(x-1)}\} \tag{7.12}$$

Similarly, we can build a GAIRE model for Layer 0 in which all subsystems are actually distributed on Layer 1.

$$G_0 = \{G_1\} \tag{7.13}$$

$$A_0 = \{A_1\} \tag{7.14}$$

$$I_0 = \{I_1\} \tag{7.15}$$

$$R_0 = \{R_1\} \tag{7.16}$$

$$E_0 = \{E_1\} \tag{7.17}$$

After obtaining all the GAIRE models for every layer, we can build the global organizational framework in two steps: (1) building a GAIRE global structure diagram and (2) building global goal and interaction diagrams.

1. Building a GAIRE global structure diagram: Assuming every subsystem consists of one aggregated goal G_i, actor A_i, interaction I_i, rule R_i, and environment E_i, all itemsets of G, A, I, R, and E form one GAIRE system. We build a global organizational structure GAIRE diagram by linking all GAIRE itemsets. The global GAIRE diagram, which also looks like the subsystem diagram, assists us with creating the global framework of an agent organization.
2. Building global goal and interaction diagrams: We build a global organizational goal diagram G capturing the relationships within the goal set $\{G_1, G_2, \ldots, G_i, \ldots, G_p\}$ and a global interaction model I capturing the interaction between the actors, goals, and environment in the subsystem set $\{S_1, S_2, \ldots, S_q\}$. The diagrams G and I can be developed based on a GAIRE system using either aggregated itemsets or detailed itemsets on demand. These two diagrams help us to grasp the overall framework of the organizational goal and interaction dynamics. Figure 7.5 shows a sample of a goal diagram.

Since open agent systems are normally very complicated, it would be very difficult for us to decompose an open agent system into individual parts. Given this consideration, one decomposition strategy that may be useful is to grasp the main impact factors in a system. A typical example is the Internet, in which some key link nodes are authorities, while other key nodes are hubs [7]. These two types of node contribute the majority of the formation evaluation of the Internet. In AOSE, for instance, some functions in the system are fundamental for functional requirements analysis; we may start with them. Similarly, some key nonfunctional qualities will be the primary and most important factors in nonfunctional requirements analysis. We will study this idea in detail and check to see whether power law

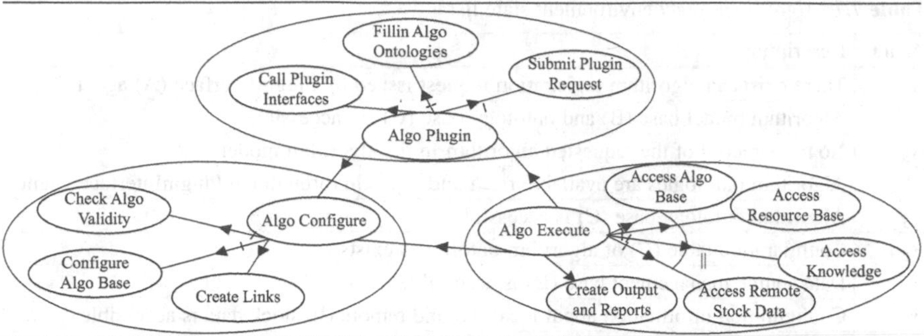

Fig. 7.5 Global goal diagram

distribution [8, 9] and scale-free distribution [10] exist in an open agent organization such as the Internet.

7.6 Organizational Dynamics Analysis

The literature review shows that the modeling of organizational dynamics is based on methods from the following fields: Markov decision process [5], science of complexity [11], dynamic system theory [12], logic-based and reactive approaches, and statistical approaches. In addition, scenarios, sequences, states, and activities are often used for visual representation of system dynamics.

Here we take the $\text{POMDP}_{\text{AEI}}$ model as an instance. In F-TRADE, the goal of the agent *AlgoPluginAgent* is to fulfill the registration of an algorithm into the system. The corresponding environment states and actions of the agent are listed in Tables 7.7 and 7.8, respectively. The state list of the *AlgoPluginAgent* environment is related to all active actors in F-TRADE which have close correlation and involvement in the goal of ConfigureAlgo. Figure 7.6 shows the Markov state chain diagram of agent *AlgoPluginAgent*. The direction of lines with arrows shows the interaction life cycle of this agent–environment interaction. In considering the interaction of environment actors with the agent, Fig. 7.6 is upgraded to Fig. 7.7, in which the simple state-action chain becomes an interaction network; the interaction network is composed of multiple state chains which link and couple all relevant agents and the environment.

Furthermore, this state-action chain can be formally specified. For instance, state transition from s_1 to s_2 under condition of a_1 is represented as follows ($t_1 < t_2$):

- $\exists$ apr : AlgoPluginRequest(apr.depender = $\text{P}_{\text{LUGINPERSON}}$ $\wedge$ apr.dependee = pluginInterface $\wedge$ Fulfilled() $\rightarrow$ [$\Diamond \leq$ t2 CheckAlgorithmValidity]$\Diamond$ $\leq$ t1 AcceptPluginRequest.Fulfilled() (7.18)

Table 7.7 *AlgoPluginAgent* environment state list

State	Description
s_1	There exists an algorithm registration request issued by PluginInterface (A) agent
s_2	Algorithm model base (B) and ontology base (C) are accessible
s_3	No track record of the requested algorithm in the algorithm model base
s_4	Algorithm ontologies are available from and typed in through the PluginInterface agent
s_5	Algorithm ontology base (C) is accessible
s_6	Configuration table (C) of algorithm ontologies exists
s_7	Data source management base (D) is accessible
s_8	Configuration information about local (E) and remote (F) stock data is accessible
s_9	Specific data sources are selected or configured through PluginInterface agent
s_{10}	No track record of the requested algorithm is available in the XML configuration files (G)
s_{11}	Connections to algorithm model base, ontology base, and data source management base are open
s_{12}	Status of registration result is available for PluginInterface agent
s_{13}	Algorithm model base, ontology base, and data source management base are closed

Table 7.8 *AlgoPluginAgent* action list

Action	Description
a_1	Receive algorithm registration request from algorithm PluginInterface agent
a_2	Open data connection to algorithm model base (B)
a_3	Query to see whether the requested algorithm is available in the algorithm model base
a_4	Receive algorithm ontologies from algorithm PluginInterface agent
a_5	Open data connection to algorithm ontology base (C)
a_6	Register and configure algorithm ontologies into algorithm ontology base (C) with help from OntologyService (H)
a_7	Open data connection to data source management base (D)
a_8	Query configuration information about data sources of local and remote stock data (EF)
a_9	Configure data sources link list (D)
a_{10}	Create configuration file for the newly registered algorithm in XML (G)
a_{11}	Return status with successful registration or not, and do exception processing if required
a_{12}	Close connections to algorithm model base, ontology base, and data source management base

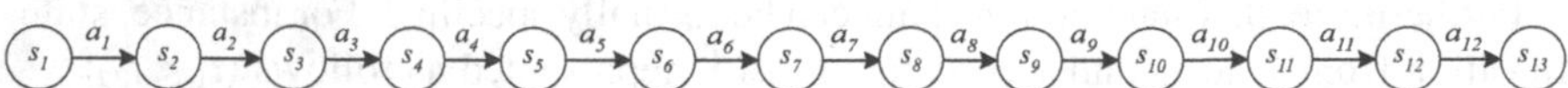

Fig. 7.6 Markov state-action chain

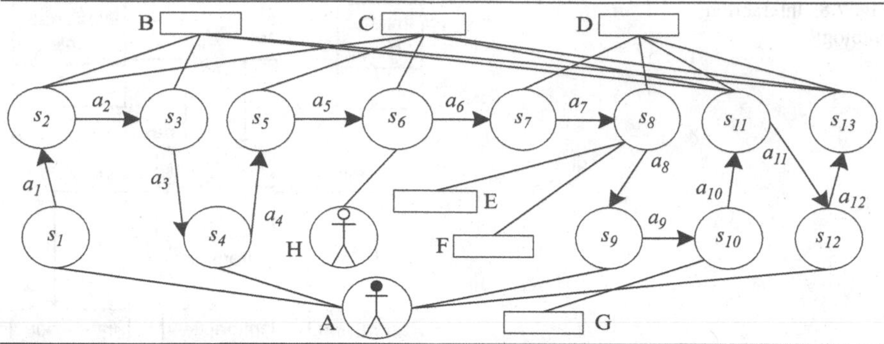

Fig. 7.7 AlgoConfigureAgent environment state list

7.7 Interaction Ontology

Interaction is a key concept in agent society which supports exchange, cooperation, coordination, and communication in order to complete tasks and finish goals. *Interaction model* (IM) identifies dependencies and relationships among various roles taken by agents and services in a MAS organization. To specify the common aspects of various IM instances, we define interaction ontology. *Interaction ontology* defines common ontological items relevant to an interaction. From the perspective of organizational metaphor, an IM of an agent organization deals with global interaction ontologies such as organizational roles, organizational protocols, organizational levels, organizational rules, and communication content.

In the process of building IM, on the other hand, the above organization-centric ontological elements are often instantiated into distinct interaction-centric terminologies. For instance, interaction protocols are utilized instead of organizational protocols; more detailed and instantiated ontological items will be specified to give a complete view of the interaction protocol. In addition, we pay more attention in building the IM to interaction ontologies such as interaction protocols, interaction patterns, interaction levels, and interaction rules, or constraints. Figure 7.8 shows the ontologies for an interaction, in which the interaction is an abstract concept.

7.7.1 Interaction Protocols

Ongoing conversations between agents often fall into typical patterns. In such cases, certain message sequences are anticipated, while at any point in the conversation, other messages are expected to follow. These typical patterns of message exchange are called *interaction protocols* [13]. In general, an agent interaction protocol is a set of rules that guides the interaction between several roles or agent services so that the goals can be achieved. A designer of agent systems has the

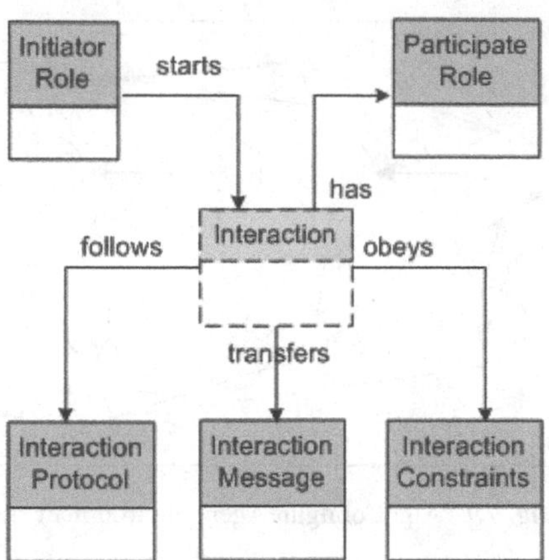

Fig. 7.8 Interaction ontology

choice to make the agents sufficiently aware of the meanings of messages, goals, beliefs, and other mental attitudes possessed by the agents. The agent planning process causes such interaction protocols choices to arise spontaneously.

Interaction protocols are usually governed by standards bodies, such as FIPA, or written up and documented in widely distributed, immutable documents. Based on this kind of protocol, agents must adhere to the order of a protocol and cannot deviate. There is usually no scope within a protocol to allow for the communication and handling of agents with differing goals. A protocol session is a mechanical process rather than a belief- or goal-driven process. In the implementation of agent systems, agents must additionally be hard coded with a protocol if they are to use it to communicate with other agents.

The above rigid protocols have a number of basic shortcomings in migrating to open agent systems, however. For instance, an open agent system is a dynamic and evolutionary system targeting certain goals rather than a play system or a program with predefined scripts and scenarios. In open systems, this reduces the flexibility of the agents, because standard protocols must be designed, approved (by a standards body), and coded into the agents to be used. This leads to the study of interaction protocols for open agent systems.

In addition, interaction protocols are similar to but also different from *communication protocols* in that the former must be adhered to for holders to fulfill their goals and tasks autonomously with much richer messages. More difference can be found in [14].

7.7.2 Organizational Patterns

A *pattern* is a proven solution to a problem in a context, or more simply, "Each pattern is a three-part rule, which expresses a relation between a certain context, a problem, and a solution" [15]. As an element in the world, each pattern encompasses a relationship between a certain context, a certain system of forces which occurs repeatedly in that context, and a certain spatial configuration which allows these forces to resolve themselves [15]. Patterns have roots in many disciplines, including literate programming and most notably in Alexander's work on urban planning and building architecture [16]. Inspired by this above work, *software patterns* [17] have been used for domains as diverse as software development organization and process, exposition and teaching, and software architecture. Instances of software patterns include interaction design patterns [18, 19] for human–computer interaction and user interface design, and patterns for role, architecture, system, society, environment, and task.

With organizational metaphor, we call all patterns in the agent organization *organizational patterns*. *Organizational patterns* are patterns in the MAS organization's behaviors, activities, effectiveness, and accumulation of choices which often create path dependencies that shape current actions. Similar to our discussion about the organization-oriented modeling of MAS, organizational patterns can also be observed from individual layers of the MAS organization. We present organizational patterns in terms of the following three layers of patterns.

1. *Organizational macro-pattern* capturing the essence of the *macrostructure* of a MAS organization: This kind of pattern is the observation of high-level *meta-patterns* independent of specific agent platforms or implementation. Organizational macro-patterns can be classified into organizational framework (architecture), interorganizational interaction patterns, evolutionary (behavior) patterns of a MAS organization, and interaction patterns in and between a MAS organization and its environment. Macro-patterns are developed to be inherited and applied by any specific pattern in designing agent systems. For instance, the organizational framework of agent systems may take the form of peer to peer, layered or hierarchical, central control, vertical integration, pyramid, joint venture, structure-in-5, bidding, hierarchical contracting, co-optation, takeover or hybrid architecture, and so forth.
2. *Organizational micro-pattern* describing the essence of the *microstructure* in a MAS organization: This type of pattern consists of mobility patterns, intraorganizational interaction patterns, security patterns, message-passing patterns, task patterns, and the like. Intraorganizational interaction patterns capture behavior modes of coordination, cooperation, negotiation, auction, and other forms of interaction in an agent organization. These may encompass direct interaction, mediator (mediation), dispatch, broker (brokerage), matchmaker (matchmaking), messenger and facilitator, contract net, monitor, embassy, wrapper, master–slave, and others. For instance, task patterns which represent how agents perform various tasks can take a master–slave or other form.
3. *Organizational design patterns* for implementing a MAS organization through specific implementation techniques: Organizational design patterns in a Java-based MAS can be instantiated and implemented as design patterns in Java [20–22].

The modeling of organizational patterns is discussed in Sect. 7.9. A template for describing patterns in natural language is given, accompanied by a sample of a contract net protocol.

7.7.3 Interaction Levels

An interaction is located on a certain organizational level; we call the organization level which supports an interaction an *interaction level*. Interaction can occur on various levels or across multiple levels in a MAS organization and its environment. This insight leads to the following forms of interaction: inter-role interactions, interagent (interservice) interactions, intra-subsystem interaction (a subsystem can be group or team of closely or loosely coupled agents), inter-subsystem interaction, and intersystem interaction (interaction between MAS organization and environment).

7.7.4 Interaction Rules

Any interaction is governed by organizational rules and cardinalities. We call organization rules which monitor interaction activities *interaction rules* or *interaction constraints*. Possible interaction constraints include communication directions, transport means of interaction, interaction time-out, and interaction exceptions. Communication between agents can be undertaken in the following ways: blackboard or message-based communication and one-way or two-way peer-to-peer, unicast, multicast, or broadcast interactions. Messages in a communication can be pulled or pushed synchronously or asynchronously between initiators and target responders. Furthermore, an interaction is only valid in a certain time period; beyond the given time slot or point, the interaction may cease with time-out exceptions. Some other exceptions related to an interaction include operation exception, protocol exception, and message exception. All these rules and constraints must be respected or addressed in the design and implementation of interactions in a MAS organization.

7.8 Interaction Protocols Engineering

Derived from communication protocol engineering, the aim of interaction protocol engineering [23] is to develop interaction protocols in which the following phases are considered:

- Analysis
- Interaction ontology
- Formal description
- Validation
- Implementation
- Conformance testing

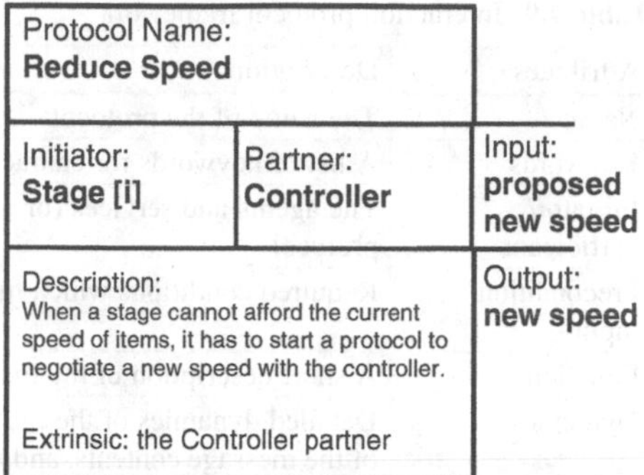
Protocol Name:
Reduce Speed
Initiator:
Stage [i]
Partner:
Controller
Input:
proposed new speed
Description:
When a stage cannot afford the current speed of items, it has to start a protocol to negotiate a new speed with the controller.
Extrinsic: the Controller partner
Output:
new speed

Fig. 7.9 GAIA protocol diagram

In this work, we introduce analysis, interaction ontology, specifications of protocols, and architecture of interaction protocols individually.

7.8.1 *Analysis*

The task of analysis is to define the requirements for a protocol. This informal phase makes use of natural language descriptions and sometimes chronograms, use cases, or algorithms to describe the requirements. The aim of this phase is to record (1) the definition of the protocol, (2) the good properties to incorporate, (3) the bad properties to avoid, (4) the type of messages and data, and (5) the message sequences.

Several templates have been developed to support the analysis of interaction protocols; for instance, Fig. 7.9 shows the GAIA *protocol diagram* [2] for analyzing and representing an interaction protocol. Comparing the framework proposed by [23] for interaction protocol engineering, we propose a more comprehensive template which is more sensitive to MAS, as shown in Table 7.9. We also believe the pattern description template discussed in Sect. 7.9. can be used for requirements analysis of the interaction protocol.

7.8.2 *Interaction Protocol Ontology*

Interaction is usually modeled in terms of diagrams such as activity and sequence diagrams, and state chart and the like, in UML, AUML, or other techniques. The interaction protocol can basically link most of the interaction ontologies. Figure 7.10 shows the *interaction protocol ontologies*. Interaction protocol ontologies can further be filled into or embodied through the protocol diagram and pattern template (discussed in Sect. 7.9) on demand.

Most interaction ontologies can be linked by interaction protocols, so we will look at the interaction ontologies of an interaction protocol. As shown in Fig. 9.3, an

Table 7.9 Interaction protocol framework

Attributes	Descriptions
Name	The name of the protocol
Keywords	A list of keywords for characterizing this protocol: auctions, secured, etc.
Initiator/ participant	The agents and services (or roles) which initiate and participate in the protocol
Precondition/ input	Required conditions which must be true for initiating the protocol
Function	A short description of the meaning and principle of this protocol
Dynamics	Detailed dynamics of the protocol with all the messages used, the meaning of the message contents, and what messages are possible for any interaction state
Message	Interaction content which will be transferred by the protocol between participants
Conditions/ constraints	Conditions and constraints which must hold during the execution of the protocol: deadlock freeness, liveliness, termination, etc.
Postcondition/ Output	Status and conditions will be true after the execution of the protocol
Termination	Normal exit interaction states; for instance, the customer has the product and the merchant has the money
Exception	Abnormal termination which may be triggered by exceptions of time-out, message, or others

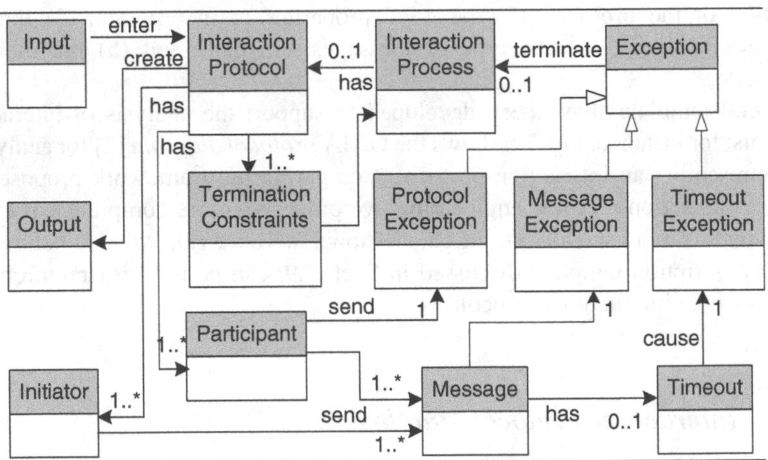

Fig. 7.10 Interaction protocol ontology

interaction process encompasses an interaction protocol which has one to many *initiators* and *participants*. An initiator, sometimes with a participant, sends a message to other participants following the conversation protocol. Each message may have a time-out constraint; if time-out occurs, it will cause a *time-out exception*. Similarly, *message exception* may be caused by a message, and *protocol exception* may be caused by a protocol. As a result, an exception will terminate the interaction process should the exception occur.

7.8.3 Specifications of Interaction Protocol

Analysis generates informal documents described in natural language. However, we need an accurate and rigorous tool—some formal description techniques—to solve ambiguities and errors, and to refine the requirements and functions which may occur during analysis. The following are some formal description techniques which are used for the specifications of interaction protocols:

- UML-based modeling languages: UML, UAML, Agent UML
- Finite state machine, automaton, state chart, FIPA PDN
- Petri net
- LOTOS, Estelle, SDL, Z, PDL, APRIL, grammar, FLBC
- Temporal logic, event calculus, POS, 3APL
- FIPA–ACL, KQML

Some of the above specifications provide protocol languages. *Protocol language* usually describes an interaction protocol diagrammatically, such as in UML or in AUML [24]. For instance, an interaction protocol can be expressed diagrammatically as a *protocol diagram*. The protocol diagram of the contract net protocol [13] is shown in AUML in Fig. 9.4. Using AUML, an interaction protocol can be presented through a three-layer framework [24]: templates and packages to present the protocol as a whole, sequence and collaboration diagrams to capture interagent dynamics, and activity diagrams and state charts to capture both intra-agent and interagent dynamics. In GAIA, a protocol is expressed informally as a protocol diagram, as shown in Fig. 7.9.

Taking a close look at individual protocol/pattern description techniques, almost every one has advantages and drawbacks in modeling open agent organizations. Similar to our discussion about integrative modeling in the early requirements analysis of agent organization, the same idea can also be used for modeling interaction protocols and patterns. There are two ways of developing modeling methodology by integrating diagrammatic modeling and logical specifications. One is to develop two separated system for visual modeling and logical modeling, respectively; another is to integrate visual and logical modeling into one description system. In Chap. 8, contract net protocol and GAIA protocol will be represented in temporal logics. More work is being conducted on developing the latter form of integrative modeling system. Paurobally [25] proposed a combined approach consisting of extended propositional dynamic logic and extended state charts for modeling interaction protocols both graphically and logically. Figure 7.11 [25] shows a combined approach with extended state chart notation over AUML and Petri nets.

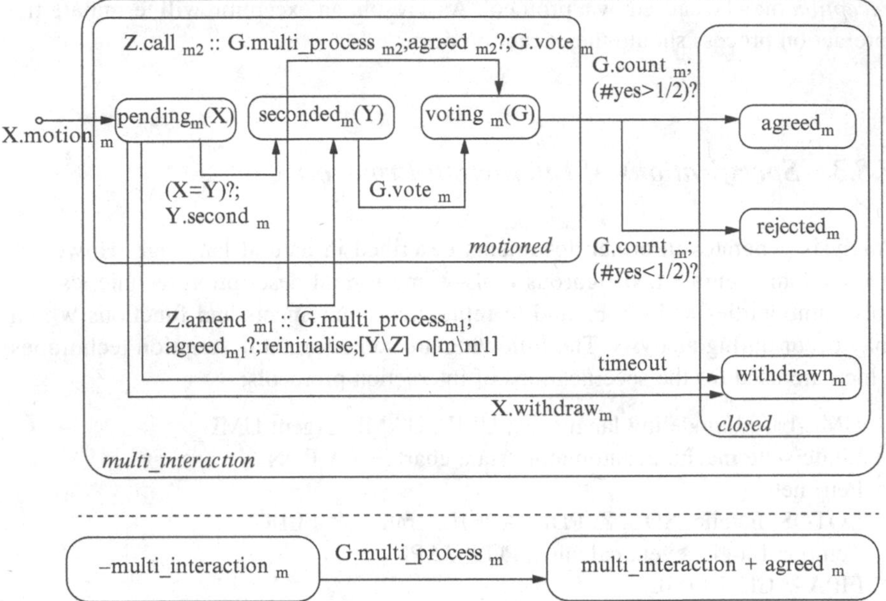

Fig. 7.11 Extended state charts notation over AUML and Petri nets for multilateral contract net

7.8.4 *Interaction Metaprotocols*

To capture the characteristics and dynamics of open agent systems, more complex interaction protocols must be embodied in the design of agent organizations. Bauer [26] and FIPA [13] define four types of macrolevel interaction protocols from the architecture perspective (called *interaction metaprotocols*) which could be instantiated or implemented by interaction protocol designers. They are (1) parameterized protocol, (2) nested protocol, (3) interleaved protocol, and (4) complex nested protocol.

A *parameterized protocol* (as shown in Fig. 7.12) is described as having one or more unbound formal parameters. It therefore defines a family of protocols; each protocol is specified by binding the parameters to actual values. Typically, the parameters represent agent roles, constraints, instances of communicative acts, and nested protocols.

A *nested protocol* (Fig. 7.13) can be defined and applied if it is used several times within the same specification. In contrast to a parameterized protocol, it is only an abbreviation for a fixed part of a protocol. Additionally, nested protocols are used to define the repetition of a nested protocol according to guards and constraints.

Interleaved protocols (Fig. 7.14) show that a protocol is performed between different agents, and to complete or progress the protocol, an agent has to perform another protocol with other agents.

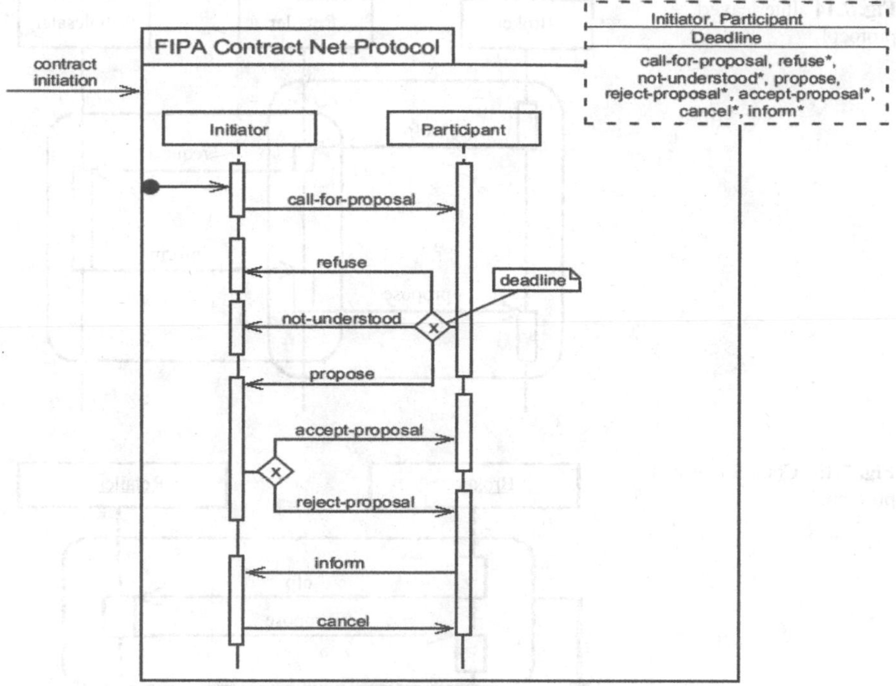

Fig. 7.12 Parameterized protocol diagram of contract net protocol in AUML

Fig. 7.13 Nested protocol

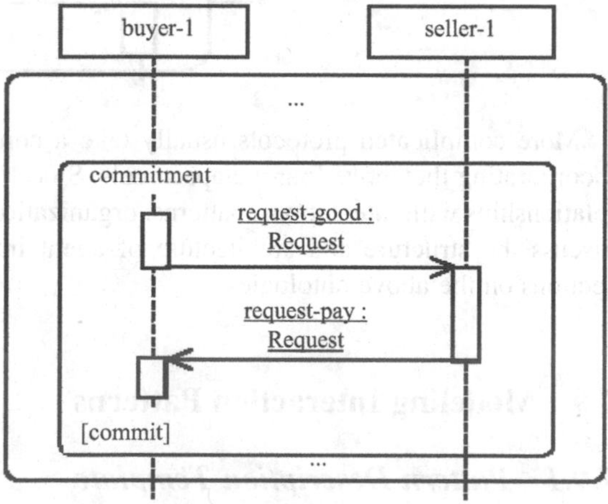

A *complex nested protocol* (Fig. 7.15) defines the parallel or decisional combination of nested protocols. It has to take into consideration the thread of interaction at the beginning and the end of the complex nested protocol. Furthermore, the start and end points within the nested protocols have to be considered.

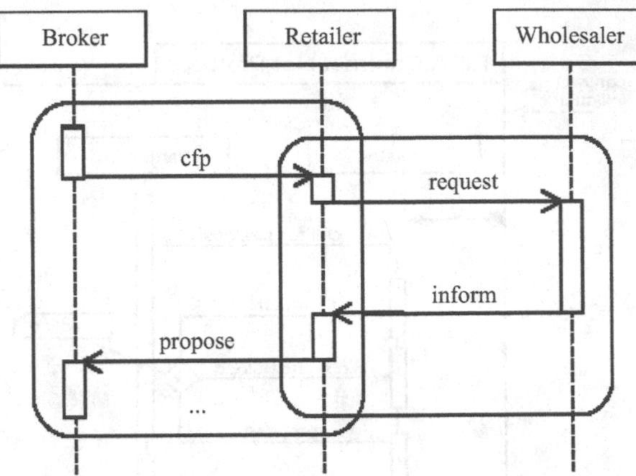

Fig. 7.14 Interleaved protocol

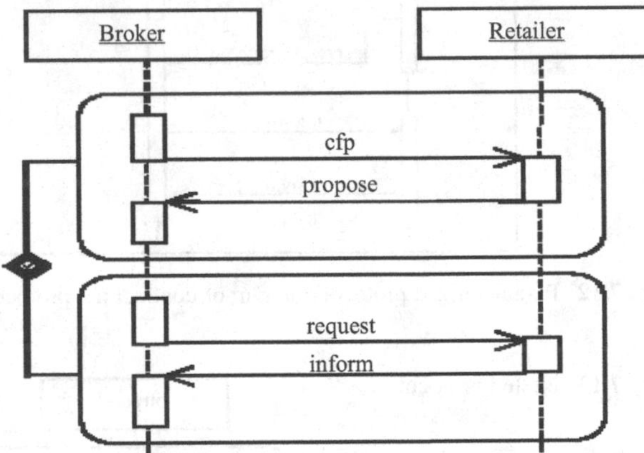

Fig. 7.15 Complex nested protocol

More complicated protocols usually take a combined or instantiated form by incorporating the above four metaprotocols. Since interaction protocols have close relationships with interactions, patterns, organizational rules, and structure, we also discuss the structure and architecture of agent interaction protocols in relevant sections on the above ontologies.

7.9 Modeling Interaction Patterns

7.9.1 Pattern Description Template

In general, a *pattern description scheme* (PDS) is utilized to model collections of attributes and properties that, when taken together, fully capture an organizational pattern. In the literature, several general purpose schemata have been proposed suggesting that a "canonical" scheme might exist that can be used for different patterns. These schemata are similar in essence. They specify key aspects of a pattern either in ontology-based specification or in diagrams, using modeling techniques such as *colored Petri nets* [27], UML [28] or AUML, and *finite state machine* [29]. However, a single, general pattern description scheme is, to some degree if not completely, inadequate for all categories of patterns, especially for specific pattern classes in MAS organizations. To this end, some remedial actions can be taken; for instance, complementary ontological elements can be added to the current PDS to capture particular features of an agent pattern. Table 7.10 lists a PDS template for agent patterns.

7.9.2 Case Study: Contract Net Protocol

This case study demonstrates how a concrete interaction protocol pattern can be instantiated using the above pattern description template. We take the well-known contract net protocol [30] as the instance. Figure 7.16 presents the contract net protocol by following the pattern description template in Sect. 7.9.1.

Other researchers have developed or are developing formal specifications for modeling organizational interaction patterns, such as those we have discussed in Sects. 7.9.1 and 8.6. A more comprehensive approach is to model an organizational pattern in an integrative specification both graphically and logically. There are several open problems about which modeling techniques can be developed, and how, to represent organizational patterns using an integrative and comprehensive modeling specification, both diagrammatically and formally. We think that the following formula could indicate a direction for the development of such integrative modeling technique:

$$\begin{aligned}\text{Integrative Modelling} = {} & \text{States (state chart)} \\ & + \text{Scenario (scenario analysis)} \\ & + \text{Actor relationship (abstract/soft entities and goals)} \\ & + \text{Logic} + \text{Sequence} + \text{Activity}\end{aligned} \tag{7.19}$$

7.10 Agent–Environment Interaction

7.10.1 What is Agent–Environment Interaction?

A multiagent system is closely coupled with its surroundings. These surroundings are termed the *environment* of the agent organization in which the agents operate and exist. The environment can be internal or external according to whether our position is as an internal or external observer, respectively. Thus, an agent environment is a relative concept which may incorporate all actors (which may take the form of human, agents, services) and relationships that are involved in the existence and evolution of an agent or an agent organization.

Table 7.10 Agent pattern description scheme

Ontological items	Descriptions
Name	A crisp name that captures the essential idea underlying the pattern
Aliases (known as)	Other well-known names for the pattern, if any
Forces/motivation	Which aspects of the problem are the forces that led to the development of the pattern? What are the prerequisites for using the pattern?
Problem/intent	What does the design pattern do? What is its rationale and intent? What particular design issue or problem does it address?
Dynamics/solution/ collaborations	Typical scenarios describing the run-time behavior of the pattern. How do the participants of the pattern collaborate to achieve their goal?
Structure	A graphical representation of the classes in the pattern
Organizational layer	In which organizational layer of the MAS organization is this pattern located? It may have additional fields that are specific for a particular category and that do not make sense for other categories
Participants	The agents or services that participate in the pattern
Dependencies	Does the pattern require any specific environment before it can be applied?
Consequences	How does the pattern support its objectives? What are the trade-offs and results of using the pattern? What aspect of the system structure can be independently varied?
Context/applicability	In which situations can the design pattern be applied? What are examples of poor design that the pattern can address? How can these situations be recognized?
Organizational roles	The organizational roles and their functionalities which are involved in the pattern
Messages	The types of message or transfer method that become involved both inside and outside of the pattern
Temporal ordering	Sequence of actions of all participants in the pattern
Known uses	Examples of real systems in which the pattern has been successfully applied
Example	A simple, abstracted example for how to use the pattern
Implementation/ pseudo-code	What pitfalls, hints, or techniques should you be aware of when implementing the pattern? Are there language specific issues? A pseudo-code may list the key rationale of the pattern
Related patterns	What design patterns are closely related to this one? What are the important differences? With which of the other patterns should this one be used?
See also	References to other patterns that solve similar problems or that can be beneficially combined with this pattern. Also, what are the potential conflicts with other patterns?

From a social science perspective, both physical and electronic environments comprise the following elements:

- *Actors* (or entities in some literature) such as all institutions, groups, other organizations, and individuals

Name: Contract-net

Aliases: none

Forces: The Contract-net protocol is a one-to-one copy of the behaviour shown by participants in market. The idea is to have the participants calculate their local cost for performing a particular task and then announcing the resulting cost to a manager who decides on the task assignment. This leads to an optimal resource allocation in the case of independent tasks. If inter-dependencies between tasks exist, the task allocation process can get caught in local minima.

Problem: Allocate a particular task to one out of several potential contractors while minimizing the local costs.

Structure: As shown in Figure 7.12

Organizational Layer: Organizational micro-pattern

Participants: For Contract Net, there are two different types of agents, an Initiator and a participant as Contractor. At any time, any one agent can be an Initiator, a Contractor or both. Contract Net creates a means for contracting as well as subcontracting tasks (or jobs), in this sense Initiators are managers and Participants are contractors.

Solution: Contract Net specifies the interaction between agents for fully automated competitive negotiation through the use of contracts. In essence, Contract Net allows tasks to be distributed among a group of agents.

Dependencies: Using the Contract-net protocol for a task allocation process requires that there exist several potential contractors with comparable abilities. For optimal performance, the Contract-net also requires independent tasks.

Consequences: If the tasks are not independent from each other, the Contract-net protocol is likely to lead to sub-optimal solutions as it can get stuck in local optima. To escape from local optima, additional mechanisms may be necessary.

Context: Automated competitive negotiation among agents

Organizational Roles: The Contract-net protocol has two roles: the manager and the bidder. The manager is responsible for putting together a task description that tells the potential bidders what to do. In closed agent societies, the task descriptions are usually given is some proprietary format; in open agent societies, some sort of standard must be applied. The second task of the manager is to determine the set of possible bidders. This can be achieved by broadcasting the task announcements to all possible receivers or by performing a pre-selection of already known candidates. While the first approach ensures that all potential contractors are reached, it is not very resource effective and can lead to unnecessary computations on the bidders side. Thus, a pre-selection of candidates is usually advantageous. The task of a bidder is to compute the cost for performing the announced task and to send an offer (if the bidder can fulfil the task at all) to the manager. If the manager grants the task to a particular bidder, the bidder guarantees task completion within the constraints provided by the task announcement.

Messages: The Contract-net has four messages: *call for bids*, *bid*, *grant* and *reject*. The *call for bids* message is issued by the manager and contains the task description. The bidders reply to this message with a bid message if they can perform the task. If not, no message is sent to the manager. After completing the bid collection phase, the manager selects the best bid and sends a *grant* to the successful bidder; the other bidders receive *reject* messages.

Temporal Ordering: The temporal ordering of the messages is shown in Figure 7.12.

Known Uses: Simulated distributed acoustic sensor network, electronic marketplace

Example: See [31]: Contract Net Protocol in Java

Implementation: [omitted]

See Also: Nested Contract Net

Fig. 7.16 Pattern for contract net protocol

– *Principles* such as relationships, dependencies, customs, norms, rules, constraints, and policies
– *Processes* such as communication and interaction under some protocols and patterns, as well as changing social and technological *forces* that are outside the organization being analyzed but that have a direct or potential impact on the organization *spatially* and/or *temporally*

For each of the above aspects, some specific concerns can be extracted. For instance, the following questions need to be answered in terms of forces:

1. What kinds of demands are environmental factors making on the organization?
2. How do environmental forces put constraints on organizational action?
3. Identify opportunities for action that the demands might create for the organization.
4. Are demands from environmental factors (including relevant constituencies/stakeholders) likely to be stable or likely to change in the near term, or over the longer term?

An agent or small agent organization on a certain level of a MAS organization is lodged in and continuously interacts with its environment temporally and spatially. Agents and their environment coexist, cooperate, and coordinate with each other in the life cycle of the MAS organization. All dialogue between agents and environment is called *agent–environment interaction* (AEI). The AEI is abstractly modeled, as shown in Fig. 7.3. This *abstract environmental model* is intended as an abstract, computational representation of the environment in which the MAS will be situated, perceived, and operated. It is made available to agents to sense the state of their environment, to enable their actions to be effected, or to consume/extract information from the environment.

Modeling agent environment and agent–environment interaction is attracting more and more attention from agent researchers. More work is being conducted on developing "scientific methods" to understand, capture, and represent the complexity of the environment and agent–environment interaction. In the AISB symposium [32], work on the measurement of behavior, quantitative description of agent–environment interaction, computer modeling, system identification, and embodied computational neuroscience has been reported.

In searching the literature, other modeling work has been found which has contributed or may also contribute to agent–environment interaction using logic-based approaches, reactive approaches, and statistical approaches. From the perspective of theoretical foundations, AEI modeling can be classified into three classes: (1) modeling based on Markov decision process, (2) modeling based on the science of complexity, and (3) dynamic system theory. In this book, we discuss all three approaches.

7.10.2 Modeling Based on Markov Decision Process

With respect to the theory of the Markov decision process (MDP), [5, 33] assumed that an agent–environment interaction could be modeled as an MDP_{AEI}. Such an MDP is a four-element tuple:

$$MDP_{AEI} = < S, A, T, R > , \tag{7.20}$$

It consists of four elements:

- A finite set of states S of environment
- A finite set of actions A that can be performed in the settings
- A probabilistic transition function $T(s_t, a_t, s_{t+1})$ representing the occurrence probability of transiting from state s_t into state s_{t+1} after taking some action
- An immediate reward function $R(s_t, a_t) : S \times A \to \mathbb{R}$ associating a value with each state

The policy

$$\pi = (s_1, s_2, \ldots, s_t, \ldots, s_n) \tag{7.21}$$

is called Markov's policy if state s_t depends only on time t but not on the previous states. The detailed process of the MDP is defined below [5]:

- The agent and environment can be modeled as synchronized finite automata.
- The agent and environment interact in discrete time t.
- The agent can sense the state of the environment $s_t \in S$ and use it to make an action $a_t \in A$.
- After the agent has acted, the environment makes a transition to a new state s_{t+1}.
- The agent receives a reward rt after performing an action, $r_t = R(a_t, s_{t+1})$.
- The goal is to find a policy in which the reward to be received is maximized according to the given input state.

The agent states form a Markov chain as shown in Fig. 7.17.

By introducing the environment observations, an AEI can be further modeled as a partially observable Markov decision process (POMDP) [5, 33], called $\text{POMDP}_{\text{AEI}}$. The $\text{POMDP}_{\text{AEI}}$ model is a six-element tuple

$$\text{POMDP}_{\text{AEI}} = < S, A, T, R, O, D > , \tag{7.22}$$

where $O(s_t, r_t)$ is the finite set of observations related to state s_t and reward r_t and D (a_t, s_{t+1}, d_{t+1}) is the probability of making observation d_{t+1} from state s_{t+1} after having taken action a_t.

Inspired by the theory of POMDP, Bouzid [34] proposed a stochastic model for dealing with the uncertainties and errors of agent sensors and effectors in situated agent systems by introducing a set of observations and probabilistic observation functions. With certain key concepts like observation, distance, and transition probability distribution, each agent occupies a state and transits from one state to another with a certain probability in a discrete environment. There are two steps in the execution of the stochastic simulation: each agent perceives its environment,

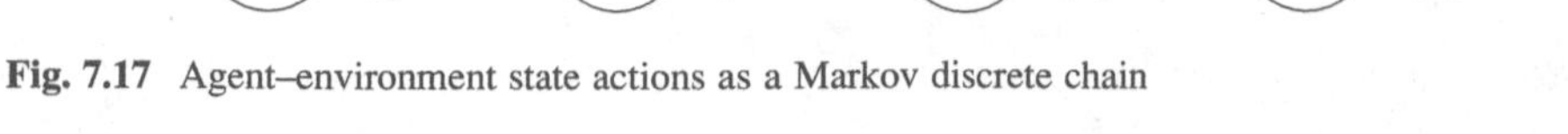

Fig. 7.17 Agent–environment state actions as a Markov discrete chain

and then an agent action is executed with a random state transition according to a given probability distribution. The total computing time $O\ (t_{\text{simul}}) \approx O\ ((N+M)^2)$, where N is the number of agents and M is the number of active objects.

Goldberg [35] developed an online, real-time method for modeling and evaluating the interaction dynamics between a situated agent and its environment. The evaluation method uses behavior-based control in conjunction with augmented Markov models (AMMs), a version of semi-Markov processes able to represent non-first-order Markovian systems in first-order form. The method utilizes AMMs to capture agent–environment interaction dynamics in terms of the history of behaviors executed while performing a task. These models provide the data that are used online and in real time to evaluate the system and suggest task-dependent, performance-improving modifications to the agent's controller/policy.

7.10.3 Modeling Based on the Science of Complexity

In traditional philosophy, *causality* is taken as the dominant explanatory principle, expressing the belief that "things happen because someone, or by extension something, makes them happen." However, causality has rarely provided an adequate account of the majority of systems because these systems consist of many interacting parts. The behavior of the system as a whole, and often of the individual parts, is a complex aggregation of the interactions of all the parts. No part controls the whole or can even control another part outside the influence of the rest of the system. In these systems, which probably govern most of the world, such as living organisms, ecosystems, and social or eco-social systems, there are no isolated controlling agencies and control hierarchies among components. These systems, contemporarily called "*complex self-organizing systems*," are said to be "*self-organizing*," and the behavior of aggregates of components is said to be "*emergent*."

The modern theory of the *self-organization* phenomenon has several roots:

- Cellular automata theory
- Cybernetics [36], which initially looked for control hierarchies, but quickly perceived more complex behavior
- Organismic biology, and philosophy, especially the Chinese traditional philosophy of "holism" that is opposed to the "reductionism" of physics
- Ecosystem theory
- The autopoiesis theory of Varela and Maturana
- Autocatalytic and cross-catalytic reaction theory in chemistry
- Mathematical ecology
- Thermodynamics and statistical physics of irreversible processes
- The mathematics and physics of nonlinear equations

It is therefore evident that many arguments and literatures on the principles of self-organization and self-organizing systems have emerged from one of the above foundations.

The principle of self-organization has maintained the following argument since the 1940s and 1950s: every isolated, determinant dynamic system obeying unchanging laws will develop organisms that are adapted to their environments. Every machine, as it enters equilibrium, performs the corresponding act of selection [37]. A system exhibits self-organization if its behavior shows increasing redundancy with the increased length of the protocol [36]. These conclusions are similar to the presently popular "*complex adaptive systems* (CAS)" [11, 38, 39], a new approach advocated by the Santa Fe Institute founded by theorists from multiple fields such as biology, physics, economics, and others. A basic introduction to the multidisciplinary work and pioneer theorists of CAS in SFI was given in [40]. The CAS approach is distinguished by the extensive use of computer simulations as a research tool, swarm [41], and an emphasis on systems, such as virtual ecologies or e-markets.

In reality, the above research into complex systems and complexity in Europe and America is just one branch of the "*science of complexity* (SoC)" which has recently become very popular. Two major streams of study in the western world for this new science are CAS, founded by American scientists, and complex systems theory in Europe, as discussed above. The European research on complex systems theory, which includes but is not limited to representative theoretical contributions on cybernetics, self-organization and equilibrium, synergetics, hidden order, and chaos, has long promoted the understanding of complex systems since the 1940s.

In spite of relatively fragmentary recognition by the world, compared to the work of American and European colleagues, Chinese scientists from multidisciplinary fields have conducted memorable works in so-called SoC based on eastern philosophy and culture. These studies probably benefit from the long-standing philosophy of "holism" in traditional Chinese culture. Representative work includes the theory of *open complex giant systems* [42–49], its *Metasynthesis from Qualitative to Quantitative* methodology, and its engineering technique *Hall for Workshop of Metasynthetic Engineering*. This work tries to integrate western methodology and Chinese traditional holism, and builds a methodology based on *systematology* [44].

Even though the new SoC, which has recently become very popular in interdisciplinary research, was originally inspired and driven by complex systems, multiagent systems research, which until now has been one of the main areas of complex systems research, has paid insufficient attention to this newly emergent science and has failed to absorb its lessons. The SoC, as the science of complex systems, has shown, and will continue to offer, strong inspiration, foundation, and deeply involved interaction to promote the research of multiagent systems, especially open complex agent systems.

Some research on MAS has been conducted in terms of SoC. Artificial markets, ant classification, and ecological systems are simulated through the swarm platform as per the theory of CAS. Self-organization has been studied in ecological systems.

The "influences and reaction" model [50] and the "proposition–action" model made a clear distinction between influences which are produced by agents' behavior and the reaction of the environment, and further took into account the chain reaction that might rise from the simultaneity of agent propositions. Lifeworld is proposed to capture conventions and invariants of interaction and to taxonomize the interaction between elements such as activity, spatial, material, temporal, role, and so on. Environment is modeled as a state machine, and the behaviors of agents are modeled as policies mapping states to actions. *Open giant intelligent systems* are under investigation using *artificial social agents* [51].

7.10.4 Dynamic System Theory

Several researchers have studied the dynamics of agent–environment interaction using *dynamic systems theory*. Kosecka [52] assumes agent systems to be *discrete-event dynamic systems*, in which the environment is modeled as a finite state machine. Horswill [53] takes the following views in modeling deliberate and reactive agent systems: each agent–environment pair will form a dynamic system with some behavior; the environment is a dynamic system with a known set of possible states.

Beer [12] applied the mathematical machinery of dynamic systems theory to the formalization of agent–environment interactions. Specifically, he proposed to view agent $\mathcal{A}$ and environment ε as two continuous-time dynamic systems:

$$x_{\mathcal{A}} = \mathcal{A}(x_{\mathcal{A}}; u_{\mathcal{A}}), \tag{7.23}$$

and

$$x_{\varepsilon} = \varepsilon(x_{\varepsilon}; u_{\varepsilon}). \tag{7.24}$$

An agent and its environment are in constant interaction. Formally, this means that $\mathcal{A}$ and ε are coupled nonautonomous dynamic systems exhibiting only convergent dynamics:

$$\dot{x}_{\mathcal{A}} = \mathcal{A}\left(x_{\mathcal{A}}; S(x_{\varepsilon}); u'_{\mathcal{A}}\right) \tag{7.25}$$

$$\dot{x}_{\varepsilon} = \varepsilon\left(x_{\varepsilon}; M(x_{\mathcal{A}}); u'_{\varepsilon}\right) \tag{7.26}$$

where $u'_{\mathcal{A}}$ and u'_{ε} represent any remaining parameters of $\mathcal{A}$ and ε, respectively, that do not participate in the coupling. With the assumption of coupled dynamic systems, the interaction between systems can be viewed as the trajectory of one large system whose variables are simply the variables of both the agent and the environment. Beer [12] further observed that "subtle interplay between sensory input and internal state is crucial to accurate discrimination," thus accounting for an arguably "cognitive" phenomenon in terms of dynamic interrelations between behavior, mechanism, and environment.

Dynamic systems theory provides a straightforward reading of the general notion of making interaction, not internal cognitive processing, the unit of analysis for AI research. Having defined things in this way, dynamic systems theory provides an extensive vocabulary for discussing the space of possibilities through which a given agent–environment system travels. A given region of the space, for example, might form a basin within which all possible initial configurations eventually settle into a stable, periodic "limit set." The interaction might be defined as being adaptive in relation to an arbitrary condition upon its trajectory.

As a highly general framework for formalizing agent–environment interactions, perhaps the principal challenge for the dynamic systems theory will be to formulate systems-theoretic definitions that capture the particular kinds of structure encountered in more general categories of agent–environment interaction. Relatively compact and comprehensible formulas should be developed as new, more appropriate conceptions of categories as basic as "representation" to capture particular properties of enormous dynamic systems. This calls for scientific methods for the interaction between agents and environment to pursue a computational understanding of agent organizations on all levels, from the neurological and mechanical to the social. Computational research on interaction and agency [54] using principled characterizations of agents' interactions with their environments can guide the analysis of living agents and the design of artificial agents.

Another challenge [54] for the relationship between AI and dynamic systems theory is to reconcile the quantitative and qualitative aspects of the various systems that AI research seeks to understand. Differential equations describe systems that can be characterized in terms of numerical variables, but symbolic systems require other types of analysis. In its broadest definition, systems theory is general enough to provide definitions of even very complex symbolic systems, but it does not follow that general results can cast useful light upon those systems. This may benefit from interdisciplinary research in other fields including AI and dynamic systems theory.

7.10.5 Case Study: Markov State Chain

In this section, we discuss how to build an AEI model using the modeling techniques discussed in the above section. We exemplify the usage of the $POMDP_{AEI}$ model in modeling interaction in F-TRADE. As shown in the MDPs, the following lists basic steps for building the $POMDP_{AEI}$ model.

Building the $POMDP_{AEI}$ model

Step 1: Extracting the target agent and its environment; the environment includes all active actors which have close involvement in interaction with the agent.

Step 2: Listing all possible states of the environment, i.e., the states of all active actors.
Step 3: Labeling each state with a number in ascending order in terms of the life cycle of the AEI.
Step 4: Listing all actions the agent is performing on its environment in the lifecycle and marking them in the same numbering system through synchronization with the environment states.
Step 5: Linking all states mediating by actions one by one in terms of the numbering system.
Step 6: Optimizing the life cycle of the state transformation and the performance of actions in the AEI.
Step 7: Checking the model for uncertainties and errors.

As a case study of the analysis of agent organizational dynamics, we have discussed the $POMDP_{AEI}$ model for the agent *AlgoConfigureAgent* in Sect. 7.6. A Markov state-action chain has been built in terms of the roles and activities of the agent in F-TRADE.

7.11 Summary

Taking an organizational metaphor, this chapter has presented the blocks for mapping a system to an organization. Visual modules and models have been introduced in this chapter to represent actors, actor roles, environment as an actor, organizational rules in a system, organizational structure, and protocols, patterns, and rules for modeling interactions.

The models presented in this chapter follow the framework called ORGANISED proposed in Chap. 6, which assumes that a system can be represented in terms of key elements: organizational rules, goals, actors, norms, interactions, structures, environment, and dynamics. Chap. 8 will further discuss the modeling of these elements in terms of formal tools and models.

References

1. Erl, T.: Service-Oriented Architecture: A Field Guide to Integrating XML and Web Services. Pearson Education, Upper Saddle River (2004)
2. Zambonelli, F., Jennings, N.R., Wooldridge, M.: Developing multiagent systems: the GAIA methodology. ACM Trans. Softw. Eng. Methodol. **12**(3), 317–370 (2003)
3. Weiss, G. (ed.): Multiagent Systems: A Modern Approach to Distributed Artificial Intelligence. The MIT Press, Cambridge, MA (2013)
4. Odell, J.J., Parunak, H.V.D., Fleischer, M., Brueckner, S.: Modeling agents and their environment. In: Proceedings of the 3rd International Conference on Agent-Oriented Software Engineering III, AOSE2002, pp. 16–31, Springer, Berlin/Heidelberg (2002)

5. Vasilyev, A.: Synergetic approach in adaptive systems. Master thesis, Transport and Telecommunication Institute, Riga, Latvia (2002)
6. Castro, J., Kolp, M., Mylopoulos, J.: Towards requirements-driven information systems engineering: the TROPOS project. Inf. Syst. **27**(6), 365–389 (2002)
7. Dai, R.W., Cao, L.B.: Internet—an open complex giant system, Science in China (Series E). Sci. China Ser. E **33**(4), 289–296 (2003) (in Chinese)
8. Albert, R., Barabási, A.L.: Topology of evolving networks: local events and universality. Phys. Rev. Lett. **85**, 5234 (2000)
9. Albert, R., Barabási, A.L., Jeong, H., Bianconi, G.: Power-law distribution of the World Wide Web. Science **287**, 2115 (2000)
10. Barabási, A.L., Albert, R.: Emergence of scaling in random networks. Science **286**, 509–512 (1999)
11. Waldrop, M.M.: Complexity: The Emerging Science at the Edge of Order and Chaos. Simon & Schuster, New York (1992)
12. Beer, R.D.: A dynamical systems perspective on agent-environment interaction. Artif. Intell. **72**, 173–215 (1995)
13. FIPA Interaction Protocol Library Specification, Foundation for Intelligent Physical Agents, **fipa.org/specs/fipa**00025/
14. Jennings, N.R.: On agent-based software engineering. Artif. Intell. **117**(2), 277–296 (2000)
15. Alexander, C.: The Timeless Way of Building. Oxford University Press, New York (1979)
16. http://hillside.net/references.html
17. Gamma, E., Helm, R., Johnson, R., Vlissides, J.: Design Patterns. Addison-Wesley, Reading, MA (1995)
18. http://www.pliant.org/personal/Tom_Erickson/InteractionPatterns.html
19. http://www.hcipatterns.org/publications.html
20. Hannemann, J., Kiczales, G.: Design pattern implementation in Java and AspectJ. In: Proceedings of the 17th AM SIGPLAN Conference on Object-oriented Programming, Systems, Languages and Applications, OOPSLA'02, pp. 161–173, ACM, New York (2002)
21. http://www.javaworld.com/columns/jw-java-design-patterns-index.shtml
22. Grand, M. (ed.): Java Enterprise Design Patterns: Patterns in Java, vol. 3. Wiley India, Mumbai (2002)
23. Huget, M.P., Koning, J.L.: Interaction Protocol Engineering in Multiagent Systems. LNCS/LNAI State-of-the-Art Survey-2650. Springer, Berlin/Heidelberg (2003)
24. Odell, J.J., Parunak, H.V.D., Bauer, B.: Representing agent interaction protocols in UML. In: Proceedings of the 1st International Workshop on Agent-Oriented Software Engineering, AOSE2000, pp. 121–140, Springer, Berlin/Heidelberg (2001)
25. Paurobally, S., Cunningham, J., Jennings, N.R.: Developing agent interaction protocols using graphical and logical methodologies. In: Workshop on Programming MAS, Second International Conference on Autonomous Agents and Multiagent Systems, AAMAS'03 Melbourne, Australia, 14–18 July 2003
26. Bauer, B., Müller, J.P., Odell, J.: Agent UML: a formalism for specifying multiagent interaction. In: Proceedings of the ICSE 2000 Workshops on Agent-Oriented Software Engineering, AOSE2000, pp. 91–103, Springer, Berlin/Heidelberg (2001)
27. de Araújo Lima, E.F., de Figueiredo, J.C.A., Guerrero, D.D.S.: Using coloured petri nets to compare mobile agent design patterns. Electron. Notes Theor. Comput. Sci. **95**, 287–305 (2004)
28. Mak, J.K.H., Choy, C.S.T., Lun, D.P.K.: Precise modeling of design patterns in UML. In: Proceedings of the 26th International Conference on Software Engineering, ICSE'04, IEEE Computer Society, pp. 252–261, Scotland, UK (2004)
29. Noriega, P.: Agent mediated auctions: the fishmarket metaphor. Ph.D. thesis, Universitat Autonoma de Barcelona (1997)
30. Smith, R.G.: The contract net protocol: a high level communications and control in a distributed problem solver. IEEE Trans. Comput. **C-29**(12), 1104–1113 (2006)

31. Sprinkle, J.M.: Model integrated program synthesis of agent interaction protocols. Master thesis, Graduate School of Vanderbilt University (2000)
32. AISB CONVENTION[S/OL]: aisb.aber.ac.uk/AgentsSchedule.html (2003)
33. Smallwood, R.D., Sondik, E.J.: The optimal control of partially observable Markov processes over a Nite Horizon [J]. Oper. Res. **21**, 1071–1088 (1973)
34. Bouzid, M.: Antoni Ligeza: temporal causal abduction. Constraints **5**(3), 303–319 (2000)
35. Goldberg, D.: Evaluating the dynamics of agent-environment interaction. Ph.D. thesis, Institute for Robotics and Intelligent Systems, University of Southern California (2001)
36. Ashby, W.R.: An Introduction to Cybernetics. Chapman & Hall, London (1956)
37. Sinha, A.K., Buckley, R.A.: Equilibrium diagram of the iron-molybdenum system [J]. J. Iron Steel Inst. **205**, 191 (1967)
38. Gell-Mann, M.: The Quark and the Jaguar [M]. Freeman, New York (1994)
39. Bonabeau, E., Dorigo, M., Theraulaz, G.: Swarm Intelligence: From Natural to Artificial Systems. Oxford University Press, New York (1999)
40. Heylighen, F.: The Evolution of Complexity. Kluwer Academic, Dordrecht (1996)
41. Eberhart, R., Shi, Y., Kennedy, J.: Swarm Intelligence. Morgan Kaufmann, San Francisco (2001)
42. Qian, X., Yu, J., Dai, R.: A new scientific field–open complex giant systems and the methodology (in Chinese). Chin. J. Nat. **13**(1), 3–10 (1990)
43. Qian, X.: Revisiting issues on open complex giant systems (in Chinese). Pattern Recogn. Artif. Intell. **4**(1), 5–8 (1991)
44. Qian, X.: Building Systematology (in Chinese). Shanghai Jiaotong University Press, Shanghai (2007)
45. Dai, R.: Qualitative-to-quantitative metasynthetic engineering (in Chinese). Pattern Recogn. Artif. Intell. **6**(2), 60–65 (1993)
46. Cao, L., Dai, R.: Open Complex Intelligent Systems (in Chinese). Post & Telecom Press, Beijing (2008)
47. Cao, L., Zhang, C., Zhou, M.: Engineering open complex agent systems: a case study. IEEE Trans. Syst. Man Cybern. C: Appl. Rev. **38**(4), 483–496 (2008)
48. Cao, L., Dai, R., Zhou, M.: Metasynthesis: M-space M-interaction and M-computing for open complex giant systems. IEEE Trans. Syst. Man Cybern. A **39**(5), 1007–1021 (2009)
49. Dai, R., et al.: Metasynthesis of Intelligent Systems (in Chinese). Zhejiang Science & Technology Press, Zhejiang (1995)
50. Ferber, J.: Multi-agent Systems: An Introduction to Distributed Artificial Intelligence. Addison-Wesley/Longman, Harlow (1998)
51. Cao, L.B., Dai, R.W.: Social abstraction for agent-based open giant intelligent systems. In: Proceedings of International Conference on Intelligent Information Technology (ICIIT-02), 22–25 Sept, pp. 47–52. Beijing, China ISBN 7-115-75100-5/0267
52. Kosecka, J., Bogoni, L.: Application of discrete events systems for modeling and controlling robotic agents. In: Proceedings of the 1994 IEEE International Conference on Robotics and Automation, IEEE, pp. 2557–2562, San Diego, CA (1994)
53. Horswill, I.: Analysis of adaptation and environment. Artif. Intell. **73**(1–2), 1–30 (1995)
54. Agre, P.E.: Computational research on interaction and agency. Artif. Intell. **72**, 1–52 (1995)
55. Puterman, M.L.: Markov Decision Processes: Discrete Stochastic Dynamic Programming. Wiley-Interscience, Hoboken (2005)

Chapter 8
Formal Modeling

8.1 Introduction

An effective step for deeply understanding the system complexities of a complex problem is to build formal models to represent the system objects, modules, relations and interactions between objects and system modules, and system properties and dynamics. This requests the building blocks for formal modeling to be developed. In this chapter, we discuss formal models for open complex systems by taking an organizational metaphor. This is built on top of first-order liner-time temporal logics as the modeling tools, which is introduced in Sect. 8.2. Temporal specification is introduced in Sect. 8.3. Sects. 8.4, 8.5 and 8.6 particularly discusses formulae for organizational elements introduced in previous chapters, including Actor, Rule, Interaction, Environment, Relationship, Int-element and Global-properties. In Sect. 8.7, FIPA ACL message specifications are introduced. Organizational goals are modeled in Sect. 8.8.

8.2 First-Order Linear-Time Temporal Logics

Formal specification techniques differ mainly in the particular specification paradigms they rely on. Formal specification paradigms can be *history-based specification*, *state-based specification*, *transition-based specification*, *functional specification*, and *operational specification* [1]. If a system is characterized by its maximal set of admissible histories (or "behaviors") over time—for instance, an agent system—then the history-based specification is suitable.

The properties of interest for history-based specification are defined by *temporal logic assertions* about system objects; such assertions involve operators referring to the past, current, and future states. The assertions are interpreted over time

L. Cao, *Metasynthetic Computing and Engineering of Complex Systems*,
Advanced Information and Knowledge Processing,
DOI 10.1007/978-1-4471-6551-4_8

structures. Time can be linear or branching. Time structures can be discrete, dense, or continuous. The properties may refer to time points, time intervals, or both. Most often it is necessary to specify properties over time bounds; real-time temporal logics are therefore necessary.

8.2.1 Formal Assertions

Formal assertions are interpreted over historical sequences of states. Each assertion is in general satisfied by some sequences and falsified by other sequences. The notation

$$(H,t)|= f \tag{8.1}$$

is used to express that assertion *f* is satisfied by history *H* at time position t ($t \in T$), where *T* denotes a linear temporal structure. The assertions *f* and *t* can be expressed in the following syntax describing the formulae in the typed first-order linear-time temporal logics [2]:

$f ::= f \wedge f \mid f \vee f \mid \neg f \mid t$ (boolean operators)
$\mid t = t \mid t < t \mid t \leq t$ (relational operators)
$\mid \forall x\text{: } sort.f \mid \exists x\text{: } sort.f$ (quantifier operators)
$\mid \circ f \mid \diamond f \mid \square f \mid f\mathcal{U}f$ (future operators)
$\mid \bullet f \mid \blacklozenge f \mid \blacksquare f \mid f\mathcal{W}f \mid f\mathcal{S}f$ (past operators)
$t ::= c \mid x \mid t.a$ (const. and var.)
| *self* | actor | depender | dependee (special terms)
| JustCreated(t) | Changed(t)
| Fulfilled(t) | JustFulfilled(t) (special predicates)

where the temporal operators can further be asserted as follows:

$\circ f$: $(H, t) \mid= \circ f$
iff (*H*, next(*t*)) $\mid= f$, *f* is true in the next state; (X)
$\bullet f$: $(H, t) \mid= \bullet f$
iff (*H*, previous(*t*)) $\mid= f$, *f* was true in the previous state; (Y)
$\diamond f$: $(H, t) \mid= \diamond f$
iff $(H, t_1) \mid= f$ for some $t_1 \geq t$, *f* will be true sometime in the future; (F)
$\blacklozenge f$: $(H, t) \mid= \blacklozenge f$
iff $(H, t_0) \mid= f$ for some $t_0 \leq t$, *f* was true sometime in the past state; (O)
$\square f$: $(H, t) \mid= \square f$
iff $(H, t) \mid= f$ for all $t \geq t_0$, *f* will be always true; (G)
$\blacksquare f$: $(H, t) \mid= \blacksquare f$
iff $(H, t) \mid= f$ for all $t \leq t_0$, *f* was always true; (H)

`fSg`: $(H, t) \mid= fSg$
iff there exists a $t_1 \leq t$ such that $(H, t_1) \mid= g$ and for every t, $(H, t) \mid= f$, f is always true *since* g became true.

`fUg`: $(H, t) \mid= fUg$
iff there exists a $t_2 \geq t$ such that $(H, t_2) \mid= g$ and for every t_1, $t \leq t_1 < t_2$, $(H, t_1) \mid= f$, f is always true *until* g becomes true.

`fWg`: $(H, t) \mid= fWg$
iff $(H, t) \mid= fUg$ or $(H, t) \mid= \Box f$, f will always be true *unless* g becomes true

The notations used above will be introduced in Sect. 8.3. In addition, we will also use the standard logical connectives $\rightarrow$ (implies), $\leftrightarrow$ (equivalent), $\Rightarrow$ (entails), $\Leftrightarrow$ (congruent), where

$$f \Rightarrow g \text{ iff } o(f \rightarrow g) \tag{8.2}$$

$$f \Leftrightarrow g \text{ iff } o(f \rightarrow g) \tag{8.3}$$

8.2.2 Real-Time Temporal Logics

Bounded versions of the above temporal operators are also introduced for handling real-time restrictions. The formulae can be in the following style [3]:

- $_{\leq d}$ (some time in the future within deadline d).
- $\Box_{\leq d}$ (always in the future up to deadline d).
- The semantics of the real-time operators can then be defined accordingly. For instance:
- $(H, t) \mid= \Diamond_{\leq d} f$ iff $(H, t_1) \mid= f$ for some $t_1 \geq t$ with $dist(t, t_1) \leq d$
- $(H, t) \mid= \Box_{\leq d} f$ iff $(H, t_1) \mid= f$ for some $t_1 \geq t$ with $dist(t, t_1) < d$

where the temporal distance function *dist* is

$dist(t, t_1)$: $\mid t_1 - t \mid \times u$, where u denotes a chosen time unit.

The terms t [2] on which the formulae are defined may be integer or Boolean constants (c) and variables (x) or may refer to the attribute's values of the class instances ($t.a$, where a can either be a standard attribute or a special attribute like *actor* or *depender*). Instances may express properties about themselves using the keyword *self.*

Special predicates can appear in temporal logic formulae. The two latter predicates are defined only for goals and dependencies:

- JustCreated(t): it holds if element t exists in this state but not in the previous state.
- Changed(t): it holds in a state if the value of term t has changed with respect to the previous state.
- Fulfilled(t): it holds if t has been fulfilled.
- JustFulfilled(t): it holds if Fulfilled(t) holds in this state, but not in the previous state.

8.3 Temporal Specification

For precise analysis and refinement of requirements, a formal specification is effective. To build formal specifications, the formal grammar must be specified first. The formal grammar consists of grammars for specification, formula, relationships, attributes, inner properties, and global properties [2].

The *specification* defines the structure of the instances together with their attributes for formal representation of an organizational actor (or entity). The basic elements describing a specification consist of *formulae* and *terms* linked by operators such as Boolean, relational, quantifier, future, past, and life-cycle operators. *Special terms* and *predicates* may also be required. The specification grammar is detailed below:

```
/* Temporal specification grammar */
<specification> :: = (<formula> "<boolean-operators>" <formula> |
    <term> | <term> "<relational-operators>" <term> | <quanti-
    fier-operators> <variable> ":" <sort> "." <formula> | <future-
    operators> <formula> | <past-operators> | <lifecycle-
    operators> | <formula>) *
<boolean-operators> ::= (∧ | ∨ | ¬)
<relational-operators> ::= (< | = | ≤)
<quantifier-operators> ::= (∀ | ∃)
<future-operators> ::= (○ | ◇ | □ | U)
<past-operators> ::= (● | ♦ | ■ | W)
<lifecycle-operators> ::= (x^* | x^+ | x^w | x.y | x|y | x+y | x∋y | [x] | [x]_y |)
<terms> :: = (<constant> | <variable> | <term> . <attribute> |
    <special-terms> | <special-predicates>)
<special-terms> ::= (self | actor | depender | dependee)
<special-predicates> ::= [JustCreated(term) | Changed(term) |
    Fulfilled(term) | JustFulfilled(term)]
```

Some of the temporal operators are interpreted in Table 8.1.

The life-cycle operators are defined in Table 8.2.

Table 8.1 Temporal operators

Operator	Interpretation
○/●	Next/previous state
◇/♦	Sometime in the future/past
□/■	Always in the future/past
$\Diamond_{\leq d}$	Sometime in the future within deadline *d*
$\Box_{\leq d}$	Always in the future up to deadline *d*
U	*fUg*: *f* is always true *since g* became true
W	*fWg*: *f* will always be true *unless g* becomes true

Table 8.2 Life-cycle operators

Operator	Interpretation
$x.y$	x *is* followed by y
$x+y$	x and y coexists
$x \mid y$	x or y occurs
x^*	x occurs 0 or more times
x^+	x occurs 1 or more times
x^w	x occurs indefinitely often
$[x]$	x is optional
$[x]_y$	x is optional if y occurs
$x \| y$	x and y interleaved
$x \ni y$	y is covered by x

8.4 Formulae for Organizational Abstraction

Supporting the organizational abstraction of agent organizations, the formulae are composed of actor, rule, interaction, environment, relationship, int-element, and global properties.

```
/* Organizational abstraction formula grammar*/
<formulae>::=(<actor>|<environment>|<rule>|<interaction>|
    <relationships>|<int-element>|<global-properties>)*
```

8.4.1 Actor

As discussed and defined in Sects. 5.2.1 and 6.1.1, an actor may take the form of a human, workspace, autonomous agent, service, entity, or resource. An entity actor may take the form of a component, belief, desire, intention, interaction, position, or self-defined entity.

```
/* Abstraction formula for Actor*/
<actor> ::=Actor<type>[attributes] [creation-properties] [invar-
    properties]
<actor_type> ::= (Human | Workspace | Agent | Service | Entity |
    Resource)
<entity> ::= (Component | Belief | Desire | Intention | Interaction |
    Position | [Self-defined]...)
<agent | service> ::=Agent | Service<role>+
```

8.4.2 Environment

As discussed and defined in Sects. 6.2.2 and 7.3, environment can be specified in terms of elements such as type, actor, resource, principle, process, and force, as shown in the following grammar. The in-actor is the object to which the environment belongs, and the out-actor is the actor which the environment encompasses. Here, the resource actor is highlighted since it is normally one of the main actors in the environment in which agents are lodged. Principle, process, and force are the principal social properties of environment.

```
/* Abstraction formula for Environment*/
<environment> ::= Environment<type><in-actor>+<out-actor>+
    [resources] [principles] [processes] [forces] [attributes]
    [invar-properties]
<environment_type> ::= (Physical | Electronic | Social | Communica-
    tion | ...)
```

8.4.3 Rule

Matching the discussion and definition of rules in Sects. 6.2.4 and 7.4, the following formula specifies the principal elements in defining an organizational rule. There are two types of rules: structural and problem-solving. Every rule is initiated by an initiator actor, and many participant actors may be involved in the operation of the rule:

```
/* Abstraction formula for Rule*/
<rule> ::= Rule<type><initiator>+<participant>+<relationships>
    [attributes] [invar-properties]
<rule_type> ::= (Structural | Problem-Solving)
<structural> ::= ([Control] [Peer] [Ownership] [Dependency])
<dependency> ::= (Single | Bidirectional | And/Or-Composition |
    And/Or-Decomposition)
<problem-solving> ::= ([Means-Ends] [Contribution] [Goal-
    Decomposition] [Iteration] [Cardinality])
<goal-decomposition> ::= (Sequence | Alternative | Concurrency |
    Hybrid)
<iteration> ::= (While-Loop | For-Loop | Interrupt | If | Pick)
<cardinality> ::= (Mandatory One | Mandatory Many | Optional One |
    Optional Many)
```

The structural and problem-solving rules actually represent certain organizational relationships in an agent system. Relationships can also be represented formally. For instance, the following grammar specifies the dependency in a structural relationship. The depended-on or dependent entities in the dependency could take the form of goal, subgoal, resource, or could be in self-defined form. Possible dependency modes could be *achieve*, *cease*, *maintain*, *avoid*, *optimize*, or *self-defined*. Other relationships such as control, peer, benevolence, and ownership could be expressed in similar grammar as dependency.:

```
/* Dependency relationship grammar*/
<dependency> ::=<type>Dependency<name><mode>Depender<name>
    Dependee<name>[attributes]  [creation-properties]  [invar-
    properties] [fulfill-properties]
<dependency_type> ::= (Goal | Subgoal | Resource | [self-defined])
<dependency_mode> ::=Mode (achieve | cease | maintain | avoid | opti-
    mize | ...)
<resource_type> ::= (XML | Data | Information | Knowledge | [self-
    defined])
```

8.4.4 Properties and Keywords

The inner properties express constraints on the lifetime of the objects, given in the above first-order linear-time temporal logics:

```
/* Inner properties grammar */
<int-element>  ::=<type><name><mode>Actor<name>[attributes]
    [creation-properties] [invar-properties] [fulfill-properties]
<creation-properties> ::=Creation<creation-property>+
<creation-property> ::=<property-category><event-category>
    <temporal-formula>
<invar-properties> ::=Invariant <invar-property>+
<invar-property> ::=<property-category><temporal-formula>
<fulfill-properties> ::=Fulfillment <fulfill-property>+
<fulfill-property> ::=<property-category><event-category>
    <temporal-formula>
<property-category> ::= [constraint | assertion | possibility]
<event-category> ::= (trigger | condition | definition)
```

Attributes are represented by facets and sort as follows:

```
/* Attributes grammar*/
<attributes> ::=Attribute<attribute>+
<attribute> ::=<facets><name> : <sort>
<facets> ::= [constant] [optional] . . .
<sort> ::= (<name> | integer | boolean | ...)
```

A set of global properties express properties on the domain as a whole:

```
/* Global properties grammar */
<global-properties> ::= Global <global-property>+
<global-property> ::= <property-category> <temporal-formula>
```

In the above grammar, some keywords are defined as follows:

Facet: represents basic properties of attributes:

- Optional: the attribute may be undefined.
- Constant: the value of the attribute does not change after its initialization.

Sort: the type of attribute; it can be either a primitive (defining the relevant state of an instance) or a class (defining references to other instances in the domain)
Mode: declares the modality of an intentional element's fulfillment.
Invariant: defines conditions that should hold throughout the lifetime of all instances.
Creation: defines conditions on the moment when an actor begins to desire a goal; it can be associated with any class.
Fulfillment: defines conditions at the moment when the goal is actually achieved; fulfillment constraints can only be associated with intentional elements and with dependencies.
Formula: is a temporal logic formal assertion, which is interpreted over historical sequences of states. Each assertion is in general satisfied by some sequences and falsified by other sequences.
Term: is one of the basic elements for defining a formula.

The *mode* (or *pattern*) of a dependency (usually a goal) identifies what is to be done with the dependency. Five modes are allowed for goals [4]. These patterns have an impact on the set of possible behaviors of the system. *Achieve* and *cease* goals generate behaviors, *maintain* and *avoid* goals restrict behaviors, and *optimize* goals compare behaviors. Some of them may even combine to identify a sequence of behaviors.

Achieve: pattern $f \Rightarrow \Diamond g$, to achieve the goal at some point in the future
Cease: pattern $f \Rightarrow \Diamond \neg g$, undo a goal at some point in the future
Maintain: pattern $f \Rightarrow gWh$, maintain a goal for some time
Avoid: pattern $f \Rightarrow \neg gWh$, prevent a goal from becoming true
Optimize: pattern maximize (objective function) or minimize (objective function)

Creation and fulfillment constraints are further distinguished by conditions for generating events. Three categories of event constraints are *trigger*, *condition*, and *definition*.

Trigger: If there are sufficient conditions for an event, then *trigger* is satisfied.
Condition: Refers to the necessary conditions for creation or fulfillment.
Definition: Refers to the necessary and sufficient conditions for creation or fulfillment.

As shown in the formal grammar, properties which may be held in the domain can also be specified. Properties include *Constraint, Assertion* and *Possibility*. Given the above formal specifications and a set of properties, a Formal Tropos specification can be verified by means of T-TOOL [2] to check whether these properties are satisfied.

Assertion: assertion properties should hold for all valid evolutions of the formal specification.
Possibility: possibility properties should hold for at least one valid scenario.

More elements of organizational abstraction will be discussed in subsequent sections with a case study from F-TRADE. With the above specifications and grammars, formal representation is able to perform precise refinement of the requirements roughly obtained from the visual modeling.

8.5 Modeling Roles

Formal grammar for actors has been discussed in Sect. 8.4.1. There, we say that an actor may take one to many roles in an agent organization. In Sect. 7.2.2, we describe the role schema from GAIA and present an example of the role PLUGINPERSON in the GAIA format. In this section, we try to make it more formal and represent it using temporal logics.

Figure 8.1 exemplifies the role PLUGINPERSON in temporal logics based on the informal GAIA format. In the specifications, relevant actors and constant attributes are listed. Three protocols have been decomposed. For each protocol, the two stakeholders are requester and responder. Based on certain inputs and rules for the interaction, the outputs of the protocol can be constant or a status. The rule of a protocol can be illustrated as follows: For protocol ApplyRegistration, the AlgorithmRegisterAgent submits PluginRequest; when it is fulfilled, PluginInterfaces will be called by the InterfaceAgent and the AlgorithmPluginRequestID will be generated as the output of this protocol. Activity FillinAlgoRegisterOntologies is performed to fill in ontological items in plugging in an algorithm; this is conducted after the fulfillment of ApplyRegistration and CheckAlgorithmValidity. Permissions on the role PLUGINPERSON allow it to call PluginInterfaces from InterfaceAgent, read AlgorithmID, and fill in AlgoRegisterOntologies. As a liveness responsibility, the AlgorithmRegisterAgent must conduct the three protocols and one activity in order from ApplyRegistration to CheckAlgorithmValidity to FillinAlgoRegisterOntologies and finally to SubmitAlgoPluginRequest. The time invariant for conducting the above protocols and activity is defined as in the safety attribute.

Role PLUGINPERSON
InformalDef *This role involves in applying registering a non-existent algorithm, typing in attribute items of the algorithm, and submitting plug in request to F-TRADE.*
FormalDef
Actor name AlgorithmRegisterAgent
Actor name ConfigureAgent
Actor name InterfaceAgent
Actor name OntologyService
Actor name DirectoryService
Attribute constant self
Attribute constant algoid: AlgorithmID
Existed: Boolean
Attribute constant ai: AttributeItems
Attribute constant reqid: AlgorithmPluginRequestID
Available: Boolean
Protocol ApplyRegistration
Requester: AlgorithmRegisterAgent
Responder: InterfaceAgent
Input: PluginRequest()
Rule:
∀ ara: AlgorithmRegisterAgent(ara.PluginRequest() ∧ Fulfilled(self)) →
(∃ ia: InterfaceAgent(ia.CallPluginInterfaces() ∧ Called()))
Output: reqid.Available()
Protocol CheckAlgorithmValidity
Requester: AlgorithmRegisterAgent
Responder: ConfigureAgent
Input: algoid
Rule:
∀ara: AlgorithmRegisterAgent(algoid.Existed() ∧ ¬algoid.Registered() →
(∃ algoid.Valid())
Output: algoid.Plugable()
Activities
∃ ara: AlgorithmRegisterAgent((∃ (ara.ApplyRegistration.Called() ∧ Fulfilled(self)) ∧
(ara.CheckAlgorithmValidity.Plugable() ∧ Fulfilled(self)) →
$\Diamond_{\leq t1}$ ara.FillinAlgoRegisterOntologies.Filled(ai))
Protocol SubmitAlgoPluginRequest
See Figure 8.3
Permissions
∀ ia: InterfaceAgent(ara.Call(ia) → CallPluginInterfaces()) →
$\Diamond_{\leq t1}$ ara.Read(algoid) →
$\Diamond_{\leq t2}$ ara.Fill(ai)
Responsibilities
Liveness:
∃ ApplyRegistration.Fulfilled() →
$\Diamond_{\leq t1}$ CheckAlgorithmValidity.Fulfilled() →
$\Diamond_{\leq t2}$ FillinAlgoRegisterOntologies.Filled() →
$\Diamond_{\leq t3}$ SubmitAlgoPluginRequest.Fulfilled()
Safety (Invariant):
$0 < t_1 < t_2 < t_3$

Fig. 8.1 Modeling role in temporal logics

8.6 Modeling Interaction

For modeling interaction, we principally talk about how to formally represent the interaction protocol and pattern. Protocol and pattern can be formally represented. The following presents the relevant formula:

```
/*grammar for protocol*/
<protocol>  ::=  Protocol <name> [function]  [Message] <initiator>
    <responder>+ <input> <rule> <output> [termination] [exception]
<exception> ::= Exception <type>
<type> ::= [message] [timeout] [protocol]
/*grammar for pattern*/
<pattern> ::= Pattern <name> [alias] <force>* [problem] <structure>
    [layer] <participant>+ <solution> <dependency>* <role>+ [mes-
    sage] [context] [known use] [example] [see also]
```

Figure 8.2 re-illustrates the contract net protocol in temporal logics [5]. In Sect. 8.5, we present a number of protocols in the role schema in temporal logics. The protocol SubmitAlgoPluginRequest mentioned in role PLUGINPERSON is shown in Fig. 8.3.

In addition, a message transferred in an interaction can also be formalized. This is addressed in the following section, where we discuss the FIPA Agent Communication Language [6].

```
start ⇒ announce(task(Name, Desc, Req, self))
(¬award(T,A)) S announce(T) ∧ competent(A,T)
    ⇔ possible(A,T)
possible(self, T) ⇒ bid(T, self)
bidded(T,X) ∧ most-preferable(T,Y) ⇒ award(T,Y)
(¬completed(T,self,X) ∧ ¬split(T,T1,T2)) S award(T, self)
    ∧ •result(T,R) ⇒ completed(T,self, R)
```

Fig. 8.2 Contract Net in temporal logics

```
Protocol SubmitAlgoPluginRequest
Requester: AlgorithmRegisterAgent
Responder: ConfigureAgent
Input: reqid, algoid
Rule:
   ∀ara: AlgorithmRegisterAgent(ara.
      FillinAlgoRegisterOntologies.Fulfilled(algoid)
      ∧ Fulfilled(algoid)) →
         ◊≤t1 Fulfilled(reqid)
Output: reqid.Successful()
```

Fig. 8.3 Protocol SubmitAlgoPluginRequest

8.7 FIPA ACL Message Specifications

8.7.1 ACL Protocol Description Language

Agent communication usually follows a pattern of interaction protocols. Messages in agent communication are transferred between agents through attachments in the interaction protocol. Formally, an IP can be described by some form of Protocol Description Language (PDL). In a PDL, performatives and parameters must be defined to capture and denote interaction ontologies. For instance, KQML [7, 8] and FIPA ACL define corresponding PDLs, respectively. In our work, we conduct FIPA-compatible analysis and design. Tables 8.3 and 8.4 list FIPA ACL message parameters [9] and performatives for communicative acts [6], respectively.

Table 8.3 FIPA ACL message parameters

sender	receiver	reply-to	content	language	encoding
ontology	protocol	conversation-id	reply-with	in-reply-to	reply-by

Table 8.4 FIPA ACL categories of communicative acts

Performative	Info passing	Info requesting	Negotiation	Actions performing	Error handling
accept-proposal			x		
agree				x	
cancel		x		x	
cfp			x		
confirm	x				
disconfirm	x				
failure					x
inform	x				
inform-if	x				
inform-ref	x				
not-understood					x
propose			x		
query-if		x			
query-ref		x			
refuse				x	
reject-proposal			x		
request				x	
request-when				x	
request-whenever				x	
subscribe		x			

8.7.2 Modeling ACL Messages

Based on the above ACL message parameters and performatives, a FIPA ACL message consists of two parts: one concerns the administration of messages, while the other is relevant to the message body. The following message template can be instantiated to present particular messages for agent interactions. This template specifies the ACL message abstract formula grammar in temporal logics:

/* *ACL Message Abstract Formula Grammar* */
<*formulae*> ::= (<*communicative act*> . <*message parameter*>)*
<*communicative act*> ::= (accept-proposal | agree | cancel | cfp | confirm | disconfirm | failure | inform | inform-if | inform-ref | inform-ref | not-understood | propose | query-if | query-ref | refuse | reject-proposal | request | request-when | request-whenever | subscribe | envelop)
<*message parameter*> ::= ([sender] $_{\text{Anonymous}}$ + receiver^{+} + [reply-to^{+}] + ([language] $_{\text{content} \ni \text{language}}$ + [ontology]) . [content] $_{\text{conversation-id | in-reply-to}}$ + [encoding] $_{\text{envelop}}$ + [protocol] $_{\text{conversation-id | ○ reply-by}}$ + conversation-id + [reply-with] + [in-reply-to] + [reply-by])

For the instantiation of ACL messages, FIPA has proposed four types of content language specification dealing with different representations of the content of ACL messages; they are semantic language (SL) (FIPA SC00008I), constraint choice language (CCL) (FIPA XC00009B), knowledge interchange format (KIF) (FIPA XC00010C), and resource description framework (FIPA XC00011B).

Figure 8.4 lists an ACL message example in the SL format by following the above abstract formula syntax. The XML ACL Representation (FIPA SC00071E) [10] specifies the representation of ACL messages in XML; Fig. 8.5 lists the format

```
(request
 : sender (: name
algomanager@datamining.it.uts.edu.au : 8080)
   : receiver (: name
      (: movingaveragebasic@datamining.it.uts.edu.au : 8080)
      (: pairstrading1@datamining.it.uts.edu.au : 8080)
                )
   : ontology F-Trade_PSO
   : language SL
   : protocol FIPA_request
   : content (action
                 (distribute-algoid
                    (: movingaveragebasic ma89221)
                    (: pairstrading1 pt78232)
                  )
                )
 )
```

Fig. 8.4 An ACL message example in SL

```
/* XML ACL Representation Syntax */
<!—Document Type: XML DTD -->
<! ENTITY %communicative-acts "accept-proposal | agree | cancel | cfp | confirm | disconfirm | failure | inform | not-
    understood | propose | query-if | query-ref | refuse | reject-proposal | request | request-when | request-whenever |
    subscribe | inform-if | inform-ref | proxy | propagate" >
<! ENTITY % msg-param "receiver | sender | content | language | encoding | ontology | protocol | reply-with | in-
    reply-to | reply-by | reply-to | conversation-id | user-defined">
<! ELEMENT fipa-message (%msg-param;)*>
<! ATTLIST fipa-message act (%communicative-acts;) #REQUIRED conversation-id ID #IMPLIED>
<! ELEMENT sender (agent-identifier)>
<! ELEMENT receiver (agent-identifier+ )>
<! ELEMENT content (# PCDATA)>
<! ATTLIST content href CDATA #IMPLIED >
<! ELEMENT language (#PCDATA)>
<! ATTLIST language href CDATA #IMPLIED>
<! ELEMENT encoding (#PCDATA)>
<! ATTLIST encoding href CDATA #IMPLIED>
<! ELEMENT ontology (#PCDATA)>
<! ATTLIST ontology href CDATA #IMPLIED>
<! ELEMENT protocol (#PCDATA)>
<! ATTLIST protocol href CDATA #IMPLIED>
<! ELEMENT reply-with (#PCDATA)>
<! ATTLIST reply-with href CDATA #IMPLIED>
<! ELEMENT in-reply-to (#PCDATA)>
<! ATTLIST in-reply-to href CDATA #IMPLIED>
<! ELEMENT reply-by EMPTY>
<! ATTLIST reply-by time CDATA #REQUIRED href CDATA #IMPLIED>
<! ELEMENT reply-to (agent-identifier+)>
<! ELEMENT conversation-id (#PCDATA)>
<! ATTLIST conversation-id href CDATA #IMPLIED>
<! ELEMENT agent-identifier (name, address?, resolvers?, user-defined*)>
<! ELEMENT name EMPTY>
<! ATTLIST name id ID #IMPLIED refid IDREF #IMPLIED>
<! ELEMENT addresses (url+)>
<! ELEMENT url EMPTY>
<! ELEMENT url href CDATA #IMPLIED>
<! ELEMENT resolvers (agent-identifier+)>
<! ELEMENT user-defined (#PCDATA)>
<! ATTLIST user-defined href CDATA #IMPLIED>
```

Fig. 8.5 FIPA ACL specifications in XML

from the FIPA in terms of the referenced specification in detail. Fig. 8.6 presents the example in Fig. 8.4 following ACL Message Representation in XML specification.

8.8 Modeling Organizational Goal

Functional (⬭) and nonfunctional (⬯) goals can be visualized through goal decomposition in an organizational goal and structure diagram. Formally, a goal can also be represented by formal grammar. For instance, the goal RegisterAlgo is expressed as follows (Fig. 8.7):

```
<!—Document Type: XML DTD -->
<! ENTITY %communicative-acts "request" >
<! ENTITY % msg-param "receiver | sender | content | language | ontology | protocol">
<! ELEMENT fipa-message (%msg-param;)* >
<! ATTLIST fipa-message act (%communicative-acts;) #REQUIRED conversation-id ID #IMPLIED>
<! ELEMENT sender (agent-identifier)>
<! ELEMENT receiver (agent-identifier+)>
<! ELEMENT content (action, action-content*)>
<! ATTLIST content href CDATA #IMPLIED >
<! ELEMENT language (#PCDATA)>
<! ATTLIST language href CDATA #IMPLIED>
<!—This file represents a fragment of a ACL message-->
<fipa-message act = %communicative-acts;, conversation-id = "">
  <sender>
        <name>algomanager@datamining.it.uts.edu.au : 8080 </name>
    </sender>
  <receiver>
        <name>movingaveragebasic@datamining.it.uts.edu.au : 8080 </name>
    </receiver>
    <receiver>
        <name>pairstrading1@datamining.it.uts.edu.au : 8080 </name>
    </receiver>
    <content>
        <action> distribute-algoid </action>
   <action-content> movingaveragebasic ma89221 </action-content>
        <action-content> pairstrading1 pt78232 </action-content>
    </content>
    <language> SL </language>
    <ontology> F-Trade_PSO </ontology>
    <protocol> FIPA_request </protocol>
</fipa-message>
```

Fig. 8.6 An ACL message example in XML

Goal RegisterAlgo

InformalDef *When an algorithm component has been coded and the algorithm isn't available from the system at the moment, this algorithm component can be registered into the system by calling plug-in interfaces, filling in algorithm registration ontologies, and upload the algorithm module.*

FormalDef

Actor AlgorithmProvider
Actor AlgoPluginAgent
Actor ConfigureAgent
Actor InterfaceAgent
Actor OntologyService
Actor DirectoryService
Mode achieve
Attribute constant ca: CodeAlgo
Attribute constant aro: AlgoRegisterOntology
Attribute constant algo: Algorithm
registered: boolean
Creation condition

- Fulfilled(ca) $\wedge \neg$ Existed(algo)

Invariant ca.actor = actor
Fulfillment condition

$\forall$ac: AlgorithmComponent (ac.algo = algo $\rightarrow$

$\lozenge_{\leq t1}$ $\exists$ cpi: CallPluginInterfaces (cpi.actor = actor $\wedge$ Fulfilled(cpi) $\wedge$ pi.Called) $\wedge$ $\lozenge_{\leq t2}$($\exists$faro: FillinAlgoRegisterOntologies (faro.depender = actor $\wedge$ Fulfilled(faro) $\wedge$ aro.Filled) $\wedge$ $\exists$ uac: UploadAlgoComponent (uac.depender = actor $\wedge$ Fulfilled(uac) $\wedge$ ac.uploaded)))

Fig. 8.7 Modeling goal in temporal logics

8.9 Summary

Formalization plays an important role in precisely representing a problem. Formal modeling is a tool that aims to formalize system complexities. Formal modeling tools complement visual modeling in terms of building more precise mechanisms and tools for "quantifying" system complexities.

In this chapter, formal models are introduced to represent actors, roles, environment, rules, properties, interactions, communications, and goals related to a complex system. These tools should be used in alignment with the visual tools discussed in Chap. 7 on visual modeling to build both a concrete and formal representation of system components, relationships, interactions, dynamics, structures, and environment in a complex system.

References

1. van Lamsweerde, A.: Formal specification: a roadmap. In: Proceedings of the Conference on The Future of Software Engineering, ICSE'00, pp. 147–159. ACM Press (2000)
2. Fuxman, A., Liu, L., Mylopoulos, J., Pistore, M., Roveri, M., Traverso, P.: Specifying and analyzing early requirements in TROPOS. Submitted to J. Requir. Eng. http://www.cs.toronto.edu/~liu/ (2004)
3. Koymans, R.: Specifying Message Passing and Time-Critical Systems with Temporal Logic. LNCS-651. Springer, Berlin/Heidelberg (1992)
4. Dardenne, A., Lamsweerde, V.A., Fickas, S.: Goal-directed requirements acquisition. Sci. Comput. Program. **20**, 3–50 (1993)
5. Fisher, M., Wooldridge, M.J.: Specifying and executing protocols for cooperative action. In: Proceedings of the 2nd International Working Conference on Cooperating Knowledge-Based Systems, CKBS94, pp. 295–306, Keele University. Springer (1995)
6. FIPA Agent Communication Language Specification, Foundation for Intelligent Physical Agents www.fipa.org/repository/aclspecs.html
7. Finin, T., Fritzson, R., McKay, D., McEntire, R.: KQML as an agent communication language. In: Proceedings of the 3rd International Conference on Information and Knowledge Management, CIKM'94, Maryland, USA, pp. 456–463. ACM Press (1994)
8. KQML Advisory Group: An Overview of KQML: A Knowledge Query and Manipulation Language www.agent.ai/doc/upload/200302/chal92_1.pdf
9. SC00067F. FIPA Agent Message Transport Service Specification, 2002/12/03, www.fipa.org/specs/fipa00067/
10. SC00071E. FIPA ACL Message Representation in XML Specification. 2002/12/03, www.fipa.org/specs/fipa00071/

Chapter 9
Integrative Modeling

9.1 Introduction

Requirements analysis (RA) [1] is arguably the most important stage of information system development. It answers the most fundamental of all design questions: "What is the system intended for?" This is the phase in which technical considerations have to be balanced against social and organizational concerns, and where the operational environment of the system is modeled. Not surprisingly, this is also the phase where the greatest number of errors, and the costliest, are introduced to a system. It is thus the critical phase of information system development.

In this chapter, we present the following integrative modeling framework for the RA:

1. Integrating functional and nonfunctional requirements.
2. Integrating visual and formal modeling.
3. Integrating business-oriented requirements. In the discussion, we take the original goal-oriented RA as an example for the visual modeling.

9.2 Integrating Functional and Nonfunctional Requirements

9.2.1 Functional Requirements Analysis

Functional requirements analysis concerns the understanding of the goals of a system by studying an organizational setting. The output of this analysis is an organizational model which includes relevant organizational actors and their respective organizational goals, global and local organizational relationships, rules, policies, and constraints.

L. Cao, *Metasynthetic Computing and Engineering of Complex Systems*,
Advanced Information and Knowledge Processing,
DOI 10.1007/978-1-4471-6551-4_9

Taking goal-oriented RA as an example, we try to adopt concepts offered by *i** framework for functional RA, such as actor (actors can be agents, positions, or roles), as well as social dependencies among actors, including goal, softgoal, task, and resource dependencies.

9.2.2 Nonfunctional Requirements Analysis

Nonfunctional requirements analysis (or quality attributes, qualities, or more colloquially "-ilities") refers to qualities of service, development objectives, or architectural constraints [2]. The most common nonfunctional requirements [3] are the global qualities of a software system, such as flexibility, maintainability, usability, and so forth. The characteristics of such requirements include the following:

1. They are usually stated only informally.
2. They are often controversial (e.g., management wants a secure system but staff desires user-friendliness).
3. They are difficult to enforce during design and implementation.
4. They are difficult to validate.

9.2.3 Analyzing Integrative Requirements

Based on the discussion in the above two sections, functional requirements essentially focus more on functional goals, while nonfunctional requirements concern nonfunctional goals. Nevertheless, functional goals and nonfunctional goals are closely related or even interwoven in a system. It is to some degree difficult to separate them from each other in a real system, even though they are often discussed as being in different situations and having different objectives. For instance, the system-to-be is described along with relevant functions and qualities within its operational environment.

On the other hand, even though we can build individual models for functional and nonfunctional requirements, respectively, the representation of these two types of model can have the same or similar diagrammatic and formal descriptions. For instance, in goal-oriented RA, both functional and nonfunctional requirements analyses can be described by modeling techniques like strategic dependency model and strategic rationale model in TROPOS [20].

An integrative model must and can encompass both functional and nonfunctional requirements. This strategy is the most practical and effective in both modeling and understanding the problem. It will further benefit the later system design and implementation.

Weaknesses of integrative modeling may include the following:

1. Too many goals will be involved in one model.
2. The diagram could be very complicated and comprehensive.

However, just as we use decomposition technique to understand a complex system by dividing it into multiple subsystems, functions and qualities can also be considered separately first and then united into an entity. Various diagrammatic notations will be used to label the respective functional and nonfunctional requirements in integrated diagrams. This can enable the two types of goal to be traced and understood according to different motivations.

9.3 Visual Modeling

Most of the existing RA methodologies take the definition and description of diagrammatic notations as a core process in analyzing requirements. Diagrammatic modeling absorbs and inherits a number of best practices, such as structured object-oriented software engineering [4]. For instance, use case [5] is used for discussing requirements with actors; *i** framework and TROPOS methodology provide graphical notations for describing actors, goals, dependencies, rationale, and so forth. We call this type of practice *visual modeling*. More formal work on visual modeling would develop visual modeling languages for system analysis and design, offering such advantages as providing an effective and concrete interaction for different stakeholders of the process.

Nevertheless, visual modeling languages normally emphasize the structural fact in a system; they often lack formal definitions of semantics and logics. This could lead to subjective models, which cannot easily be refined in a straightforward way in a system design. Other weaknesses in building a conceptual model include knowing when the refinement should be stopped and how to check for conflict and inconsistencies. In the following, we take TROPOS visual modeling as an example to illustrate the visual modeling.

9.3.1 Goal-Oriented Visual Modeling

The *i** (pronounced "i-star", stands for "distributed intentionality") framework [6–9] and the later TROPOS (TROPOS adopts concepts for actor and goal from the *i** framework) have been developed based on goal-oriented RA. They help support process engineering and requirements engineering. They focus on modeling intentional and strategic relationships between actors. Stakeholders are represented as actors who depend on each other for goals to be achieved, softgoals to be satisfied, tasks to be performed, and resources to be furnished. The intentional elements consist of softgoals, goals, tasks, and resources. Nonfunctional and functional aspects of process description and design reasoning can be supported in this framework.

The *i** framework focuses on the modeling of strategic relationships. It considers such relationships to be "strategic" in the sense that each party is assumed to be concerned with opportunities and vulnerabilities, and seeking to protect or further its interests. The *strategic dependency* (SD) model and *strategic rationale* (SR) model are developed in *i** framework. The SD model describes the network of relationships which are held among actors. The SR model describes and supports the reasoning that all actors go through concerning their relationships with other actors to accomplish their goals and tasks. These two models assist the modelers in understanding the existing processes and generating alternatives in order to reach a new process design that better addresses the organization's objectives.

9.3.1.1 Strategic Dependency Model

An SD model is a diagram involving *actors* who have *strategic dependencies* with one another [10]. An actor can be an agent, a role, or a position. An agent is a concrete, physical actor or subsystem. A role is a function in an organization. A position is a group of roles that are institutionalized in an organization. There are three entities which constitute a dependency. A dependency describes an "agreement" (or *dependum*) between two actors: the *depender* and the *dependee*. The *depender* is the depending actor; the actor who is depended upon by another is called the *dependee*; the object on which the dependency relationship centers is called the *dependum*.

The type of dependency describes the nature of the agreement. There are four types of dependency, namely, *goal*, *task*, *resource*, and *softgoal*, according to the nature of the dependum involved. Yu and Mylopoulos [11] gave a detailed introduction to these dependencies. In a (hard) goal dependency, the depender depends on the dependee to bring about a desired state. The dependee is free to decide how to achieve the goal on his own. A task dependency specifies that the depender depends on the dependee to complete a certain task by carrying out certain activities, i.e., how the task is to be performed, but not why. A resource dependency specifies that the depender depends on the dependee for the availability of information or a physical resource. *Softgoal* dependencies are similar to goal dependencies, but their fulfillment cannot be defined precisely (for instance, the appreciation is subjective, or the fulfillment can occur only to a given extent), i.e., a goal that can be achieved to variant degrees, and needs to be optimized. A softgoal is different from a (hard) goal dependency in that there is no a priori, cut-and-dried criterion for what constitutes meeting the goal.

In a dependency, the depender may be vulnerable if the dependee fails to bring about the dependum. There are three degrees of strength of dependencies in the SD model: open (uncommitted), committed, and critical [11].

9.3.1.2 Strategic Rationale Model

The SD model focuses on the intentional dependencies among actors beyond the usual understanding based on entity flows and activities. It helps the modeler to obtain a deeper understanding of a process and identify what is at stake and for whom, and what impacts are likely if a dependency fails. However, the SD model only describes why a process is structured in a certain way. It does not really support the process of suggesting, exploring, and evaluating alternative solutions. The strategic rationale (SR) model of *i** addresses this; Chung [8] gave the following description of the SR model.

The SR model is a graph consisting of four main types of nodes, goals, tasks, resources, and softgoal nodes, and two main types of links, namely, means–ends links and task-decomposition links. Task-decomposition links describe how a task can be decomposed into subtasks or subgoals. Means–ends links specify how a goal may be achieved. They provide information about why an actor would perform a task, pursue a goal, need a resource, or want a softgoal. From the softgoals, the modeler can tell why one alternative may be chosen over others.

*i** framework has been related to different application areas, including requirements engineering, software processes, and business process reengineering. In particular, *i** has been transformed into a tool of agent-oriented early requirements analysis; it has been used in some agent-oriented methodologies (e.g., the TROPOS in [9]) and requirements engineering (e.g., [12]).

In the following work, conception in the *i** framework and TROPOS will be adopted for the visual modeling of the infrastructure F-TRADE. We will explicitly extend these frameworks to support organization-oriented analysis in terms of organizational metaphor.

9.4 Formal Specifications

Formal specifications complement some of the weaknesses of visual modeling. They present mechanisms for defining models with precise semantics. The disadvantages of formal specifications are that:

1. They are difficult to understand and utilize in modeling without strong skills in these specifications.
2. They are not good at visualizing interactions between stakeholders.

Lamsweerde [13] gave a nice roadmap for formal specifications. A *formal specification* is generally an expression in formal language, at a level of abstraction of a collection of properties which the modeled object should satisfy. "Formal" is often confused with "precise" (the former entails the latter, but the reverse is not true). Generally speaking, a specification is *formal* if it is expressed in a language made of three components:

1. Rules for determining the sound grammatical formation of sentences (the *syntax*).
2. Rules for interpreting sentences in a precise, meaningful way within the domain considered (the *semantics*).

3. Rules for inferring useful information from the specification (*inference*). Formal specification techniques essentially differ from semiformal techniques (such as dataflow diagrams, entity relationship diagrams or state transition diagrams) in that the latter do not formalize the assertion part.

A specification must thus satisfy a higher-level specification and be satisfied by lower-level specifications as a whole. *Formal language* is required for expressing formal specification. Each construct in the language normally has a two-level generic structure [14]: an *outer semantic net layer* [15] for *declaring* a concept, its attributes, and its various links to other concepts, and an *inner formal assertion layer* for *formally defining* the concept.

We introduce the logic foundation, the specifications, and the grammars for the formal modeling in detail in Chap. 8.

9.5 Integrative Modeling Framework

Requirements engineering has undergone a process from incomplete to complete. The previous diagrammatic RA developed informal box and arrow notations with minimal syntax and has evolved to become visual modeling by adding ontologies. Formal specification, on the other hand, used to be described by formal notations with ontology, syntax, and semantics. Nowadays, a formal conceptual model can be deployed for formal modeling with an assertion language for specifying rules and constraints [16]. However, a more systematic trend is to build *an integrative model by either proposing techniques for visual modeling and formal modeling, respectively, or combining these techniques to build an integrative modeling approach.*

The normal analysis of RA focuses on *goal-oriented requirements*. Goal-oriented RA emphasizes and extracts requirements from a technical point of view; however, *business-oriented requirements* indicate the need to provide more intuitive and understandable information to business persons. Business-oriented requirements are all the social and organizational policies, laws, orders, norms, constraints, and capabilities that the system is expected to satisfy in order to fulfill business objectives and processes in the organization. Business-oriented requirements can be generally divided into two categories: *business-oriented functional requirements* and *business-oriented nonfunctional requirements*. We hope the add-ons of the explicit discussion of business-oriented requirements will help us understand the business problem and operationalize the system analysis and design.

9.5.1 Business-Oriented Functional Requirements

Functional requirements capture all the functions and activities the proposed system is expected to serve. For the proposed F-TRADE, it is necessary to develop the following functional requirements discussed in the relevant sections above: services supporting trading and mining, data services, system services, and algorithm services.

9.5.2 Business-Oriented Nonfunctional Requirements

Nonfunctional (or technical) requirements [1] are all the operational objectives related to the environment, hardware, and software of the organization. For the proposed F-TRADE, the following objectives, discussed in the related sections above, are key objectives which the infrastructure must attain: component plug-and-play, enterprise application integration, and formation of a knowledge portal for financial decision supports in trading and mining.

9.5.3 Integrative Modeling

It can be seen, therefore, that *integrative modeling* is a kind of modeling methodology:

1. It combines functional and nonfunctional requirements.
2. It integrates goal-oriented and technical requirements in business fashion.
3. It tries to systematically integrate formal specifications and informal diagrammatic notations.

Visual modeling equipped with formal languages can interpret the problem and handle requirements engineering in a concrete and visualized way, which aids discussion with stakeholders; it can also support deductive reasoning, which assists with precise and explicit understanding of the problem and modeling procedures. Furthermore, integrative modeling covers not only goal-oriented requirements but also business-oriented requirements. Figures 9.1 and 9.2 show the evolution of requirements engineering and a framework of integrative modeling, respectively.

In Chaps. 7 and 8, we propose and detail the individual techniques for visual modeling and formal modeling. For visual modeling, several effective concepts from *i**, TROPOS, GAIA, and other technologies can be adopted and extended on demand to build the organization-oriented analysis of functional and nonfunctional requirements. At the same time, first-order linear-time temporal logic [17–19], also presented in formal TROPOS [13], can be utilized to develop formal grammar and specifications to complement the above visual analyses for specifying goals and softgoals completely and precisely. Explicit formal specifications are very important for refinement and later design and implementation.

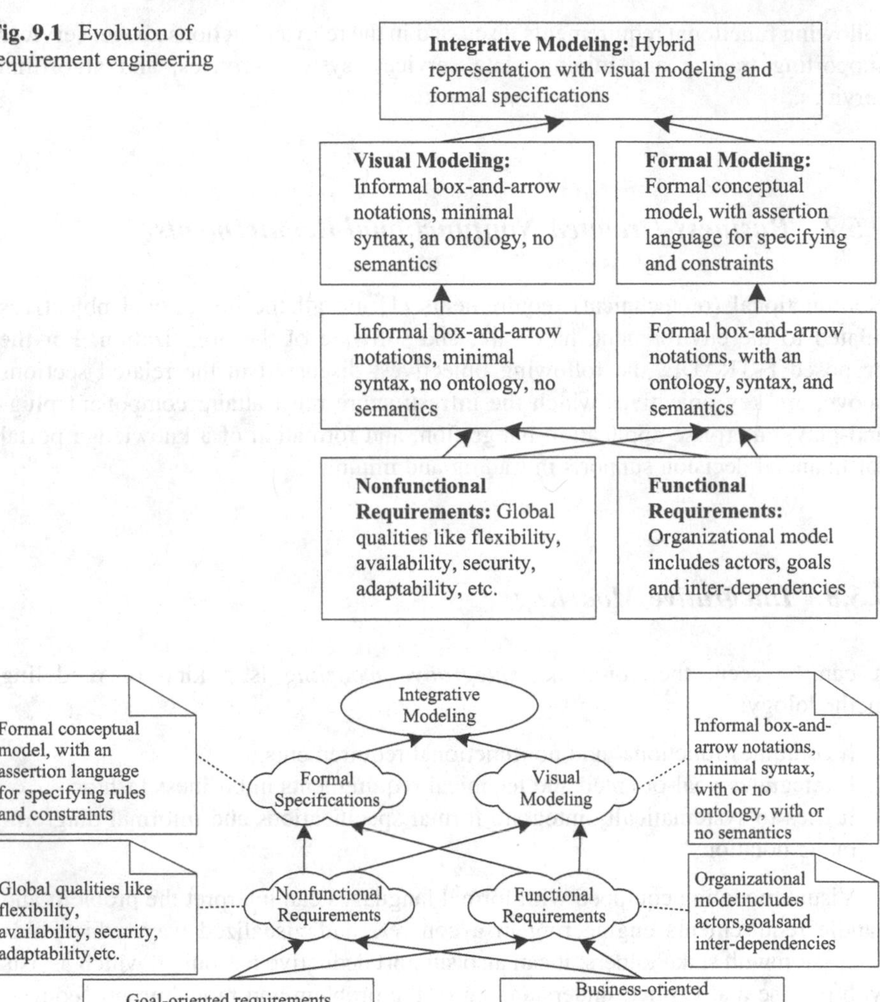

Fig. 9.1 Evolution of requirement engineering

Fig. 9.2 Framework of integrative modeling

9.6 Summary

In this chapter, integrative modeling for agent-oriented requirements analysis has been investigated. The main conclusions and work in the integrative analysis of agent-oriented requirements are summarized below:

1. Analyzing agent-oriented early requirements in terms of techniques and experience obtained in goal-oriented requirements engineering
2. Undertaking goal-oriented requirements engineering in terms of organizational metaphor when discussing agent-oriented requirements

3. Extending *i** framework to organization-oriented visual modeling of agent-based systems
4. Building formal specifications to refine and clarify requirements extraction in visual modeling

Conflicts also often emerge in system modeling, and modeling mechanisms for conflict resolution and responsibilities refinement are therefore essential for the sound and robust capture and development of system requirements. For instance, conflict resolution and responsibility refinement can be performed through scenario-based analysis, sequential analysis, and state transition analysis.

References

1. Satzinger, J.W., Jackson, R.B., Burd, S.D.: Systems Analysis and Design in a Changing World, 2nd edn. Course Technology (Thomson Learning), Boston (2002)
2. van Lamsweerde, A.: From system goals to software architecture. In: Bernardo, M., Inverardi, P. (eds.) Formal Methods for Software Architectures. LNCS-2804, pp. 25–43. Springer, New York (2003)
3. Mylopoulos, J., Chung, L., Yu, E.: From object-oriented to goal-oriented requirements analysis. Commun. ACM **42**(1), 31–37 (1999)
4. Kruchten, P.: The Rational Unified Process: An Introduction, 3rd edn. Addison-Wesley, Reading/Boston (2000)
5. Leffingwell, D., Widrig, D.: Managing Software Requirements: A Use Case Approach, 2nd edn. Addison-Wesley, Boston (2003)
6. Yu, E.S.K.: Modelling Strategic Relationships for Process Reengineering. Ph.D. thesis, University of Toronto, Department of Computer Science (1995)
7. Yu, E.S.K.: Towards modeling and reasoning support for early-phase requirements engineering. In: Proceedings of the 3rd International Symposium on Requirements Engineering, RE'97, IEEE Computer Society, pp. 226–235
8. Chung, L., Nixon, B.A., Yu, E., Mylopoulos, J.: Nonfunctional Requirements in Software Engineering. Kluwer Academic, Boston (2000)
9. Goal oriented requirement language. At http://www.cs.toronto.edu/km/GRL (2001)
10. Castro, J., Kolp, M., Mylopoulos, J.: Towards requirements-driven information systems engineering: the TROPOS project. Inf. Syst. **27**(6), 365–389 (2002)
11. Yu, E.S.K., Mylopoulos, J.: Understanding "Why" in software process modelling, analysis, and design. In: Proceedings of the 16th International Conference on Software Engineering, ICSE'94, IEEE Computer Society, pp. 159–168, Los Alamitos, CA (1994)
12. Perini, A., Pistore, M., Roveri, M., Susi, A.: Agent-oriented modeling by interleaving formal and informal specification. In: AOSE 2003, Melbourne, Australia (2003)
13. van Lamsweerde, A.: Formal specification: a roadmap. In: Finkelstein, A. (ed.) The Future of Software Engineering, pp. 147–160. ACM Press, New York (2000)
14. van Lamsweerde, A., Darimont, R., Letier, E.: Managing conflicts in goal-driven requirements engineering. IEEE. Trans. Softw. Eng., Special Issue on Managing Inconsistency in Software. Development. **24**(11), 908–926 (1998)
15. Brachman, R.J., Schmolze, J.G.: An overview of the KL-ONE knowledge representation system. Cogn. Sci. **9**(2), 171–216 (1985)
16. http://www.cs.toronto.edu/~jm/2507S/Notes04/FormalRML.pdf
17. Manna, Z., Pnueli, A.: The Temporal Logic of Reactive and Concurrent Systems. Springer, New York (1992)

18. Koymans, R.: Specifying Message Passing and Time-Critical Systems with Temporal Logic. LNCS, vol. 651. Springer, New York (1992)
19. Dardenne, A., Lamsweerde, V.A., Fickas, S.: Goal-directed requirements acquisition. Sci. Comput. Program. **20**, 3–50 (1993)
20. Fuxman, A., Liu, L., Mylopoulos, J., Pistore, M., Roveri, M., Traverso, P.: Specifying and Analyzing Earlý Requirements in TROPOS. Submitted to J. Requir. Eng. http://www.cs.toronto.edu/~liu/ (2004)

Chapter 10
Agent Service-Oriented Architectural Design

10.1 Introduction

Architectural design is a crucial stage in software design. It aims to build an overall picture of what a problem-solving system for tackling a complex problem looks like by focusing on architectural aspects. These aspects involve basic computing unit definition and functionalities; design patterns; architectures for integrating related resources and applications; a system architecture for integrating and supporting computing units, resources, applications, and their interactions; integration strategies; and the communication, coordination, and management of computing units, resources, and applications.

Following the OSOAD methodology discussed in Chap. 6, this chapter specifically discusses agent service-oriented architectural design. We introduce agent models and service models to capture basic computing units in a complex system. Agent architecture patterns and structural and functional service patterns are presented as design patterns for agent services. Agent service-oriented integration architectures include integration levels and techniques, as well as architectures for application integration. Different agent service-oriented integration strategies are discussed, such as multiagent + Web services and multiagent + service-oriented computing. Agent service management, communication, and coordination are discussed. Case studies are presented to show corresponding functions and systems.

10.2 Agent Service Model

The design of agent service models seeks to specify patterns for agent architectures, intra-agent activities, and interagent communications. We define agent model and service model, respectively, and this section introduces the architectural design of agent models and service models.

L. Cao, *Metasynthetic Computing and Engineering of Complex Systems*,
Advanced Information and Knowledge Processing,
DOI 10.1007/978-1-4471-6551-4_10

10.2.1 Agent Model

Agent models specify agent architectures and the definition of agent classes. Agent architectures and agent classes are detailed individually below.

An *agent architecture* defines specific commitments of the internal structure and operation of an agent or a collection of agents. There are four main categories of agent architectures; deductive agent architecture, reactive agent architecture, belief–desire–intention agent architecture, and hybrid agent architecture.

- *Deductive agent architecture* [1, 2]: in which decision-making is realized through logical deduction; see Fig. 10.1.
- *Reactive agent architecture* [1–3]: in which decision-making is implemented in the form of direct mapping from situations to actions; see Fig. 10.2.
- *Belief–desire–intention agent architecture* [1, 2]: in which decision-making depends upon the manipulation of data structures representing the beliefs, desires, and intentions of the agent (Fig. 10.3).
- *Hybrid agent architecture*: in which decision-making is achieved via a combinatorial mechanism embedding two of the above three decision-making principles.

An *agent class* describes a type of computerized object that performs roles and actions situated in an environment to meet its design objectives. It is usually defined in terms of a set of attributes in a specification. Some *agent instances* are instantiated from these classes. There may be a one-to-one or many-to-one correspondence between roles and agent types; a number of closely related roles can be mapped into the same agent class for the purpose of convenience or efficiency (Fig. 10.4).

Attributes used for defining an agent class may include agent name, type, locators, owner, roles, address, messages, input variables, preconditions, output

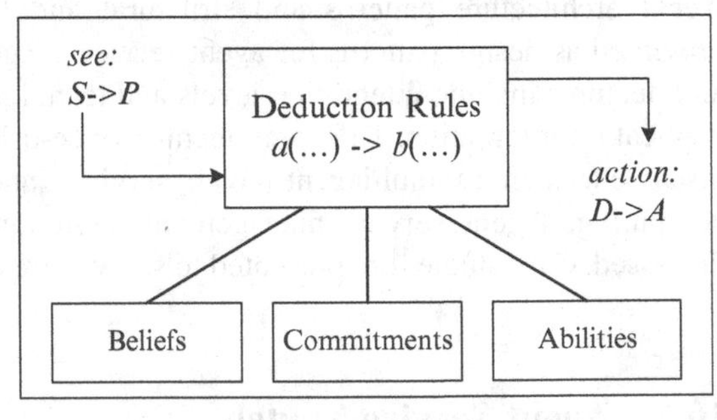

Fig. 10.1 Deductive agent architecture

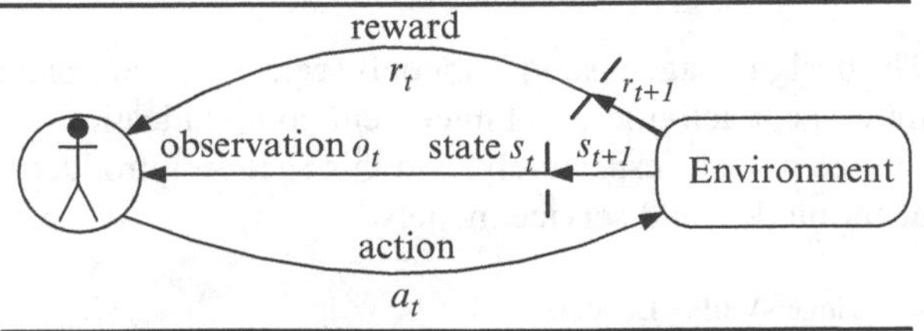

Fig. 10.2 Reductive agent architecture

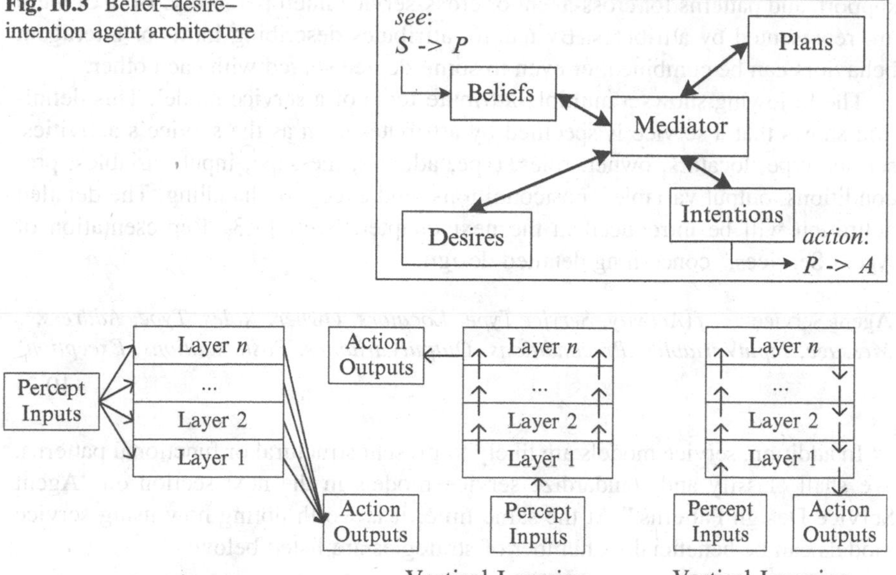

Fig. 10.3 Belief–desire–intention agent architecture

Fig. 10.4 Hybrid agent architecture

variables, postconditions, and exception handling. The detailed definition will be introduced in the next chapter, Sect. 11.3 "Representation of Agent Services," concerning detailed design.

$$\text{Agent Service} ::= f(\mathit{Name}, \mathit{Type}, \mathit{Locators}, \mathit{Owner}, \mathit{Roles}, \mathit{Address}, \mathit{Message}, \mathit{InputVariables}, \mathit{Preconditions}, \mathit{OutputVariables}, \mathit{Postconditions}, \mathit{Exception}) \tag{10.1}$$

10.2.2 Service Model

A *service* is an application that exposes its functionality through an API. In other words, a service is a resource that is designed to be consumed by software rather than by humans. The service models described here are typical of those involved in agent activities and cross-agent communication. Their roles are:

1. Intended as coherent blocks of activities that are required to realize the intra-agent or intraservice roles and their properties
2. Centered around enabling and managing interagent or interservice communications in scenarios of legacy integration and enterprise integration

The definition of service models fulfilling the first role above describes the basic attributes of agent service that represent the intra-agent or intraservice roles and properties. On the other hand, the definition of services that manage interagent and interservice communications describes the design interaction and communication support, and patterns for cross-agent or cross-service interoperability. Both of them are represented by attributes. By nature, attributes describing intra- or interagent behaviors can be combined, or even to some degree shared with each other.

The following shows a multiple-attribute tuple of a service model. This definition shows that a service is specified by attributes such as the service's activities, service type, locators, owner, roles, type, address, message, input variables, preconditions, output variables, postconditions, and exception handling. The detailed definition will be introduced in the next chapter, Sect. 11.3 "Representation of Agent Services," concerning detailed design.

$$\text{Agent Service} ::= f(\textit{Activity}, \textit{Service Type}, \textit{Locators}, \textit{Owner}, \textit{Roles}, \textit{Type}, \textit{Address}, \textit{Message}, \textit{InputVariables}, \textit{Preconditions}, \textit{OutputVariables}, \textit{Postconditions}, \textit{Exception}) \tag{10.2}$$

In addition, service models are likely to present structural or functional patterns. We shall classify and standardize service models in the next section on "Agent Service Design Patterns." At the same time, it is worth noting how using service models can be beneficial. A number of strategies are listed below:

- The model types represent the overall roles and positions of services within an agent service-oriented architecture.
- Service models can be aligned with enterprise business processes, policies, workflow, and security standards and specifications.
- Specific design standards can be applied to different service models.

10.3 Agent Service Design Patterns

Agent service design patterns refer to certain repeatable structures, architectures, services and applications in agent services. Here we focus on agent architecture patterns and service patterns.

10.3.1 Agent Architecture Patterns

The following section introduces the four classes of agent architectures. Let P be the set of perceptions, A be the set of all possible actions, Bel be the set of all possible beliefs, Des be the set of all possible desires, and Int be the set of all possible intentions.

10.3.1.1 Deductive Agent Architecture

Deductive agent architecture [1, 2] defines a symbolic representation (logical formulae) of an agent's environment and its desired behavior based on traditional symbolic AI. In this architecture, a *deliberate* agent [4] is viewed as a theory prover in which decision-making is realized through logical deduction. The deduction rules have the form:

$$\phi(\ldots) \rightarrow \varphi(\ldots) \tag{10.3}$$

The main task in building logic-based agent architecture is to present the agent's goals and reactive behaviors through the definition and refinement of a logically executable specification. In such agents, the internal state is assumed to be a database of formulae of classical first-order predicate logic. The relevant work and discussion include *deliberate agents* [4], *situation calculus* [5], *situation automata* [6], and the like.

Let ϕ be an executable specification (directly executed to produce the agent's behaviors), L be the set of sentences of classical first-order logic, $D = \wp(L)$ be the set of L databases, d be the members of D, φ be the concluded specification with any variables instantiated, and $Do(\ldots)$ be the predicate prescribing an action a. Then, a deliberate agent's perception function *see* has the form:

$$see : S \rightarrow P \tag{10.4}$$

The perceived information is mapped with a database and a percept and further mapped to a new database. The function *next* has the form:

$$next : D \times P \rightarrow D \tag{10.5}$$

When an agent perceives information, it may select an *action* in terms of its reduction rules, e.g., $d \vdash_{\rho} \phi$ means that the formula ϕ can be proven from the database d using only the deduction rules ρ. The *action* function takes the form:

$$action : D \rightarrow A \tag{10.6}$$

An action of a logic-based agent can be described in pseudo-code as follows:

```
function action(d : D) : A
begin
    for each a ∈ A do
        if d ⊢ρ Do(a) then
            return a
        end if
    end for
    for each a ∈ A do
        if d ⊢ρ ¬Do(a) then
        return a
    end if
end for
return null
end function action
```

10.3.1.2 Reactive Agent Architecture

According to the theory on reactive agents [1–3], intelligent behavior is not disembodied but is an emergence of the interaction maintained by the agent with its environment. In such an agent, decision-making is implemented in the form of direct mapping from situation to action. This is sometimes referred to as a *behavioral* or *situated* approach. In the establishment of reactive agents [7], behaviors are often implemented as rules of the following form, which simply map perceptual input directly to actions:

$$\wp(P) \rightarrow \wp(A) \tag{10.7}$$

Subsumption architecture [8–10] is arguably the best-known reactive agent architecture, which has two defining characteristics. The first is that an agent's decision-making is realized through a set of task-accomplishing behaviors. The second is that multiple behaviors can "fire" simultaneously in a cooperative manner. However, there are several fundamental, unsolved problems with reactive architecture, not only with subsumption architecture. Research on these problems further leads to creating evolutionary or artificial life agents [11] as one solution.

Let $c \in P$ be a set of percepts called the condition and $a \in A$ be an action; a behavior (c, a) will fire when the environment is in state $s \in S$ iff $see(s) \in c$. Given an agent's set of behavior rules $R \subseteq \{(c, a) \mid c \in P \text{ and } a \in A\}$, there is a binary *inhibition relation* $\prec \subseteq R \times R$[1] on the set of behaviors. This relation is assumed to be a total ordering on R. Furthermore, the decision function *action* of a creative agent can be defined as follows:

```
function action(p : P) : A
var fired : ℘(R)
var selected : A
begin
    fired := {(c,a) | (c,a) ∈ R and p ∈ c}
    for each (c,a) ∈ fired do
      if ¬(∃(c',a') ∈ fired such that (c',a') ≺ (c,a)) then
          return a
      end if
 end for
 return null
 end function action
```

[1] $b_1 \prec b_2$ if $(b_1, b_2) \in \prec$, read this as "b_1 inhibits b_2", that is, b_1 is lower in the hierarchy than b_2, and will hence have priority over b_2.

10.3.1.3 Belief–Desire–Intention Agent Architecture

Belief–desire–intention (BDI) architectures [1, 2] have roots in the philosophical tradition of understanding *practical reasoning* [12]. Practical reasoning involves two important processes: deciding *what* goals we want to achieve (deliberation) and *how* we are going to achieve them (means–ends reasoning).

The process of practical reasoning in a BDI agent is summarized as seven main components [1]. The state of a BDI agent at any given moment is a triple (B, D, I), where $B \subseteq Bel$, $D \subseteq Des$, and $I \subseteq Int$:

- A set of current beliefs *Bel*, representing information the agent has about its current environment
- A belief revision function (*revise beliefs, brf*), which takes a perceptual input and the agent's current beliefs and, on the basis of these, determines a new set of beliefs

$$brf : \wp(Bel) \times P \to \wp(Bel) \tag{10.8}$$

- An option generation function (*generate options*), which determines the options available to the agent (its desires), on the basis of its current beliefs about its environment and its current intentions

$$options : \wp(Bel) \times \wp(Int) \to \wp(Des) \tag{10.9}$$

- A set of current *options*, representing possible courses of action available to the agent
- A *filter* function, which represents the agent's deliberation process and which determines the agent's intentions on the basis of its current beliefs, desires, and intentions

$$\begin{gathered} filter : \wp(Bel) \times \wp(Des) \times \wp(Int) \to \wp(Int) \\ \text{where, } \forall B \in \wp(Bel), \forall D \in \wp(Des), \forall I \in \wp(Int), filter(B, D, I) \subseteq I \cup D \end{gathered} \tag{10.10}$$

- A set of current *intentions*, representing the agent's current focus—those states of affairs that it has committed to trying to bring about
- An action selection function (*execute*), which determines an action to perform on the basis of current intentions

$$\begin{aligned} &execute : \wp(Int) \to A \\ &action : P \to A \end{aligned} \tag{10.11}$$

The action of a BDI agent is further defined by the following pseudo-code:

```
function action(p : P) : A
begin
   B := brf(B, p)
   D := options(D, I)
   I := filter(B, D, I)
   return execute(I)
end function action
```

There are also other forms of agent architectures. These include the blackboard system and various organizational structures.

10.3.2 Structural and Functional Service Patterns

Structural service patterns extract and specify generic service architectures. Here, we list some service models for legacy integration and enterprise integration.

Service models used for cross-agent or service integration include [13] proxy service, wrapper service, adapter service, intermediary service, service interceptor, and coordination service [14]. Other services include matchmaking service, mediation service, brokerage service, gateway service, negotiation service [15], auction service [16], orchestration engine, and so forth. Table 10.1 lists their definitions and relevant context scope.

The above discussion attempts to classify and standardize service types. By nature, every service is unique, but many services ultimately undertake similar functions and exhibit common features. In light of this reality, we can further categorize service models and customize them to whatever extent necessary. Paul B. Monday [17] introduces 15 service patterns by illustrating their usefulness in the Web service environment. They are service-oriented architecture, architecture adapter pattern, service directory pattern, business object pattern, business object collection pattern, business process (composite) pattern, asynchronous business process pattern, event monitor pattern, observer pattern, publish/subscribe pattern, physical tiers pattern, Faux implementation pattern, service factory pattern, data transfer object pattern, and partial population pattern.

10.4 Agent Service-Oriented Integration Architectures

Application integration is when two or more applications are combined in terms of business requirements, data share, data transfer, business logics, or business processes. This section first discusses general integration levels and strategies; agent service-oriented architectures for legacy integration and enterprise integration are then outlined.

10.4.1 Integration Levels and Techniques

Enterprise application integration (EAI) and integrated enterprise are currently very popular for developing enterprise-wide applications. They are the combination of processes, applications, standards, and databases. They support a seamless integration of two or more enterprise systems and allow them to operate as one, so that

Table 10.1 Structural service patterns

Service name	Service definition	Context scope
Proxy service	A proxy service generates a service interface that mirrors or resembles that of an existing application component. Generally resides alongside its corresponding component	Supports low-to-medium usage volume with auto-generated interface but lacks a quality service interface; legacy integration
Wrapper service	A wrapper service creates a service interface that exposes specific parts of existing applications. It is often designed to encapsulate the functionality exposed by an accompanying legacy adapter	Customized, suitable for low-to-medium volume; limited to the exposed functionality; legacy integration
Adapter service	A specialized piece of software that connects heterogeneous applications by providing an open interface (API) to all proprietary APIs	Could be used as wrapper; scalable and varying versions; legacy integration
Coordination service	Follows a predefined assembly model providing functions such as context creation, registration for coordination, and protocol selection to enable atomic transactions or handle conditions related to the execution of overall business process activities	As atomic transaction coordinators enabling ACID-like transaction, or business activity coordinator handling execution conditions; legacy or enterprise integration
Matchmaking service	Is generally accomplished through a market or by providing middle agents, or is based on a broker, middle agent or agents that act as a directory for services available in the system	For one-to-many or many-to-many negotiation mechanisms in e-markets [24]; may cover gateway and broker
Gateway service	Enables data to flow between different applications or networks through one interface	Can be an application or an interface
Brokerage service	Facilitates the advertising and discovery of agent services	Usually acts as a role of SOA for matchmaking or mediation
Mobile agents	*Mobile agents* are programs that can migrate from host to host in a network, at times, and to places, of their own choosing. The state of the running program is saved, transported to the new host, and restored, allowing the program to continue where it left off	Differ from process migration systems and applets; act as secure brokering and many other applications [25]
Negotiation service	Two or more parties competitively jointly search a space of possible solutions with the goal of reaching a consensus	With auction, distributive or integrative negotiation [26], contract net [27]
Auction service	Specifies bidding or other forms of convention for negotiating prices and offers based on a mediator, either in an auction house or not	Auction [28] or auction house [29, 30], combinatorial auctions
Orchestration engine	Encapsulates and executes business process logics that involve a variety of applications and resources. May contain new business rules, exception handling, and transaction management	Invokes existing logic or broker component, integrates with other applications; EAI integration

information and business processes across applications can be shared. From the perspective of integration levels, there are usually four solutions for building the above applications:

- *Presentation Integration* or User Interface Integration: This allows the integration of new software through the existing presentation of the legacy software. It is typically used to create a new user interface, but may be used to integrate other applications.
- *Data Integration*: This allows the integration of software through sharing and access to the data that is created, managed, and stored by the software, without the involvement of business logic; it is typically used for the purpose of reusing or synchronizing data across applications.
- *Functional Integration* or Method Level Integration: This permits the integration of software for the purpose of invoking existing functionality from other new or existing applications. This integration is normally performed interfacing with and invoking programming logic from one application to another. It is also called application level integration.
- *Business Process Integration* or Process Level Integration: Sharing data or business logic facilitates a new automated business process, or the merging of two or more existing processes. In EAI-oriented business process integration, a broker component and a separate orchestration module are used to enable cross-application communication.

On the other hand, from the point of view of technologies and integration strategies, application integration may take one of the following forms:

- *Middleware-Based Integration*: Multiple applications are combined via a middleware technique, for instance, JDBC-based data access; remote invocation protocols such as RPC, DCOM, and CORBA-based distributed technologies; and adapters such as connectors, gateways, and mediators. Message queue and messaging-centered middleware, such as IBM MQSeries and MSMQ, are the most recent EAI solution. More progressive middleware applications are server-based applications that provide server-side application hosting and component-based brokerage such as broker, adapter, gateway, and connector.

Middleware-based EAI has several drawbacks. Most EAI solutions aim to integrate comprehensive applications in an enterprise to improve overall communication using a specific middleware. This makes the implementation and deployment of EAI costly, inflexible, and inefficient.

- *Web Service-Oriented Integration*: A Web service-based integration layer is established by exposing legacy APIs via Web services for sharing data and business logic. Web service-based EAI requires less time and money and facilitates inter-application communication.

It supports the step-by-step integration of applications and allows the enterprise to break applications into separate business logic units in terms of integration flexibility and business process requirements. Each unit can be packaged as a separate Web service. Web service-based integration is typically used for internal integration. The weakness of Web service-based EAI stems from the limitations of Web services. It might not provide quality of service such as autonomy, flexibility, security, reliability, performance, and uptime to the level required by Internet-based integration.

- *Agent Service-Oriented Integration*: Applications are packed (or wrapped) as agents or services via a mediation layer or integration layer that supports the automated integration of data and business logic. Agent service-oriented integration can deal with some limitations in Web service-based integration. It provides capabilities such as autonomy, flexibility, and efficiency. In the remainder of this chapter, we shall focus on agent service-oriented integration.

10.4.2 Architectures for Application Integration

Agent service-oriented legacy integration can utilize agent services of proxy, wrapper, adapter, coordination for atomic transactions, gateway for data access, broker, mediator, or matchmaker on demand. On the other hand, broker, mediator, matchmaker, mobile agents, negotiation, auction, and orchestration are more suitable for enterprise integration.

The basic strategy here is to plug-in some form of agent services; they are used to wrap each of the legacy or enterprise applications, or act as middleware by exposing APIs and establishing messaging-based communication between two applications. In the abovementioned agent services, proxy, wrapper, and coordination can be used for wrapping legacy applications, while gateway, broker, mediator, matchmaker, mobile agents, negotiation, auction, and orchestration are mainly used to build a middle integration layer between legacy and enterprise applications. Figures 10.5 and 10.6 illustrate these two types of integration architectures.

The instantiation of the above two solutions may require specific integration architectures. For instance, Fig. 10.7 illustrates an agent service-enabled hub and spoke architecture, where a centrally located server hosts the integration logic that controls the orchestration and brokering of all inter-application communication. Application 1 is integrated after being wrapped as a service; Application 2 interacts with a wrapper server via an agent adapter; both of them interact with the server

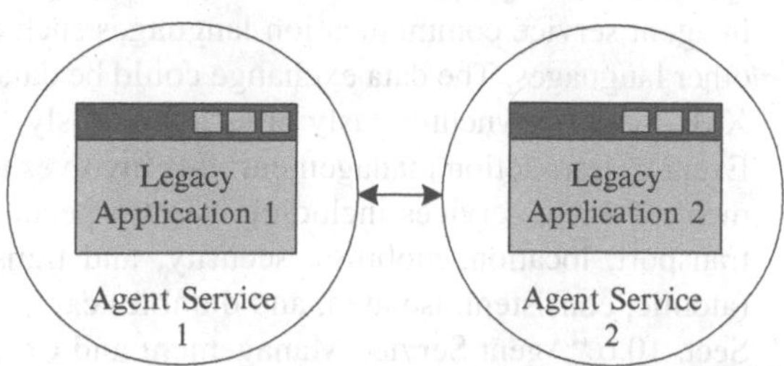

Fig. 10.5 Agent service-based wrapping integration

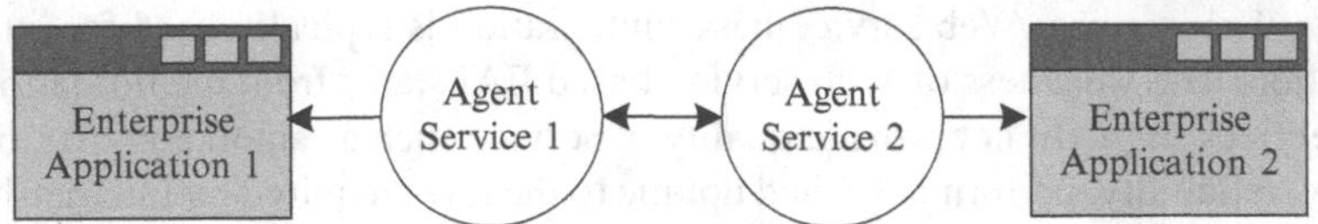

Fig. 10.6 Agent service-based middleware integration

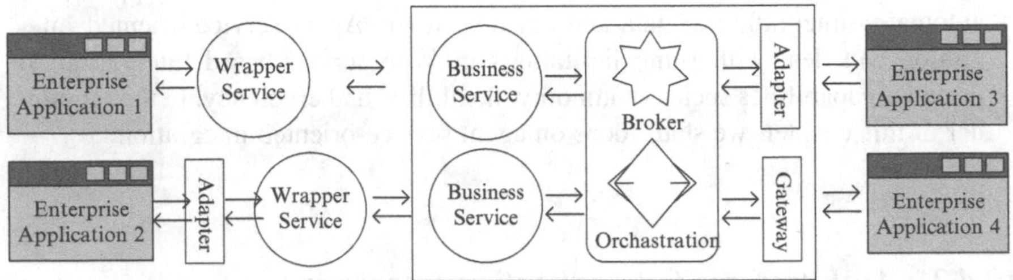

Fig. 10.7 Agent service-enabled hub and spoke integration

through business services. In addition, Applications 3 and 4 are combined with the server via an agent adapter and a gateway agent, respectively.

In the implementation of agent service-oriented legacy integration and EAI, the following issues may need to be considered: integration level, integration type, interface design, message passing, event or transaction management, and exception handling.

- Integration level: this involves the level at which the two applications are integrated, and can be on the level of presentation, data, method, or process.
- Integration type: this is to determine that the inter-application communication is based on either wrapping or middlewaring integration.
- Interface design: agent services integrating legacy applications or components may generate agent service interfaces automatically, mirror the legacy applications, expose application APIs, or implement a predefined specification programmatically.
- Message passing: this defines the communication protocol and transport method between enterprise information systems. The communication protocol could be RPC, socket, HTTP, ATP, SOAP, or other specific protocols embedded in specialized integration techniques. The dialogue between applications could be in agent service communication languages such as KQML, KIF, FIPA ACL, or other languages. The data exchange could be database or document centric (e.g., XML-based) asynchronously or synchronously.
- Event or transaction management: this involves the administration and management of agent services including event aspects such as dispatching, directory, transport, location, mobility, security, and transaction support such as ACID (atomic, consistent, isolated, and durable) capabilities. These will be detailed in Sect. 10.6 "Agent Service Management and Communications."

- Exception handling: this is essential for capturing all integration exceptions in the execution of the integrated system. Exceptions in the most important activities of agent service communications and transaction processing must be recorded and handled immediately; corresponding solutions should be triggered automatically.

10.5 Agent Service-Oriented Integration Strategies

Agent service-oriented integration can be undertaken in terms of various strategies and techniques. One method is to combine multiagent and Web services technologies, which is a Web service-centric method. Another direction is to utilize service-oriented computing to establish integration architecture while implementing intelligent and automated process integration via multiagent technology. These two directions are discussed below.

10.5.1 Multiagent + Web Services

This is a new research direction in the dialogue between AAMAS and *XML-driven Web service* (hereafter referred to as *Web service*). There is some existing work on this topic; for instance, agent-based dynamic service integration, agent-based automatic negotiation in Web services, conversations with Web services, agents for Web service composition, etc. However, the dialogue between multiagent and Web service which we advocate here is designed to bridge the gaps between them in inter-application communication.

In this approach, Web services are taken as being the foundation of building integrated applications. We call this "Web service-driven agent systems." In the process of building agent service-oriented applications, Web service is recommended to build the enterprise integration infrastructure. We advocate the usage of second-generation Web services. Figure 10.8 [13] illustrates all the main elements of second-generation Web service specifications and the relationships between them.

The basic work for the Web service-based infrastructure includes:

- Providing a service description that, at minimum, consists of a WSDL document
- Being capable of transporting XML documents using SOAP over HTTP

Mutiagent technology plays a significant role in other aspects beyond the above basic infrastructure linkage. The following lists the main functionalities a multiagent may serve in the "multiagent + Web service" approach:

- Management, dispatching, conversation, negotiation, mediation, and discovery of Web services: For these types of issue, multiagents can provide great advantages for flexible, intelligent, automated, and (pro-) active operations. For

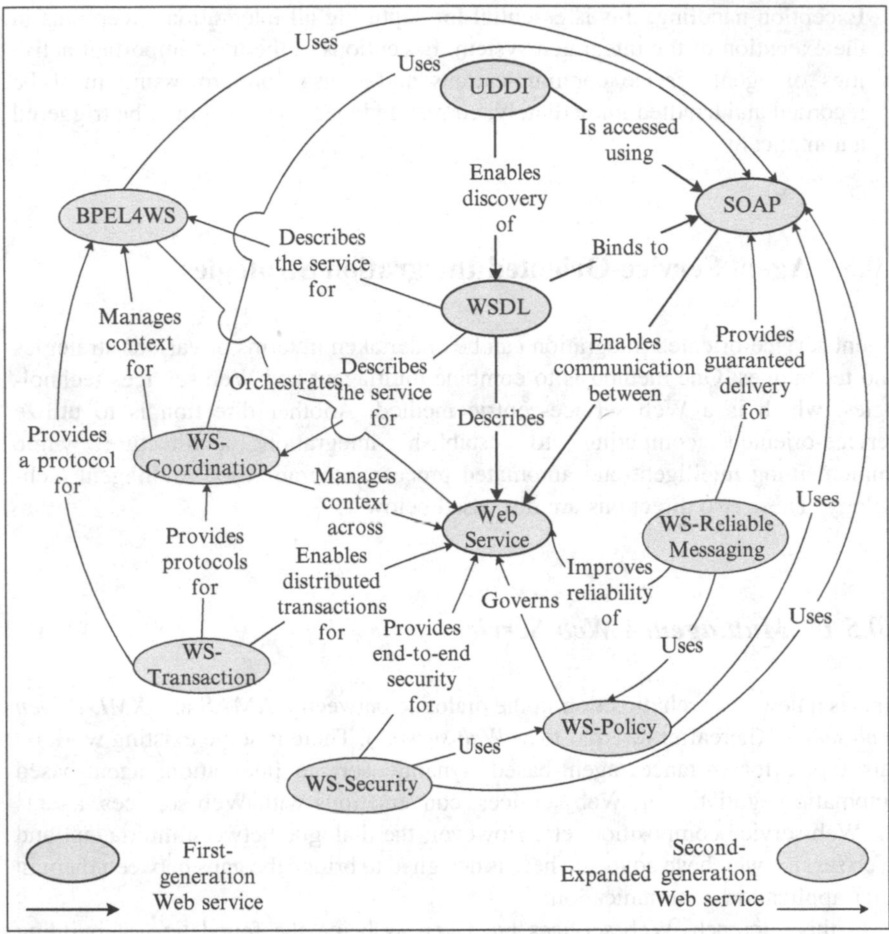

Fig. 10.8 High-level relationships between first- and second-generation Web service standards

instance, a gateway agent located at a remote data source listens to requests and extracts data from the source after receiving data request messages; it generates and dispatches an agent to deliver the extracted data to the requester after completion. This can reduce the network payload and enhance the flexibility of remote data access.

- Agent-based services: Services for middlewaring and business processes such as adapter, connector, matchmaker, mediator, broker, gateway, and negotiator can be customized, personalized, and specialized as agents. For instance, an agent-based coordinator taking *partial global planning* [18, 19] has the capability to decide its own goals and generate short-term plans in order to achieve them; it may further alter local plans in order to better coordinate its own activities. Agent-based services can enhance the automated and flexible decision-making of these services.

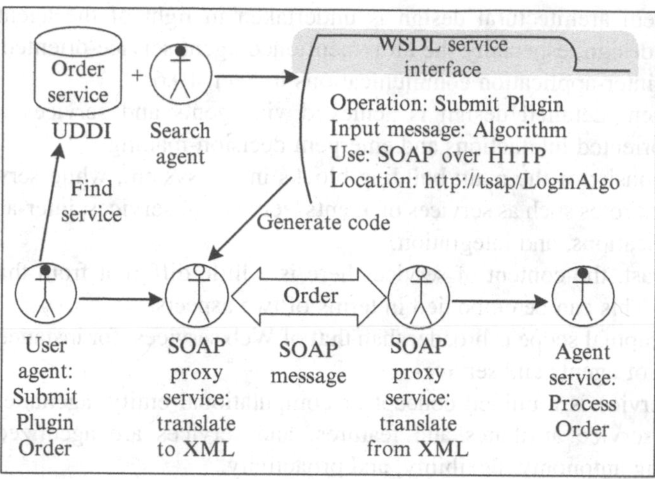

Fig. 10.9 Web service-driven agent systems for EAI

- Agentized components: Applications and application components in integration enterprise can be agentized on demand. This can follow the normal method of building multiagent systems. However, the agentized components are mainly used to establish specific functional components rather than a stand-alone multiagent system; for instance, agents for human–computer interaction, resource access dispatching, management of Web services, and enterprise business logic. The problem here may include how to link and administer these agents in the Web service-centric environment.

Figure 10.9 describes the architecture of a Web service-driven agent system. In this system, the client defines and submits an algorithm plug-in order via the user agent. The user agent finds the order service through the UDDI and search agent. A SOAP proxy service is generated using WSDL. This proxy service is accepted as a bridge to talk to the process order agent service via SOAP message, or to translate the message.

10.5.2 Multiagent + Service-Oriented Computing

This strategy is based on the combination of multiagent and service-oriented computing. It is not necessary for service-oriented computing to be Web service-driven or service-centric. In this approach, multiagents act as the main building blocks of the system infrastructure. A system following this approach is called a "multiagent-driven service system". The fundamental theory of the multiagent-driven service system is as follows:

- The system analysis is performed in terms of agent-oriented early and late requirements analyses, in particular the organization-oriented analysis previously discussed in Chap. 9.

- The system architectural design is undertaken in light of the agent service-oriented design, especially the aforementioned agent service-oriented architecture and inter-application communications in Sect. 10.6.
- The system detailed design is achieved via agents and services and agent service-oriented interactions and emergent decision-making.
- Agents consist of the main building blocks in the system, while services play significant roles such as services of agents, services of services, inter-application communications, and integration.
- By contrast, the content of services here is a little different from that in Web services. This can be embodied in terms of two aspects:
- The conceptual scope is broader than that of Web services; for instance, services could be of agents and services.
- Agent service is a unified concept or computational entity, agents encompass explicit service attributes and features, and services are agentized through supporting autonomy, flexibility, and proactivity.

The above discussion constitutes the main theory of "agent service-oriented computing" that we advocate and emphasize in this book. In fact, the system analysis and design of open complex agent systems and the implementation of F-TRADE covered in previous chapters, as well as in subsequent chapters, all refer to and detail this approach. Details about implementing the "multiagent + Web service" approach are beyond the scope of this book, however, and are not dealt with here.

With regard to the above theory, Figs. 10.6, 10.7, and 10.8 illustrate a number of solutions for agent service-based legacy and enterprise integration. In Sect. 10.8 "Case Study," we shall exemplify the agent service-oriented architecture via the case study system F-TRADE. In addition, agent service-oriented detailed design will be explained in Chap. 11.

10.6 Agent Service Management and Communications

The main issues related to the management and communications of agent services include naming, directory, communication, transport, mediation, and discovery of agent services in the integrated enterprise applications. The main tasks and objectives of these aspects are sketched below. More information is given in Chap. 11 "Agent Service-Oriented Detailed Design":

- *Naming of agent services*: Establishing a common namespace and a naming specification is a key to fostering the unified view and management of agent services. A good way to establish a standard vocabulary within an organization is to put together a glossary. Some industry-standard glossaries are available, but it is recommended that they should be customized to relate key terms to existing business processes and systems within the organization. To this end, ontology is required.

- *Directory of agent services*: The directory of agent services provides a location for agent services to register directory entries and a consistent means by which agents and services can discover services.
- *Communication of agent services*: This involves the communication model, communicative act, and communication control in interagent service communications.
- *Transport of agent services*: This performs the transport of messages between agent services and deals with the transport protocol for the delivery of messages.
- *Mediation of agent services*: This involves mechanisms for both local mediation in a subsystem and global mediation in an agent society.
- *Discovery of agent services*: This concerns the following two aspects—querying an agent service with which to communicate, and searching for a message.
- *Security of agent services*: This is a hard problem in both Web service- and multiagent-driven integration. For Web services security, the main work would be to implement XML and Web services security specifications. For multiagents, it depends on specific implementation techniques, but this is still an open problem for any multiagent theory.

10.7 Agent Service Coordination

Coordination is "the glue that binds separate activities into an ensemble" [20]. It has been a hot research topic in service-oriented computing and agent-based computing. Coordination concerns the composition and communication of agent services. Agent service coordination is a life cycle of connecting multiple agent services into a composite business logic or application through communication between them. Agent service coordination researches the development of a coordination method, coordination model, QoS-enabled coordination mechanisms, and so on. This section introduces them respectively.

10.7.1 Coordination Methods

Several coordination methods feature in studies in the literature. Typical methods include workflow-based coordination, business logic-based coordination, and process-based coordination. Emerging methods consist of organization-based coordination and goal-based coordination. Agent service-based systems are dependency communities with all relevant organizational factors such as agent services, workflow, business processes, and business logics coexisting in a highly integrated manner. As a result, workflow-based service coordination must naturally involve other organizational aspects such as business processes, business models, and/or business logics intra- or inter-organization.

There are thus two ways to study agent service coordination strategies. The first is to split one of the organizational members from the dependency community and study coordination methods based on this organizational factor. The second is to study the coordination dynamics in terms of the total agent service organization rather than local factors. We call the first a specific *organizational member-level service coordination* or *dimension-level coordination* method, while the second is referred to as *organization-level service coordination*.

10.7.1.1 Organizational Member-Level Coordination

Definition 10.1 Organizational member-level service coordination studies coordination dynamics and supporting mechanisms in terms of a specific organizational member in an agent service-based society.

Many organizational members may be involved in the composition and coordination of agent services. They are basically segmented into *static* and *dynamic* organizational members. Static organizational members include the member actor, role, rule, goal, norm, and environment as discussed in the ORGANISED framework. In coordination dynamics analysis, more attention is paid to dynamic organizational members, such as system workflows, business processes, business tasks, business models, and business logics. In addition, the events and life cycle of coordination, all possible scenarios in a coordination dynamics analysis, and so on, also deserve attention in the local or global *behavior* representation of coordination dynamics. Figure 10.10 presents a multiple-dimensional coordination analysis model.

Organizational member-level service coordination often observes and deals with coordination dynamics from the perspective of one member of one dimension, rather than from a comprehensive aspect. In some research, multiple members from

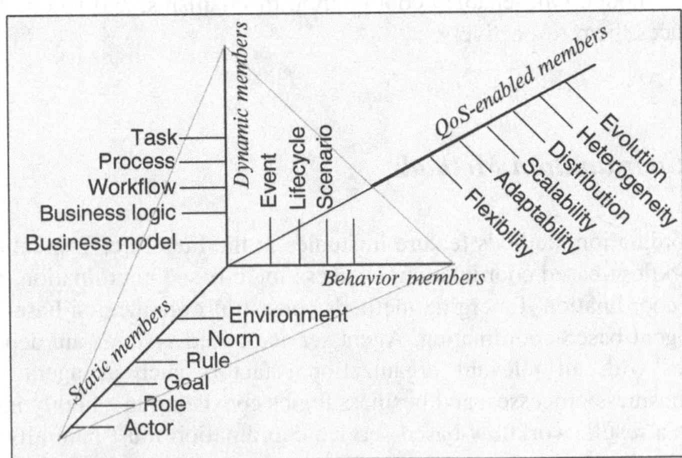

Fig. 10.10 A multi-dimensional coordination analysis model

one dimension or across multiple dimensions may be studied. At present, organizational member-level coordination methods have largely become the focus of the research frontier; for instance, the well-researched workflow-based coordination and a combined research with process-based coordination. However, not all members in the three dimensions are studied. This is partly due to the fact that static members are fundamental organizational factors of behavior and dynamic dimensions. Some emerging coordination methods focus on event-based coordination, life-cycle-based coordination, and business logic-based coordination. We believe a multidimensional and multilevel combination of the above organizational members is the best direction for coordination method analysis.

Workflow-based coordination is a typical type of service coordination method. It adopts the concept of workflow to specify the flow of relevant business logic, control, and data among agent services. Workflow-based coordination involves one operational system. In the scenario of distributed software systems and virtual organizations, it supports one workflow or links different workflows spanning multiple participating organizations. To handle such situations, cross-organizational workflow-based service coordination can be an effective approach.

In practice, member-level coordination has to involve one or multiple organizational members since they are essential for the coordination dynamics analysis and are closely coupled in an agent organization. If this kind of method is observed in a systematic manner, it moves into the second category, namely, organization-level service coordination.

10.7.1.2 Organization-Level Coordination

Definition 10.2 Organization-level coordination takes a comprehensive view of the coordination of agent services in an agent organization. It involves all or most of the fundamental organizational members distributed on multiple dimensions of the coordination analysis model.

Ideally, a more comprehensive and practical manner for coordinating agent services is to consider possibly all the organizational members related to coordination dynamics in an agent service society. We call this *organization-level coordination* or *integrated coordination* because it observes and deals with agent service coordination dynamics in terms of the global agent organization rather than as a part of local organizational factors. Organization-level service coordination looks for a systematic coordination strategy which glues together workflow, processes, business models, and logics in the form of behavior aspects, such as the life cycle of agent service systems across all relevant participating organizations. A key research issue for organization-level coordination is to establish an appropriate driving force or abstraction metaphor to bind those distributed organizational members into a seamless entity and develop modeling techniques to depict the coordination. In the next section, we will introduce a coordination organization model which may be useful for designing and representing organization-level coordination.

QoS-enabled members, on the other hand, depict the collective nonfunctional performance aspects of agent service coordination in an agent organization. They reflect the organizational quality of services resulting from a coordination strategy. QoS-enabled members are indicative of organizational coordination performance such as flexibility, adaptability, scalability, distribution, heterogeneity, and evolution. This is definitely an important aspect in coordination studies. Scant research has been undertaken in this area, for instance, into dynamic and adaptive service composition, flexible coordination, and the evolution of service processes.

10.7.2 Coordination Modeling and Patterns

Coordination of agent services is represented by coordination models. Coordination models are developed to further analyze, represent, and refine service coordination in an agent organization. The modeling of agent service coordination can basically be conducted in terms of individual members in the abovementioned multidimensional coordination analysis model, i.e., static and dynamic coordination members, behavior aspects, and QoS-enabled factors. A series of static organizational members, for instance, actors in the coordination, the roles and goals of actors, coordination rules (or transformation rules) and norms in organizing and controlling the coordination, and the environment surrounding the coordination activities, must first be abstracted. Further, the dynamic representation of coordination activities, for instance, the workflow, business process, and business logic of the agent service coordination, in addition to the static organizational members, should be analyzed. Coordination modeling can also be focused on behavior aspects such as the life cycle, the events, and scenarios of coordination. Lastly, QoS-related performance factors warrant consideration in a comprehensive modeling of coordination.

The above member-based modeling focuses on individual members and factors in the coordination process. A more comprehensive and ideal manner is to model the coordination in terms of all relevant members in the multidimensional analysis model. Although it is challenging to capture all members and build a coordination model that covers all factors in a coordination task, it may be feasible to observe major members and factors from a multidimensional perspective. Following this idea, a coordination organizational model may be established.

A coordination organizational model tries to capture and present a picture of relevant major organizational members and factors in a coordination task. Figure 10.11 shows the meta-model of coordination based on the interaction between static, dynamic, behavioral, and QoS-related aspects. It illustrates a multilevel view of analyzing the coordination of agent services. The meta-model is mapped to the abovementioned multidimensional analysis model. Following this view, the coordination organizational model represents a coordination task through dynamic collective presentation and behaviors implemented by static organizational members. QoS-enabled design targets performance implementation through static, dynamic, and behavior interaction.

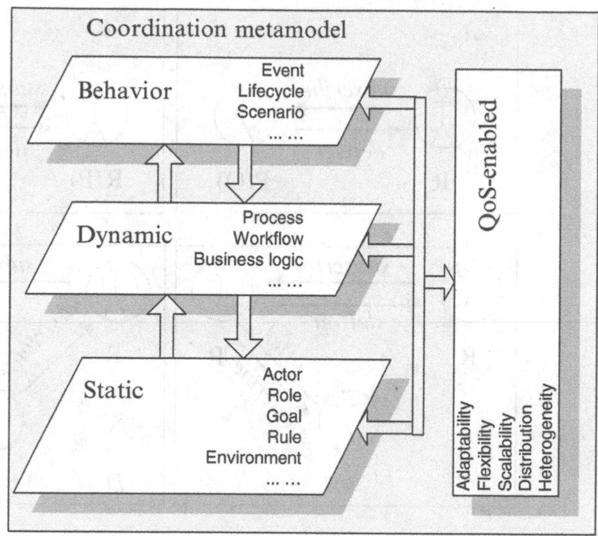

Fig. 10.11 Coordination interaction meta-model

The presentation of the coordination organization model can be performed in terms of some or all of four dimensions, namely, static, dynamic, behavioral, and QoS-enabled aspects, which depict different aspects of a coordination activity and its process. In general, static and dynamic factors are more often under consideration than aspects of behaviors and QoS-enabled performance. However, aspects of behavior and QoS-enabled aspects also deserve attention. In practice, every member in one of the four dimensions should be expanded and modeled in detail in terms of specific modeling techniques. For instance, workflow is studied specifically by some researchers using XML or UML-based modeling languages:

```
/*define a coordination*/
<coordination> ::= <static> <dynamic> [behavior] [QoS-enabled]
<static> ::= (Actor | Role | Goal | Rule | Environment | ...) +
<dynamic> ::= (Process | Workflow | Business Logic | ...) +
<behavior> ::= (Event | lifecycle | Scenario | ...) +
<QoS-enabled> ::= (Adaptability | Scalability | Distribution | Flex-
ibility | ...) +
```

Further, we present it in XML as follows:

```
<!ELEMENT Coordination (Static | Dynamic | Behavior | QoS-enabled)+>
<!ELEMENT Static (Actor | Role | Goal | Rule | Environment | ...)+>
<!ELEMENT Dynamic (Process | Workflow | Business Logic | ...)+>
<!ELEMENT Behavior (Event | lifecycle | Scenario | ...)+>
<!ELEMENT QoS-enabled (Adaptability | Scalability | Distribution |
Flexibility | ...)+>
```

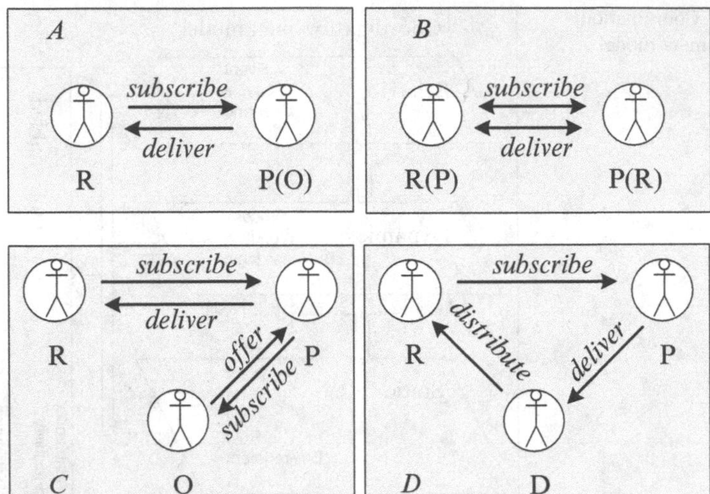

Fig. 10.12 Service coordination meta-patterns

The establishment of a coordination model is often based on a certain modeling language. Coordination modeling languages include XML-based, UML-based, OWL-S based, first-order logic-based, rule-based, and so on. We introduce concrete coordination modeling in Chap. 11 on detailed design.

In studying service coordination, the aforementioned cross-dimensional members are often categorized into different coordination patterns. A coordination pattern abstracts certain interactions between cross-dimensional members from a higher level of perspective. For instance, in Fig. 10.12, we show generic coordination patterns. Coordination patterns can be very complicated and may be presented as multilevel and cross-workflow. Complex advanced coordination patterns are composed of simple coordination meta-patterns. Some coordination meta-patterns exist, for instance, requester–provider model, peer-to-peer model, requester–distributor–provider, and owner–provider–requester. Here, *owner* refers to services that generate and offer support that can be used by others. *Requester* refers to services that subscribe to services from service providers. *Provider* refers to services that subscribe to and deliver services on behalf of service owners. *Distributor* refers to agent services that mediate or distribute services between service requesters and providers. Figure 10.12 illustrates each of these, respectively.

During detailed design, coordination patterns are further delineated as more concrete and specific coordination structures. For example, a requester–provider model may be instantiated into consumer-contract-provider model. We will discuss the detailed design of coordination in the next chapter.

10.8 Case Study

The system functional architecture of F-TRADE [21–23] is shown in Fig. 10.13. In F-TRADE, all functionality is categorized and organized into five function centers: Administration Center, Algorithm Center, Control Center, Business Services Center, and User Center. Direct application or application components of F-TRADE are agentized or instantiated as services in F-TRADE. Atomic transactions and processes are usually agentized, while inter-application communications are more likely to be implemented as services. These agents and services must easily follow predefined specifications of organizational framework and design patterns, and must implement particular APIs for corresponding agent services.

Services in F-TRADE are predefined and customized as system functional building blocks of the infrastructure. Services are mainly developed for inter-application interaction and integration, and for resource management and operations; for instance, data services dealing with data source management and operations, ontology services handling ontological management and operations, subscription services for algorithm subscriptions, system services for interfaces, agents' plug-in, system reconstruction, etc.

Nevertheless, agents are mainly used as system business building blocks. They are categorized and specified in terms of naming, activity, location, directory, and relationships in the organization. For instance, all pluggable algorithms are wrapped as algorithm agents if they implement the algorithm API. For the flexible administration, system mechanisms are provided to plug agents into F-TRADE through the Control Center. Interfaces for user agent interaction are generated according to user profile and agent ontology.

Six types of users including algorithm providers and subscribers can interact with corresponding user interface agents lodged in the above centers. User requests will be mediated and directed to target functional agents by mediator agents. The target agents will act on the requests by cooperating with other functional agents

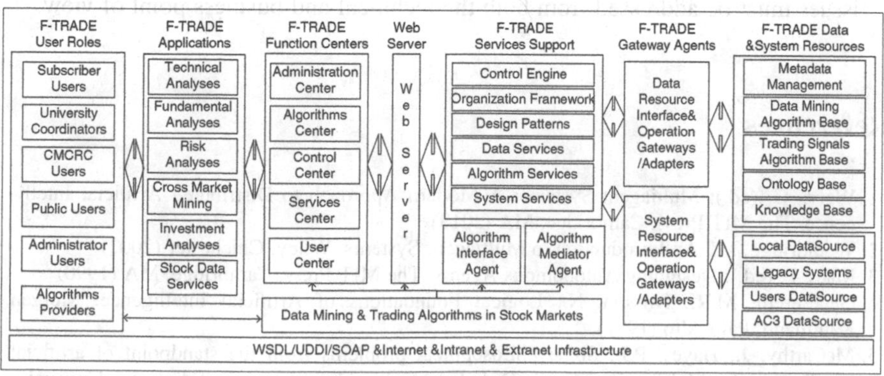

Fig. 10.13 Architecture of F-TRADE

and services as required, or by delivering requests to other relevant agents for further action.

Heterogeneous data resources, including both local, remote and multiform data formats, are managed by data resource gateways. Data source interface gateway agents deal with the registration and maintenance of data source drivers, link information, and data. Requests for data extraction, preprocessing, transportation, and operations on data sources are mediated by data source operation gateway agents; further actions are taken by functional agents as required.

The system database, algorithm base, and knowledge base of F-TRADE are maintained by different types of system resource gateway agents according to the types of information and storage. Particular agent services are designed to operate and manage user profiles, the ontology of agents, and the configuration information of the organizational framework and user permissions.

10.9 Summary

In this chapter, we have introduced agent service-oriented architectural design. The key points for this architectural design are summarized as follows:

- There are multiple candidate solutions for building agents and services in terms of different methodologies.
- We have listed valuable design patterns of agent services which constitute significant building blocks in agent service-oriented design.
- There are two solutions for implementing agent service-oriented enterprise: one is "multiagent + Web services", which aims to build Web service-driven agent systems, and the other is "multiagent + service-oriented computing", which targets multiagent-driven service systems.
- We advocate and detail the multiagent-driven service systems for agent service-oriented computing.
- In the process of architectural design, many administration and communications issues must be addressed from both the technical and business point of view.

References

1. Weiss, G. (ed.): Multiagent Systems: A Modern Approach to Distributed Artificial Intelligence. The MIT Press, Cambridge, MA (2013)
2. Wooldridge, M.: An Introduction to Multiagent Systems. Wiley, Chichester (2002)
3. Maes, P. (ed.): Designing Autonomous Agents. The MIT Press, Cambridge, MA (1990)
4. Genesereth, M.R., Nilsson, N.: Logical Foundations of Artificial Intelligence. Morgan Kaufmann, Palo Alto (1987)
5. McCarthy, J., Hayes, P.J.: Some philosophical problems from the standpoint of artificial intelligence. In: Meltzer, B., Michie, D. (eds.) Machine Intelligence 4. Edinburgh University Press (1969)

6. Rosenschein, S.J., Kaelbling, L.P.: A situated view of representation and control. In: Agre, P. E., Rosenschein, S.J. (eds.) Computational Theories of Interaction and Agency. The MIT Press, Cambridge, MA (1996)
7. Ferber, J.: Multi-agent Systems: An Introduction to Distributed Artificial Intelligence. Addison-Wesley/Longman, Harlow (1998)
8. Brooks, R.A.: A robust layered control system for a mobile robot. IEEE J. Robotic. Autom. **2** (1), 14–23 (1986)
9. Brooks, R.A.: Intelligence without reason. In: Proceedings of the 12th International Joint Conference on Artificial Intelligence, IJCAI-91, 569–595, Morgan Kaufmann (1991)
10. Brooks, R.A.: Intelligence without representation. Artif. Int. **47**, 139–159 (1991)
11. Langton, C. (ed.): Artificial life. Santa Fe Institute Studies in the Sciences of Complexity, Addison-Wesley, Reading (1989)
12. Bratman, M.E.: Intentions, Plans, and Practical Reason. Harvard University Press (1987)
13. Erl, T.: Service-Oriented Architecture: A Field Guide to Integrating XML and Web Services. Pearson Education, Upper Saddle River (2004)
14. www.specifications.ws
15. Rosenschein, S.J., Zlotkin, G.: Rules of Encounter: Designing Conventions for Automated Negotiation Among Computers. MIT Press, Cambridge, MA (1998)
16. Cassady Jr., R.: Auctions and Auctioneering. University of California Press, London (1967)
17. Monday, P.B.: Web Service Patterns: Java Edition. Apress, New York (2003)
18. Durfee, E.H., Lesser, V.R.: Using partial global planning to coordinate distributed problem solvers. In: Proceedings of the 10th IJCAI, pp. 875–883 (1987)
19. Durfee, E.H.: Coordination of Distributed Problem Solvers. Kluwer Academic Press, Boston (1988)
20. Gelernter, D., Carriero, N.: Coordination languages and their significance. Commun. ACM **35** (2), 97–107 (1992)
21. Cao, L.B., Wang, J.Q., Lin, L., Zhang, C.Q.: Agent services-based infrastructure for online assessment of trading strategies. In: Proceedings of the 2004 IEEE/WIC/ACM International Conference on Intelligent Agent Technology (IAT'04), pp. 345–349. IEEE Computer Society Press (2004)
22. F-Trade user manual. http://www-staff.it.uts.edu.au/~lbcao/ftrade/manual.pdf (2003)
23. http://www-staff.it.uts.edu.au/~lbcao/ftrade/ftrade.htm, Dec 2014
24. Veit, D.: Matchmaking in Electronic Markets: An Agent-based Approach. LNAI-2882, Springer, Berlin/Heidelberg (2003)
25. Lange, D.B., Oshima, M.: Seven good reasons for mobile agents. Commun. ACM **42**(3), 88–89 (1999)
26. Lewicki, R., Saunders, D., Minton, J.: Essentials of Negotiation, Mcgraw-Hill College, 2nd edn. (2000)
27. Smith, R.: The contract net protocol: a high level communications and control in a distributed problem solver. IEEE. Trans. Comput. **C-29**(12), 1104–1113 (2006)
28. Wolfstetter, E.: Auctions: An introduction. J. Econ. Surv. **10**(4), 367–420 (1996)
29. Smith, V.L.: Auctions. In: Eatwell, J., Milgate, M., Newman, P. (eds.) The New Palgrave: A Dictionary of Economics. Palgrave Macmillan, London (1987)
30. McAfee, R.P., McMillan, J.: Auctions and bidding. J. Econ. Lit. **25**, 699–738 (2006)

6. Bolenderdu, S.E., Kaashboek [illegible]: A [illegible] view of representation and content. In: Ande. [illegible] Bi. Stenning [illegible] (eds.) Computational [illegible] of [illegible] and [illegible]. The MIT Press, Cambridge, MA (1996)
[illegible] Bayes [illegible] systems: an introduction [illegible] Artificial Intelligence [illegible] (1998)
Brooks, R.A.: [illegible] Artificial Intelligence [illegible] 139–159
8. Brooks, R.A.: [illegible] without reason. In: Proceedings of the 12th International Joint Conference on Artificial Intelligence (IJCAI-91) [illegible] Australia (1991)
9. Brooks, R.A.: [illegible] Cambridge, MA [illegible]
10. [illegible] Studies in the Sciences of Complexity [illegible]
Brown, [illegible] Harvard University Press [illegible] (1971)
18. [illegible] Field guide to integrating [illegible] Web Services [illegible] Upper Saddle River [illegible]
[illegible]
13. [illegible] S. [illegible] Designing [illegible] for Automatic [illegible] MIT Press, Cambridge, MA (1994)
[illegible] Automata [illegible] University Press [illegible] (1987)
17. [illegible] Web Services [illegible] New York (2003)
18. [illegible] partial parse [illegible] distributed [illegible] In: Proceedings of the [illegible] (1996)
[illegible] of [illegible] Kluwer Academic Press [illegible]
[illegible]
20. [illegible] (1991)
21. [illegible] based [illegible] In: Proceedings of the [illegible] International Conference on [illegible] Press [illegible]
22. [illegible] (2014)
23. [illegible] (2014)
24. [illegible] Agent-based Approach [illegible] Springer, Berlin [illegible]
25. [illegible]
[illegible]
[illegible]
[illegible]
28. W[illegible]
[illegible] A [illegible]
[illegible]

Chapter 11
Agent Service-Oriented Detailed Design

11.1 Introduction

The previous chapter introduced agent services-oriented architectural design. In this chapter, detailed design is addressed. We first describe agent service ontology, which is key to the representation of agent services aiming for generic and consistent application. Endpoint interfaces for agent service are also described. Management work on directory, communication, transport, mediation, discovery, and other issues are detailed based on the discussion in Chap. 10.

11.2 Agent Service Ontology

The motivation for designing the ontology of agent services is to specify, organize, and manage the organizational structure of all agents and services in the problem domain. This can be dealt with in two steps:

1. Extraction of problem-solving ontologies from a domain problem
2. Development of agent service ontologies for detailed design based on the problem-solving ontologies extracted

11.2.1 Extracting Problem-Solving Ontology

Problem-solving ontology is extracted from the problem domain for designing the problem-solving system. The problem-solving ontologies that exist in a problem domain can be extracted through the analysis and buildup of a conceptual model, extended entity relationship model, or agent class diagram [1, 2]. These

L. Cao, *Metasynthetic Computing and Engineering of Complex Systems*,
Advanced Information and Knowledge Processing,
DOI 10.1007/978-1-4471-6551-4_11

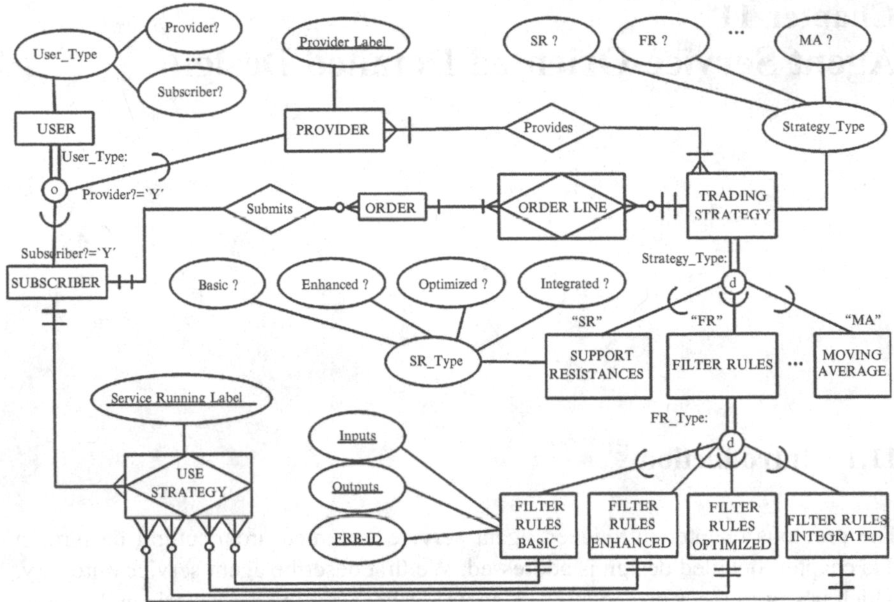

Fig. 11.1 An excerpt of problem-solving ontologies in F-TRADE in EER

models assist us to capture and indicate elements of agents and services and the relationships they embody. Figure 11.1 shows an excerpt of the problem-solving ontologies in F-TRADE in the extended entity relationship (EER) [3]. The relationships between ontological members can be classified into three high-level categories: associated ontologies, aggregated ontologies, and generalized hierarchical ontologies.

11.2.1.1 Associated Ontologies

There are a number of very common relations among ontologies in which one ontology is related to another, or many, in a meaningful and interesting connection, such as the semantic relationships of similar_to, relate_to, disjoin_with, and overlap_to, discussed in Sect. 12.3 "Ontological Semantic Relationships". The degree of an association relationship may be unary, binary, ternary, or n-nary, and is context dependent. Furthermore, the relational constraints existing in the two sides may be optional or mandatory, and are often combined in a relationship pattern in the real world, as discussed in Table 4.6. For instance, a *TradingStrategy* service can only be registered by one specific *TradingStrategyProvider* agent, while the provider agent can optionally contribute zero to many specific strategy services.

11.2.1.2 Aggregated Ontologies

Aggregated ontologies can also be called composite ontologies. These ontologies interlink through semantic relationships such as part_of, as discussed in Sect. 12.3 "Ontological Semantic Relationships". This relationship characterizes two ontologies offered by agents in a strong whole-part or assembly-part (*part_of*) relationship, i.e., one ontology is an encapsulation of its component ontologies. The composed parts are called part or component ontologies. For instance, the *Filter Rule* strategy package consists of strategies like *Filter Rule Basic*, *Filter Rule Enhanced*, *Filter Rule Optimized*, and *Filter Rule Integrated.*

11.2.1.3 Generalized Hierarchical Ontologies

These ontologies are linked through a semantic relationship such as subclass_of, as discussed in Sect. 12.3. This ontological relationship takes the form of a kind of (*is_a*) hierarchy, in which one ontology represents a more general concept, while the other represents more specialized concepts. There are thus super-type ontology and subtypes coexisting in the domain, and sometimes a subtype is again a super-type. Furthermore, there may be semantic constraints like *overlapping, disjoint, complete, or incomplete* specializations lodged in the hierarchy. For instance, a trading strategy can be a *Filter Rules*, *Support Resistances*, or *Channel Breakouts*, but cannot, say, consist of more than one strategy in any given moment. However, an agent can play a role as both a trading strategy provider and a strategy subscriber who can utilize other providers' strategies after service subscription.

As shown in Fig. 9.1, the business rules in a realistic assessment system of trading strategies often take the form of a combination of association, aggregation, and generalization. After obtaining the organizational structure of the trading strategy services, we construct a directed hierarchical graph which shows the conceptual structure (including nodes as agents and services, and edges for relations among them) between agents and services, after omitting attributes and operations.

11.2.2 Developing Agent Service Ontology

In developing agent service ontology for problem-solving, the basic idea is to decompose and classify the problem-solving ontologies extracted from the domain problem on demand and reorganize and formulate them in terms of the attributes and properties in the agent service model. The algorithm for developing agent service ontologies is as follows:

1. Extracting actors of agents and services and their goals from the problem-solving ontology diagram
2. Specifying the roles, activities, and types of agents and services

3. Analyzing relevant environment objects of an agent service and relationships to other agent services
4. Designing the agent service ontological diagram to specify and describe all agents and services
5. Refining the developed agent services through reverse reengineering and scenario-based analysis
6. Presenting the agent service ontology

Based on the agent service ontological diagram (for instance, in Fig. 11.1), we can extract the GAIRE elements for design objectives according to the organized framework. For instance, the main roles include algorithm provider, subscriber, and user; the main entity actors are composed of trading strategies and order; the main goals of the system are plug in, subscribe, evaluate, use, and optimize, which are implicitly embedded in the diagram.

11.3 Representation of Agent Services

11.3.1 Agent Service Specification

Representation of agent services is undertaken in terms of ontological engineering and agent-oriented methodology. An agent service is specified in terms of ontologies. An agent or a service consists of multiple ontological items. Each ontological item atom describes one attribute of the agent or the service.

Definition 11.1 *Agent Service Ontological Item Atom.* An ontological item atom is a basic unit which defines a type of attribute the agent or service must hold; the attribute can be any item of attributes and relationships related to an agent or a service.

An agent service ontological item is represented as a key–value tuple (KVT). A KVT consists of a set of unordered key–value pairs (KVP). Each KVP has two elements; the first element, the *key*, is a pair-element drawn from an administered namespace. A KVP is expressed in the form < *A* > k: v</*A*>; here *A* refers to one of the attributes.

All keys defined in the infrastructure are drawn from a universal and uniform ontological namespace. This makes the transparent and flexible organization of internal ontologies, agents, and services possible.

The type of *value*, which is the second element of the key–value pair, depends on the *key*. In many cases, it is another pair-element, a text string drawn from the same or another namespace. In some cases, the value is a constant or an expression of a specific type.

For the exact position and description of an agent service ontological item in a hierarchical namespace, a pair-element is often presented as an ordered tuple of tokens. This tuple denotes a name within a hierarchical namespace following a type of semantic relationship, in which the first token in the tuple is at the highest level in the hierarchy and the rightmost is the leaf. The natural encoding of a pair-element is as a sequence of text strings separated by dots.

Example 11.1 An agent service ontological item atom—*the name*—of an algorithm Moving Average Basic can be represented in terms of KVP as follows:

$$< A > Name\_MovingAverageBasic : String < /A > \quad (11.1)$$

where *A* indicates that this KVP refers to the attribute as inputs.

It is equivalent to and will naturally be encoded as follows in implementation:

$$\{\{\text{ftrade.algorithm.tradingrule.movingaverage.basic.MovingAverageBasic}\} : \{\text{java.lang.String}\}\} \quad (11.2)$$

in the Java-based F-TRADE system.

For each item atom, there may be zero, one, or many candidate items as required; they are also called an item list. For instance, an item list of *j* inputs is a list of j ($j \geq 1$) items as $< A > k_1{:}v_1, k_2{:}v_2, \ldots, k_j{:}v_j </A>$. Some of these items are mandatory (e.g., some service inputs must be specified), while others are not required (optional) for the service to be provided, but their provision may increase the efficiency of the provided service or the quality of the service. The multiplicity of constraint properties of items in item atoms is defined in Tables 7.3 and 7.6.

Definition 11.2 *Services of Agent.* Here, *services of agent* means the services an agent can provide or perform. Besides those static attributes an agent has, offering certain service is an important aspect of an agent. Services of agent consist of the dynamic functions and activities an agent can perform.

Definition 11.3 *Services of Service.* Here *services of service* means the services a service can provide or perform. In addition to those static attributes a service has, offering certain service is an important aspect of a service. *Services of service* consist of dynamic functions and activities a service can perform.

We separate services from agent and service, so that we can explicitly express the dynamic actions an agent or a service can or must perform; this is especially important for developing an agent service-oriented system.

Definition 11.4 *Agent Service Specification.* An agent service is described by multiple attributes which are presented by agent service ontological item atoms. Basic item atoms are as follows. Each agent service has a globally unique logical type of the component.

Attributes and properties of an agent service include *activity*, *service type*, *locators*, *owner*, *roles*, *type*, *address*, *message*, *input variables*, *preconditions*, *output variables*, *postconditions*, and *exception*.

1. *Activity* (*SA*) describes the function of the agent service.
2. *Locators (SL, RL) of either sender or receiver indicate location of agent services.*

3. *Owner (SO) is the one who holds the agent or service.*
4. *Roles (SR) are held by an agent or a service.*
5. *Attributes related to service transportation are type (TT), address (TA), and message (TM).*
6. *Input variables (I) and output variables (O) are in/out parameters.*
7. *Preconditions (IC) and postconditions (OC) define constraints on executing an agent service.*
8. *Cardinality (IO) defines the cardinality property of each attribute. All such properties can be expressed in the form KVP.*
9. *Exception (E) defines various unexpected events, messages, system operations, etc.*

For all the above item atoms, the constraint properties on them are as follows: $\{\{SA, \text{MO}\}, \{SL, \text{MO}\}, \{RL, \text{MO}\}, \{SO, \text{MO}\}, \{TT, \text{MO}\}, \{TA, \text{MO}\}, \{TM, \text{OM}\}, \{I, \text{OM}\}, \{IC, \text{OM}\}, \{O, \text{OM}\}, \{OC, \text{OM}\}, \{E, \text{OM}\}\}$where MO (Mandatory One) and OM (Optional Many) are defined in Table 7.6.

11.3.2 Case Study: Algorithm Registration Agent Service

In this section, we exemplify how to use the above agent service specifications to build an agent service for F-TRADE. We take the service of registering a trading strategy algorithm as an instance. The agent service RegisterAlgorithm is expressed in the formats of KVT and an informal format from first-order logic.

Example 11.2 The agent AlgoPluginAgent has services *RegisterAlgorithm* to fulfill the registration of a trading strategy algorithm. The agent service is expressed in KVT as follows:

```
<S>Service: RegisterAlgorithm
    <SA>Activity: register(ParaArrayList)</SA>
    <SO>ServiceOwner: PLUGINPERSON</SO>
    <ST>ServiceType: Action</ST>
    <SL>SenderLocator: AgentTransport.getSender()</SL>
    <RL>ReceiverLocator: AgentTransport.getReceiver()</RL>
    <I>InputVariables: AlgoInterface.getParameter()</I>
    <IC>Preconditions: AlgoInterface.getInConstraint()</IC>
    <O>OutputVariables: AlgoInterface.getResult()</O>
    <OC>Postconditions: AlgoInterface.getOutConstraint()</OC>
    <IO>CardinalityConnective: OO</IO>
    <TT>TransType: AgentTransport.getTransportType()</TT>
    <TA>TransAddress: getTransportAddress()</TA>
    <TM>TransMessage: AgentTransport.getTransportMessage()</TM>
    <E>Exeception: FTradeException</E>
</S>
```

Table 11.1 Partial agent services

Agent service	Verb	Noun-term
ManageAlgorithm	Manage	Algoid(agent_algotype)
RegisterAlgorithm	Register	Algoid(serviceroot_algo)
DeleteAlgorithm	Delete	Algoid(serviceroot_algo)
UpdateAlgorithm	Update	Algoid(agent_algotype)
AlgorithmExecution	Execute	Algoid(agent_algotype)
OntologySearch	Search	Ontologyid(ontodomain_target)

The above service can also be expressed informally, in which service item atoms have been instantiated for practical system implementation.

This service registers a globally unique trading strategy algorithm with three mandatory inputs and the creating date of one optional algorithm. All inputs and outputs of the trading strategy are separated by semicolons and stored into an array in Java. One new trading strategy class is generated if registration is successful. Accordingly, we can define some of the basic agent services (as partially shown in Table 11.1) needed in F-TRADE to support trading and mining.

11.4 Agent Service Endpoint Interfaces

The agent service-oriented integration architecture benefits standardizing interoperability, but in no way guarantees a quality data sharing environment. The efficiency and usability of an agent service framework relies on individual agent service interfaces.

11.4.1 Designing Agent Service Interfaces

The main principles discussed here include the responsibility for designing interfaces and rules to create enterprise-wide generic and consistent interfaces.

Typical responsibilities of designing an agent service interface include the following:

- Agent service interface standard and naming conventions
- Interface implementation documents
- Agent service message standard and specifications
- Message documents
- Agent service interface consistency, clarity, and extensibility

It is worth ensuring the consistency of an agent service-oriented enterprise infrastructure; to this end, the endpoint interface design should be the first task in the system design. This is to establish generic, consistent enterprise-wide naming

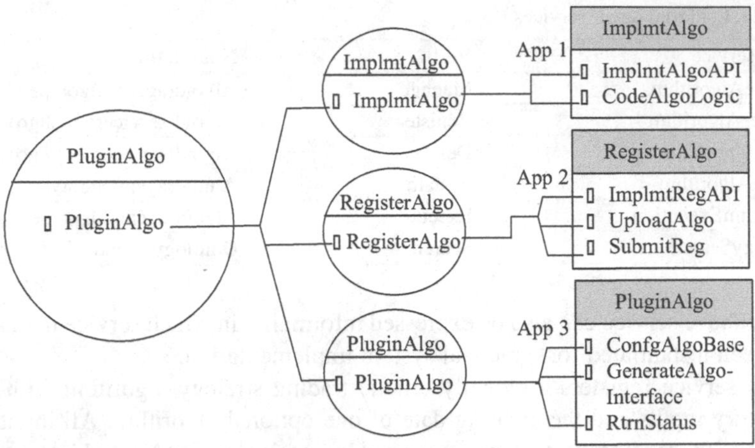

Fig. 11.2 An agent service including multiple operations via combination of multiple agents' methods

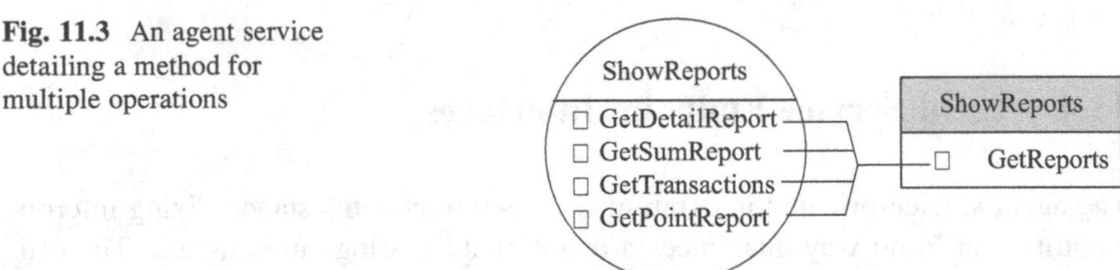

Fig. 11.3 An agent service detailing a method for multiple operations

conventions and interface design standards; an integration layer provides standard interfaces to disparate applications. Then, we can encapsulate business logics using agent services.

It is beneficial to make both the interface naming convention and the functionality of the interface more generic. These can be instantiated in the following manner:

- An agent service provides a generic operation (service interface) that represents multiple agent service methods, as illustrated in Fig. 11.2, in which the operation Implementing Algorithm (ImplementAlgo) is performed via methods Implementing Algorithm API (ImplemtAlgoAPI) and Coding Algorithm Logics (CodeAlgoLogic).
- An agent service splits a coarsely grained agent service method into multiple service operations. As shown in Fig. 11.3, the method GetReports is split into four agent service operations such as GetDetailReport, GetSumReport, GetTransactions, and GetPointReport.
- An agent service includes one to many service operations that combine various pieces of legacy or enterprise business logics, as shown in Fig. 11.3.

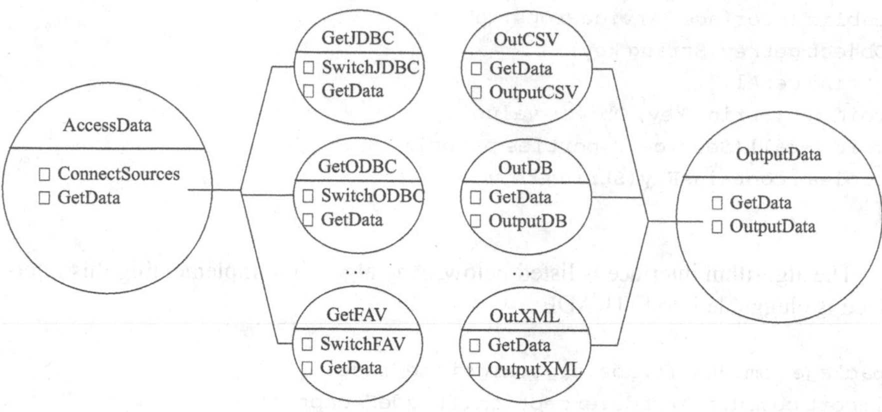

Fig. 11.4 A data service supporting multiple input types or offering multiple output formats

- An operation of an agent service encapsulates functionality from multiple business agent services; each of these agent services assembles one to multiple methods representing legacy or enterprise applications. As in Fig. 11.2, the service Plugin Algorithm (PluginAlgo) is divided into three operations—Implementing Algorithm (ImplmtAlgo), Register Algorithm (RegisterAlgo), and Plugin Algorithm (PluginAlgo) performed by three agent services. These operations are conducted through eight agent methods held by the three agents.
- A data service supports multiple data input or output formats. As illustrated in Fig. 11.4, an agent service managing data sources offers access to multiple data types, while a data output service provides multiple output formats.

11.4.2 Case Study: Algorithm Service Interface

This section illustrates the algorithm service interface. We introduce interfaces for service and service properties and then present the algorithm interface.

The following defines a generic service in terms of a set of properties.

```
public interface Service{
ServiceProperties getServiceProperties()throws ServiceException,
  ServiceFailure;
void     setServiceProperties(ServiceProperties     props)throws
  ServiceException, ServiceFailure;
}
```

Service properties describe the properties of a generic service. As discussed in Sect. 11.3 "Representation of Agent Services," each property consists of a key–value pair.

```
public interface ServiceProperties{
Object getKey(String key);
string getAll();
void set(String key, Object value);
void setall(ServiceProperties props);
boolean containsKey(String key);
}
```

The algorithm interface is listed below. Any algorithm implementing this interface is pluggable into F-TRADE:

```
package com.uts.ftrade.algo.algointerface;
import com.uts.ftrade.exception.ftradeException;
public interface AlgoInterface {
  /**set algorithm parameters
  * @param parametername
  * @param value
  * @throws ftradeException
  */
 public void setParameter(String parametername, String value) throws
  ftradeException;
/**set data source driver
*/
 public void setDataSourceDriver(String driver) throws
  ftradeException;
/**set data source URL
*/
 public void setDataSourceUrl(String url) throws ftradeException;
/**set data source user name
*/
 public void setDataSourceUser(String user) throws tsapException;
/**set data source access password
*/
 public void setDataSourcePws(String pws) throws tsapException;
/**algorithm body
*/
 public void execute() throws ftradeException;
/**get algorithm results
*/
 public String getResult(String outputString) throws
  ftradeException;
}
```

11.5 Directory of Agent Services

The basic role of the directory of agent services is to provide a location at which agent services register directory entries, and a consistent means by which agents and services can discover services. The agent registers with the directory service, passing to it a directory entry. Services can register their service descriptions as service directory entries. The agent service directory involves agent directory service [3, 4] and service directory service [3, 4]. With agent directory service and service directory service, agents and services can search directory service to locate services appropriate to their needs. Other agents can search the directory entries to find agents with which they wish to interact.

With regard to the directory of agent service, one issue is to define, present, locate, and discover the directory services of agents and services. Definition is closely related to the representation of directory service. To this end, a directory namespace and specifications for directory service are required. Location and discovery of agent services is undertaken through directory services by utilizing the representation and organization of agent directory service and service directory service.

Specification of directory service specifies what attributes will be brought into directory entry for positioning agents and services, and how these attributes should be presented in the directory entry.

On the basis of the obtained organizational ontology structure, our further work is to define an agent service directory namespace of trading strategy agents and services. F-TRADE employs attributes represented as key–value tuples to define fields in both agents and services. The directory service namespace is drawn from the domain rules of the financial trading strategy assessment according to the discussion in the agent service ontology, and also follows the generic specifications of FIPA namespace [3, 4].

Correspondingly, attributes in the namespace include common attributes and specifics which indicate the hierarchical structure required in the assessment system of trading strategies as follows:

```
/* Common agent service attributes in F-TRADE */
ftrade.agent.sender-locator
ftrade.agent.receiver-locator
ftrade.agent.owner
ftrade.agent.roles
ftrade.agent.transport-type
ftrade.agent.transport-address
ftrade.agent.transport-message
ftrade.agent.directory-entry
ftrade.agent.view
ftrade.agent.model
ftrade.agent.action
ftrade.agent.valueobject
ftrade.agent.exception
```

As a snapshot of the above service attributes, the agent locator interface is exemplified as follows:

```
public interface AgentLocator{
int hashCode();
String getType();
String getAddress();
setType(String type);
void setAddress(String address);
}
```

The following is part of the interface of agent service directory:

```
public interface AgentDirectory extends Directory{
AgentDescription[] getAgentDescription();
Vector getDirectoryEntry();
void register(AgentDescription ad) throws DirectoryFailure;
void update(AgentDescription ad) throws DirectoryFailure;
void delete(AgentDescription ad) throws DirectoryFailure;
void execute(AgentDescription ad) throws DirectoryFailure;
AgentDescription[] search(AgentDescription ad) throws
  DirectoryFailure;
void setDirectoryEntry(Vector de);
}
```

The algorithm specific agent service attributes in F-TRADE are exemplified as follows:

ftrade.agent.algorithm
ftrade.agent.algorithm.tradingstrategy.tradingstrategy-type
ftrade.agent.algorithm.tradingstrategy.filterrules.filterrule-senhanced.display-name
ftrade.agent.algorithm.tradingstrategy.filterrules.filterrule-senhanced.identifier-name
ftrade.agent.algorithm.tradingstrategy.filterrules.filterrule-senhanced.provider-indentifier
ftrade.agent.algorithm.tradingstrategy.filterrules.filterrule-senhanced.sender-locator
ftrade.agent.algorithm.tradingstrategy.filterrules.filterrule-senhanced.receiver-locator
ftrade.agent.algorithm.tradingstrategy.filterrules.filterrule-senhanced.input-messages
ftrade.agent.algorithm.tradingstrategy.filterrules.filterrule-senhanced.output-messages
ftrade.agent.algorithm.tradingstrategy.filterrules.filterrule-senhanced.execption-messages

The following trading strategy algorithm interface must be implemented for all algorithms, including trading strategies and data mining algorithms, for them to be executed after online plug into F-TRADE.

```
public interface AlgoInterface {
void setParameter(String parametername, String value) throws
  ftradeException;
void execute() throws ftradeException;
String getResult(String outputString) throws ftradeException;
void setDataSourceDriver(String driver) throws ftradeException;
}
```

The agent service ontology and naming space information among agents and services are kept in the body of agents and services (parts of them are agent locator and service locator) as a part of the knowledge of the domain services. Address management and transport resolution of agents and services is organized by agent/service locators. As part of agent directory entry, agent locators contain agent transport descriptions and transport type, and the specific address and properties of transport type, which can be used to locate an agent or an agent service. Technically, a service locator may contain references to the interfaces of a trading strategy system by services accessed programmatically, or may be called by a message transport service to select a transport for communicating with an agent via message passing.

The agent can access agent directory services to discover and locate the target agent and services with which the source agents want to communicate. In the process of searching for the destination agent, the agent will technically retrieve the agent/service ontology library and directory services for an agent/service directory entry in the form of KVT; this agent/service directory entry encloses transport information which can be parsed to locate the target agents and services according to business rules defined in the agent service directories.

A small example could be as follows. In F-TRADE, a trading strategy user clicks a trading strategy, for instance, *VWAP Execution Strategy*, in which he/she is interested, at the *Algorithm* (i.e., trading strategy) *Center*. This request will be directed to the algorithm service library, where all trading strategy and data mining algorithms are organized according to a uniform naming and hierarchical ontology structure, and the following agent directory entry will be queried. If found, it will be passed in a message to the user agent by which the request is activated.

$$\{\{\text{algorithm.execute} - \text{service}\} : \{\text{ftrade.algorithm. tradingstrategy.vwap.vwapexecutionstrategy.execute}\}\} \tag{11.3}$$

where the value of the algorithm.execute-service is the execution (execute()) service of the VWAPExecutionStrategy agent.

11.6 Communication of Agent Services

Agent service communication deals with communication model, communicative act, and communication control in agent and service communication. Message-based communication model is commonly used for communication of agent services. In this model, message structure will be discussed in terms of related message parameters; this can be called the agent service message model.

Agent service message model [3, 5–7] is essential for message packing and delivery before discussing the transport of the message. This involves which additional properties will be abstracted and packed into a message body. It is also necessary to find expression method to describe the message content, which could be a kind of communication language. For the expression of agent messages, agent ontology will also be involved.

Before the delivery of an agent service message, other communication settings, such as communicative act, conversation partners' setting, conversation type, and so forth must be defined. In Sect. 8.7, discussed message parameters, communicative acts, and how to model the message of agent service in terms of FIPA ACL specifications. That work actually builds up an agent service message ontological diagram, which is quite significant in the communication of agent services.

11.7 Transport of Agent Services

Agent service transport mainly deals with the transport of messages between agents and a transport protocol that can be used for the delivery of messages. Message transport [3, 8, 9] supports the sending and receiving of messages between agents. Transport type deals with what type of communication protocol can be used for delivering messages across networks.

In the delivery of messages between agent services, messages will usually be transformed into, or wrapped, as a secure packet, probably with transport address and verification information. This message packing will be further embedded into the agent message model.

An agent service communication channel will be set up between agent services. In message-based communication, the following issues need to be studied—transport interfaces, delivery (sending and receiving) and handling of messages, quality, and exception handling of communication.

Another issue is how to represent transport service. This includes what attributes can be selected to describe the transport service and how to represent the transport service of agents and services.

The following shows an agent service transport interface:

```
public interface AgentTransport extends Transport{
AgentLocator getSender();
AgentLocator getReceiver();
```

```
String getTransportType();
Locator getTransportAddress();
Message getTransportMessage();
void setSender(AgentLocator sender);
void setReceiver(AgentLocator receiver);
void setTransportType(String ttype);
void setTransportAddress(Locator taddress);
void setTransportMessage(Message tmessage);
}
```

11.8 Mediation of Agent Services

The mediation and management [10–15] of agent services is very important for a large agent service system to work effectively. With the organizational metaphor, an agent space can be viewed as consisting of multiple organizational branches, and there are likely to be multiple organizational hierarchies existing in the agent society. For instance, these organizational branches could be relatively isolated functional centers, or they could be activity centers, where relevant agents and services work together to fulfill actions or support services in those centers.

The mediation of agent services in the agent society can therefore exist at two levels. One supports local mediation in a subsystem; the other provides global mediation in the agent society. To meet these objectives, multi-tier mediation strategies must be designed to deal with management at different levels. Mediation protocols must be considered in the negotiation, and mediation logic must be clarified from the ask request to the reply response.

In our agent service-based trading strategy assessment system, the mediation [2] of agent services is supported by the following two-level matchmaking strategy (shown in Fig. 11.5), in which all agents and services are named, organized, and routed in a uniform service naming and addressing space. Note that the *M* in the figure is a backup of the root mediator in case the M is out of order, in which case the mediator agent *M* would start up and immediately act as M.

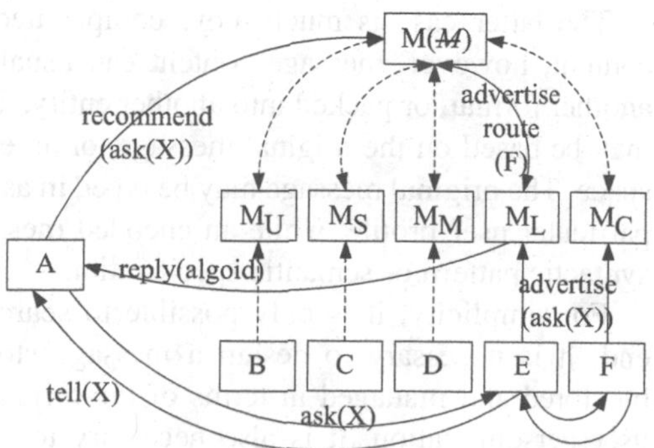

Fig. 11.5 Agent service mediation

This mediation structure is adaptive to the organizational structure of the agent-based trading strategy assessment platform. In F-TRADE, all the user actors (e.g., a trading strategy subscriber initiating an *ask(X)* request from agent *A*) interact with the system console portal, in which there are five large functional centers [12–17]: *User Center*, *Service Center*, *Administration Center*, *Algorithm Center*, and *Control Center*, each of which coordinates and administers most issues related to the specific center. Correspondingly, five center-level mediator agents, M_U, M_S, M_M, M_L, M_C, are dispatched and lodged in the background of each of these five centers, respectively.

A center-level mediator agent, for instance, the M_L for the *Algorithm Center*, accepts and organizes advertising information including functions (e.g., *ask(X)* performative) and location (parsing from the value of service attribute key, for instance, *ftrade.algorithm.tradingstrategy.vwap. vwapexecutionstrategy.execute*) of its affiliated agents and services and further reports the parsed location data to the high-level root mediator *M*. The root mediator *M* organizes and maintains a global resource location list and, in the above case, is asked by agent *A* to recommend a mediator (for instance, M_L) which locates an appropriate agent who has advertised the service requested by *A*. Once the mediator M_L learns that agent *E* is willing to accept *ask(X)* performative, it replies to *A* with the name and location information of *E*. *A* is then free to initiate a dialog with *E* to request and respond to queries.

11.9 Discovery of Agent Services

The discovery [3, 18, 19] of agent services can take two forms: the first might be to query an agent service with which to communicate; the second might be to search for a message.

For the former, an agent service can search another agent service for the requested actions or information, according to what that service advertises. This can be located by the agent/service directory service; the request may be contained in a property in the directory entries. The search matches the queried value with the enclosed property value in the target entry item.

The latter case is much more complicated. The query comes from message content; however, message content can usually be transformed or encoded into another format, or packed into another entity. This means the query of the message may be based on the original message, or an encoded message in the whole agent space. The original message may be typed in as a user personification according to a particular user profile, while an encoded message may be in the form of another syntactic pattern or semantic specification.

For simplicity, it is only possible to search for an original message. To this end, it is necessary to design a message store structure, so that messages can be stored and managed in terms of the original format and business ontology as user personification. It is also necessary to support the semantic transformation from user profile ontology to encoded message ontology. This is what ontology semantic transformation does.

11.10 Modeling Coordination

Modeling agent service coordination is performed in terms of modeling languages; however, there are at present no standard techniques for modeling coordination. Rather, the modeling of agent service coordination presents the following characteristics:

1. Many researchers are apt to develop specific modeling techniques for their coordination methods.
2. Specific modeling techniques are often based on generic modeling languages, for instance, XML-based modeling language, UML, use case, and ER diagram.
3. Ontology-based modeling is emerging as an effective technique for describing both syntactic and semantic aspects of coordination.
4. Formalization of coordination plays an increasingly important role in exactly presenting the logics and scenarios of coordination activities; for example, rule-based modeling and first-order logic-based coordination modeling.

XML-based languages are widely studied and used to model coordination, for instance, BPEL4WS and WSFL. Typical ontology-based modeling includes OWL-S and its multiple extension versions for Web services and service-oriented computing.

Coordination modeling is a process of extending and instantiating coordination meta-models and meta-patterns, discussed in Sect. 10.7. For instance, the CrossFlow project [20] proposed a dynamic provider–contract–requester-based cross-organizational architecture to support workflow management across multiple organizations. The architecture extends the meta-pattern provider–requester by adding a contract-based coordination mechanism between service requesters and providers.

In the following, we take the algorithm registration and subscription as an example to illustrate how to model agent service coordination. Figure 11.6 shows the workflow of algorithm (trading strategies or data mining algorithms) registration and subscription. In this case, the coordination is to recognize and distinguish the goals and tasks of the algorithm provider and subscriber. Algorithm providers register algorithms into F-TRADE, while algorithm subscribers buy licenses of algorithms of interest. Coordination services include user management service, algorithm availability checking service, and algorithm library service. The user management service first identifies the user, either provider or subscriber. If it is provider, its main task is to register a new algorithm. Otherwise, it would be a subscriber, who would like to buy algorithm services. For a provider, once an algorithm has been edited, the algorithm availability checking service checks whether this algorithm exists in the library or not. If it is not there, an algorithm

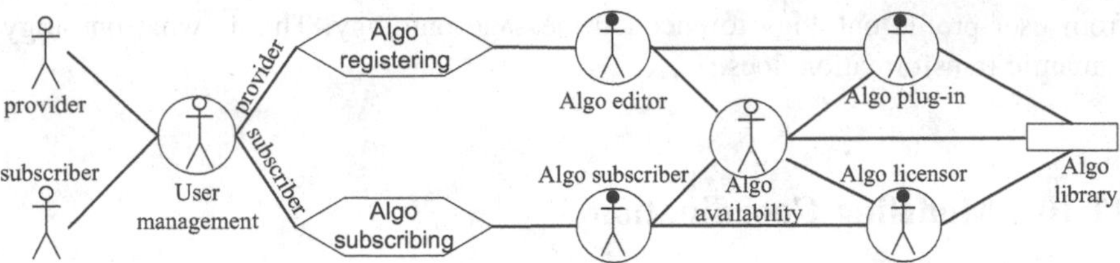

Fig. 11.6 Workflow of algorithm registration and subscription

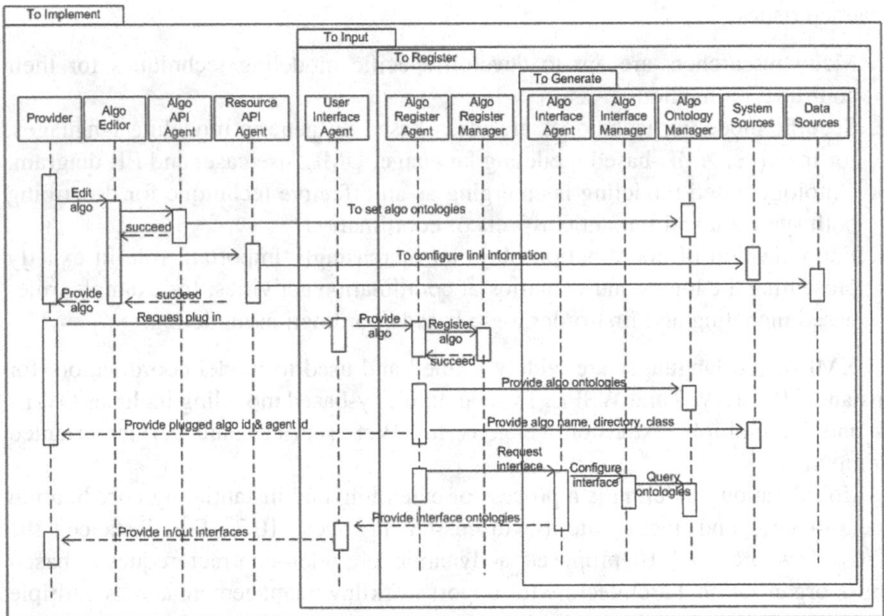

Fig. 11.7 Sequence and life cycle of algorithm registration and subscription

plug-in agent will register and plug in the algorithm to the algorithm library. With respect to algorithm subscription, the information entered through the algorithm subscriber agent is transferred to the algorithm availability checking service to check whether this algorithm exists in the library or not. If it does, the algorithm licensing agent will authorize the use of the algorithm by the subscriber.

Figures 11.7 and 11.8 specifically illustrate the sequence and life cycle of the algorithm registration. There are four embedded packages: *To Implement*, *To Input*, *To Register*, and *To Generate*, as shown in the figure. They present the agent service interactions in the plug-in activities. The package *To Implement* programs an algorithm in the AlgoEditor service by implementing the agent services AlgoAPIAgent and ResourceAPIAgent. The *To Input* package types in the agent ontologies of the programmed algorithm agent and asks to plug in the algorithm.

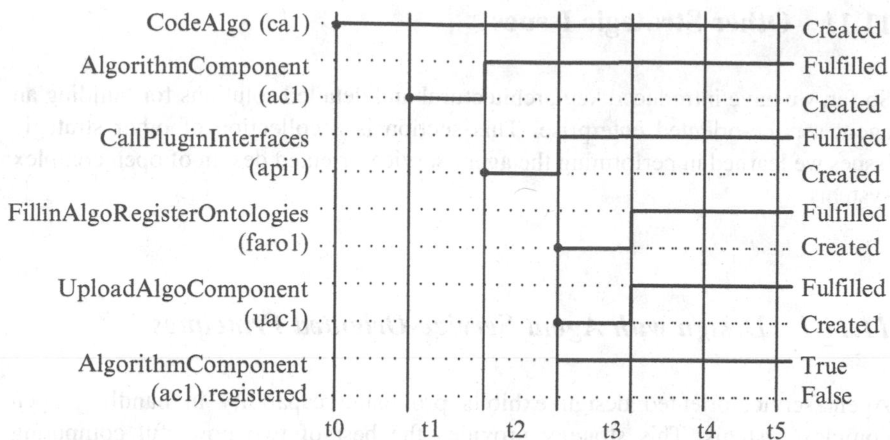

Fig. 11.8 Scenario analysis of registering and subscribing algorithms

The real registration of the algorithm agent is done by the package of *To Register*; the naming, directory, and class of the algorithm agent are stored in the algorithm base; and linkage to data resources are also set. Lastly, the input and output user interfaces for the algorithm are managed by the package *To Generate*.

```
<!ELEMENT WorkFlow (Activity | Transition )+ >
<!ELEMENT Activity (Name, Description?, Split?, Join?) >
<!ATTLIST Activity ActID ID #REQUIRED>
<!ELEMENT Name (#PCDATA) >
<!ELEMENT Split (TransRef*) >
<!ATTLIST Split Type (AND|XOR) "AND">
<!ELEMENT Join EMPTY >
<!ATTLIST Join Type (AND|XOR) "AND">
<!ELEMENT Description (#PCDATA) >
<!ELEMENT Transition (Name, Description?, To, From, Condition?)>
<!ATTLIST Transition TransID ID #REQUIRED>
<!ELEMENT To (ActRef) >
<!ELEMENT From (ActRef) >
<!ELEMENT Condition (#PCDATA | ParamRef |ActRef)*>
<!ATTLIST Condition type (parameter|workflow.state |workflow.event |
  workflow.time | CDATA) #REQUIRED>
<!ELEMENT ParamRef EMPTY>
<!ATTLIST ParamRef ParamID IDREF #REQUIRED>
<!ELEMENT ActRef EMPTY >
<!ATTLIST ActRef ActID IDREF #REQUIRED>
<!ELEMENT TransRef EMPTY >
<!ATTLIST TransRef TransID IDREF #REQUIRED>
```

11.11 Other Strategic Issues

So far, we have introduced key architectural and detailed solutions for building an agent service-oriented enterprise. This section is a collection of other strategic issues we learned in performing the agent service-oriented design of open complex systems.

11.11.1 Design with Agent Service-Oriented Principles

Agent service-oriented design exhibits promising capability in handling open complex systems. This strategy provides the best of two powerful computing paradigms. In addition, the combination of them learns from the strong aspects available in one another to offset the weaknesses of a particular paradigm.

- Plan or adjust business process models to incorporate agent service-oriented principles and instantiate these changes.
- Think of automated integration and logically partition application functionality into relatively autonomous components.

11.11.2 Create a Custom Ontological Directory

Establishing a common ontological directory within an organization is essential to foster a unified view and a generic infrastructure for future add-ons. This custom directory encompasses existing process and instantly reflects changes to business and technical issues on demand. The custodian of this ontology database has to grasp existing and emergent elements in two worlds; prevent redundancy and duplication; maintain normalization, integrity, and updates; and coordinate these vocabularies with all requesters in the organization.

11.11.3 Define a Schema Management Strategy

Closely related to the above ontological database, XML schemas play a significant role in establishing a vocabulary, associating data with a logical domain, creating validation boundaries, and encapsulating business rules. A management strategy is required to manage, maintain, and evolve XML schemas, just as for database schemas.

11.11.4 *Always Relate XML to Data*

XML unifies heterogeneous data into common data formats and bridges disparate applications as a common data representation technology. Maximizing XML utilization in the data access increases the programmer's control over data management and enterprise integration. XML brings about a number of benefits via incorporation within the programmatic logic behind data access. These benefits are embodied in such aspects as data access logic, application management logic, and system configuration logic.

11.12 Summary

In this chapter, we have discussed how to perform agent service-oriented detailed design. We have first introduced how to extract and develop agent service ontologies, and specifications have been developed for the representation of agent services. We have discussed how to design generic and consistent agent service interfaces, which is important for an agent service-oriented enterprise. Continuing the discussion in Chap. 10, directory, communication, transportation, mediation, and discovery of agent services have been detailed in this chapter. Finally, we have collected and presented other strategic issues we have learned in performing agent service-oriented design of open complex systems.

References

1. Hoffer, J.A., Venkataraman, R., Topi, H.: Modern Database Management, 11th edn. Prentice Hall, Englewood Cliffs (2012)
2. Finin, T., Fritzson, R., McKay, D., McEntire, R.: KQML as an agent communication language. In: Proceedings of the 3rd International Conference on Information and Knowledge Management, CIKM'94, pp. 456–463. ACM Press (1994)
3. FIPA SC00001L. FIPA Abstract Architecture specification (2002)
4. JSR 87: Java Agent Services: http://www.jcp.org/aboutJava/communityprocess/review/jsr087/
5. SC00071E. FIPA ACL Message Representation in XML Specification, 03 Dec 2002
6. SC00085J. FIPA Agent Message Transport Envelope Representation in XML
7. SC00084F. FIPA Agent Message Transport Protocol for HTTP Specification, 03 Dec 2002
8. SC00067F. FIPA Agent Message Transport Service Specification, 03 Dec 2002
9. XC00093A. FIPA Messaging Interoperability Service Specification
10. Veit, D.: Matchmaking in Electronic Markets: An Agent-Based Approach Towards Matchmaking in Electronic Negotiations. Springer, New York (2003)
11. SC00023J. FIPA Agent Management Specification, 03 Dec 2002
12. Cao, L.B., Luo, C., Luo, D., Liu, L.: Ontology services-based information integration in mining telecom business intelligence. In: Proceedings of PRICAI04, pp. 85–94. Springer, Berlin/Heidelberg (2004)

13. Cao, L.B., Wang, J.Q., Lin, L., and Zhang, C.Q.: Agent services-based infrastructure for online assessment of trading strategies. In: Proceedings of the 2004 IEEE/WIC/ACM International Conference on Intelligent Agent Technology (IAT'04), pp. 345–349. IEEE Computer Society Press (2004)
14. Cao, L.B., Zhang, C.Q., Luo, D., Chen, W.L., Zamani, N.: Integrative early requirements analysis for agent-based systems. In: The 4th International Conference on Hybrid Intelligent Systems, IEEE Computer Society, Kitakyushu, Japan, 5–8 December 2004. http://doi.ieeecomputersociety.org/10.1109/ICHIS.2004.63
15. Cao, L.B., Ni, J.R., Wang, J.Q., Zhang, C.Q.: Agent services-driven plug and play in the F-trade. In: Webb, G.I., Xinghuo Yu (eds.) 17th Australian Joint Conference on Artificial Intelligence: AI 2004. LNAI, vol. 3339, pp. 917–922
16. F-Trade user manual. http://www-staff.it.uts.edu.au/~lbcao/ftrade/manual.pdf (2012)
17. http://www-staff.it.uts.edu.au/~lbcao/ftrade/ftrade.htm
18. FIPA. FIPA Agent Discovery Service Specification. Version 1.2e, 20 Oct 2003
19. FIPA. FIPA JXTA Discovery Middleware Specification. Version 1.2, 20 Oct 2003
20. Grefen, P., Hoffner, Y.: CrossFlow—Cross-Organizational Workflow Support for Virtual Organizations, RIDE, 99, 90 (1999)

Chapter 12
Ontological Engineering

12.1 Introduction

From the discussion of integrative modeling and organization-oriented analysis, we have seen that conceptualization and declarative knowledge have been deeply involved in these processes. In practice, sharing, transferring, and transforming a conceptualization system from system analysis to design and implementation are critical issues in building a deployable system of software engineering. Precision, usability, continuability, deployability, and scalability are key objectives in developing the sharable conceptions and declarative knowledge in the whole life cycle of OADI. This is what *ontological engineering* [1] can do.

In this chapter, we shall discuss ontological engineering from the following aspects. Ontology profiles will be discussed in Sect. 12.2, so that concepts and relationships between the problem domain and the problem-solving system can be well understood. It is necessary to abstract the domain-specific ontologies [2] in order to understand the problem domain (capital market and its behaviors) and set up corresponding supports for the problem in the system. In addition, as a problem-solving system, it is also essential to extract agent and service ontologies [3], so that management and operations of agent services can be handled systematically and consistently.

On the other hand, understanding and maintenance of semantic relationships [4, 5] between ontology concepts is also very important for efficient and effective operations of ontologies. Semantic relationships are also fundamental for querying, transformation, and discovery of ontologies. Therefore, Sect. 12.3 will explicitly define and investigate semantic relationships between the domain-specific ontologies and also between problem-solving ontologies.

Semantic aggregation and ontology transformation [6] will be discussed in Sect. 12.5 in terms of an explicit definition of semantic relationships. Some transformation rules will be discussed in terms of different combinations of agent

L. Cao, *Metasynthetic Computing and Engineering of Complex Systems*,
Advanced Information and Knowledge Processing,
DOI 10.1007/978-1-4471-6551-4_12

ontologies. Lastly, ontological semantic aggregation and transformation across domains are discussed in Sect. 12.5.

12.2 Ontology Profiles

This section will discuss ontology profiles. The objective of this section is to enhance the understanding of the problem and problem-solver through ontological analysis. To this end, domain-specific ontology profiles are introduced in the problem domain, while problem-solving ontology profiles will be described in relation to the problem-solving system. The domain-specific ontology and the problem-solving ontology are closely related and coexist in a two-level ontology space. In this two-level ontology space, ontology semantic rules and formal formulae are abstracted for associations and matching of elements from different profiles.

12.2.1 From Ontology to Ontological Engineering

The word *ontology* is taken from philosophy, in which it means a systematic explanation of being. In the last decade, ontology has become increasingly relevant to knowledge engineering and software engineering. Many definitions of ontology have been generated and evolved over the years in different disciplines. One of the first definitions was given by [7], who defined an ontology as follows:

> Ontology defines the basic terms and relations comprising the vocabulary of a topic area as well as the rules for combining terms and relations to define extensions to the vocabulary.

A definition frequently quoted by the ontology community is *An ontology is an explicit specification of a conceptualization* [8]. Here, "conceptualization" refers to an abstract model of a phenomenon in the world which identifies the relevant concepts of that phenomenon. Guarino [9] collected seven definitions as follows, which cover comprehensive disciplines and motivation in using ontologies:

1. *Ontology as a philosophical discipline*
2. *Ontology as an informal conceptual system*
3. *Ontology as a formal semantic account*
4. *Ontology as a specification of a conceptualization*
5. *Ontology as a representation of a conceptual system via a logical theory:*
   - *Characterized by specific formal properties*
   - *Characterized only by its specific purposes*
6. *Ontology as the vocabulary used by a logical theory*
7. *Ontology as a (meta-level) specification of a logical theory*

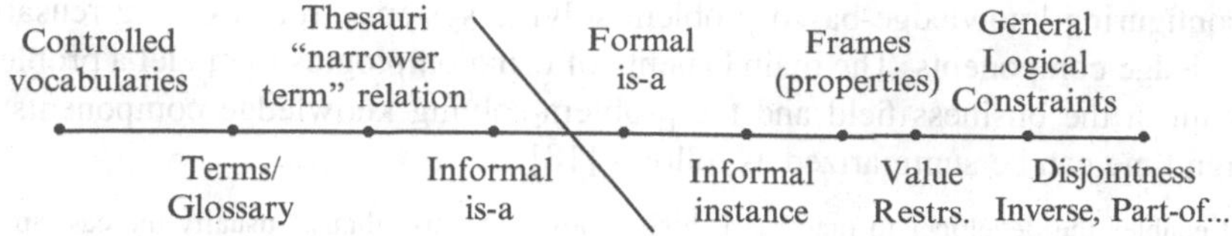

Fig. 12.1 Categorization of ontologies according to internal structure

The ontology community distinguishes ontologies that are mainly taxonomies from ontologies that model the domain in a deeper way and provide more restrictions on domain semantics; they are called *lightweight* and *heavyweight* ontologies, respectively. *Lightweight ontologies* include concepts, concept taxonomies, relationships between concepts, and properties that describe a concept. *Heavyweight ontologies* add axioms and constraints to lightweight ontologies to clarify the intended meaning of the terms gathered on the ontology.

Ontologies can be classified into categories in terms of the richness of the internal structure or the subject of the conceptualization. Lassila [31] categorized both lightweight and heavyweight ontologies in a continuous line as presented in Fig. 12.1. Ontologies can also be classified according to subjects of conceptualization into the following classes [1]: *general* [10] or *common ontologies* [11], *top-level* or *upper-level ontologies* [12], *knowledge representation ontologies* [13], *domain ontologies* [11, 13], *task ontologies* [11, 14], *domain-task ontologies* [15], *method ontologies* [14, 15], and *application ontologies* [13].

The work on ontology has concerned engineering processes such as approaches, methods, and techniques. This leads to the emergence of *ontological engineering* [1]. *Ontological engineering* refers to the set of activities that concern the ontology development process, the ontology life cycle, and the methodologies, tools, and languages for building ontologies. Modeling ontologies can take place in different languages using different approaches, some highly informal when expressed in natural language, some semiformal if expressed in a restricted and structurally defined language like Ontolingua [16] or OWL [17], or rigorously formal if represented by defined terms with formal semantics, theorems, and proofs of properties such as soundness completeness. From the above definitions of ontologies, we can see that engineering ontologies is becoming more and more explicit and formal. In later sections, we will discuss the ontological engineering of F-TRADE. The following aspects will be involved: ontology categorizations, ontology semantic relationships, ontology transformation and mapping, naming, and representation of ontologies for agent service.

12.2.2 Domain-Specific Business Ontology

Declarative knowledge modeled by ontologies in the problem domain and the problem-solver enables customers to understand the problem-solving system smoothly from the perspective of the business world. It also benefits developers in configuring knowledge-based problem-solving systems from existing reusable knowledge components. The main benefits of using ontologies to model a problem domain in the business field and the problem-solving knowledge components at design time can be summarized as follows [18]:

> It enables the developer to practice a 'higher' level of reuse than is usually the case in software engineering (i.e. knowledge engineering instead of software reuse). Moreover, it enables the developer to reuse and share application domain knowledge using a common vocabulary across heterogeneous software platforms. It also enables the developer to concentrate on the structure of the domain and the task at hand and protects him from being bothered too much by implementation details.

As a first step, a *domain-specific ontology profile* will be created to extract the essence of things in the problem domain. Domain ontologies are reusable vocabularies of concepts and relationships within a domain, activities taking place in that domain, and theories and elementary principles governing that domain [1]. In this section, we will discuss representative ontologies in the domain of capital markets in which F-TRADE works . Entities and industrial requirements discussed in Sect. 10.8 are the main elements in building domain ontologies. Here, visual models will be built to represent the ontologies of the capital markets. The visual models present ontological structure and relations between ontologies of the problem domain.

An excerpt of the ontology structure relevant to the project requirements is illustrated in Fig. 12.2. This ontology structure refers to financial information from the market microstructure [19], financial ontology from Teknowledge [20], and information from SIRCA [21]. In capital markets, subclasses of exchange type include futures exchange, index exchange, stock exchange, option exchange, and OTC exchange. Stock exchange consists of the stock market, in which there are auction markets and dealer markets. In the dealer market, the instrument type includes classes of foreign exchange, index, fixed income, stock, futures, money market, and options. For instrument stock, the financial order can be limit order, market order, or stop order. For market order, the order operation can be one of five types: amend, enter, trade, delete, and cancel. For every trade, it includes information about price, dealer, date, time, volume, and so forth.

Ontology elements in the capital market ontology base can be further refined to low-level ontologies. Figure 12.3 shows some of the refined ontologies in relation to ontology elements in Fig. 12.2. For instance, index consists of stock index and other indices. Stock index can be instantiated as NASDAQCompositeIndex for NASDAQ or other indices for individual markets. Price can be instantiated as bid price, ask price, open price, or other prices.

Synonymity and analogy are very common taxonomic phenomena in the business world. For instance, the item "long term" in the stock market is also called "long run" in foreign exchange. To express this situation in the ontological taxonomy, we name a term that is commonly used as a "*leading item*," and all other words sharing the same or similar meaning are called "*substitute items*." For instance, the leading item "closing price" may be substituted by "close price" or "daily price" in the capital markets. Relationships between domain ontologies and corresponding items in the problem-solving system will be dealt with in Sect. 12.3. The taxonomy and representation of domain ontologies will be discussed in Sect. 12.4.

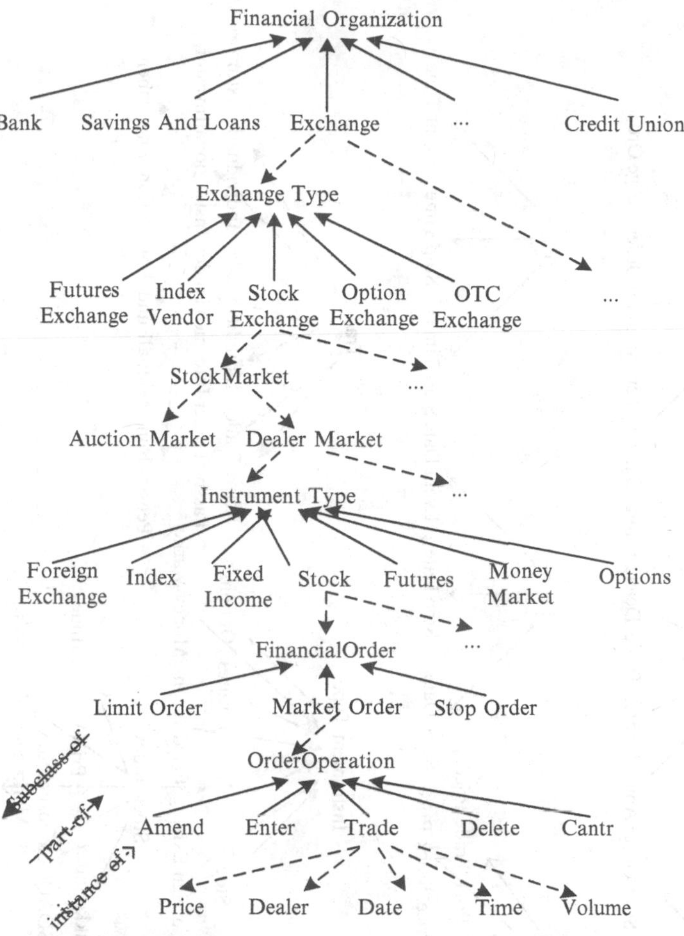

Fig. 12.2 Excerpt of domain ontologies in capital markets

12.2.3 Problem-Solving Ontology

Problem-solving methods (PSM) [22] are proposed to share and reuse knowledge and reasoning behavior across domains and tasks for the purpose of solving the problem. In the PSM, we explicitly define all ontologies that contribute closely to problem-solving methods as the *problem-solving ontology* (PSO) profile. Problem-solving ontologies capture knowledge in the following aspects:

1. Task ontologies
2. (Problem-solving) Method ontologies from the problem-solving perspective

To solve a problem, task ontologies and method ontologies are proposed to describe tasks and related methods for fulfilling those tasks. *Task ontologies* [11] provide a systematic vocabulary of the terms used to solve problems associated

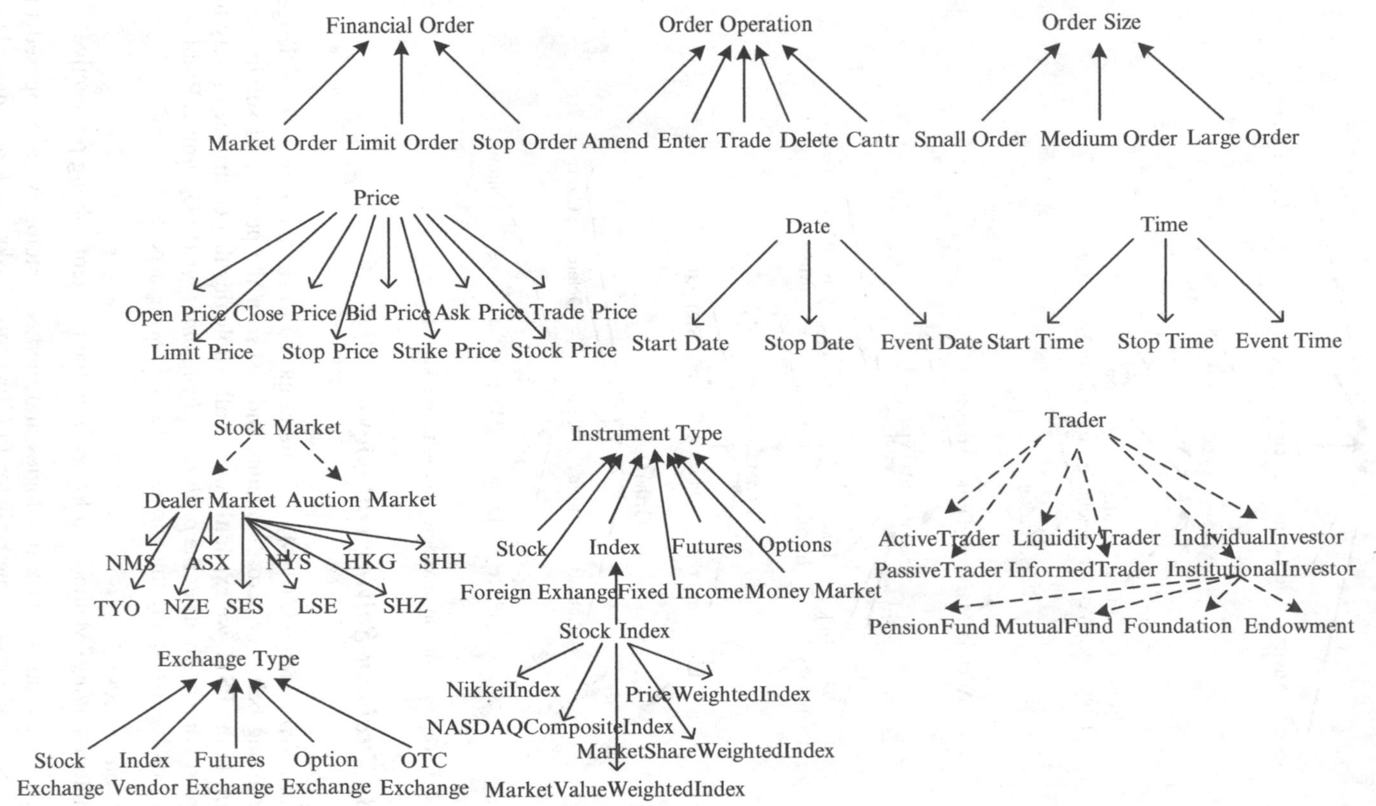

Fig. 12.3 Excerpt of domain ontologies according to internal structure

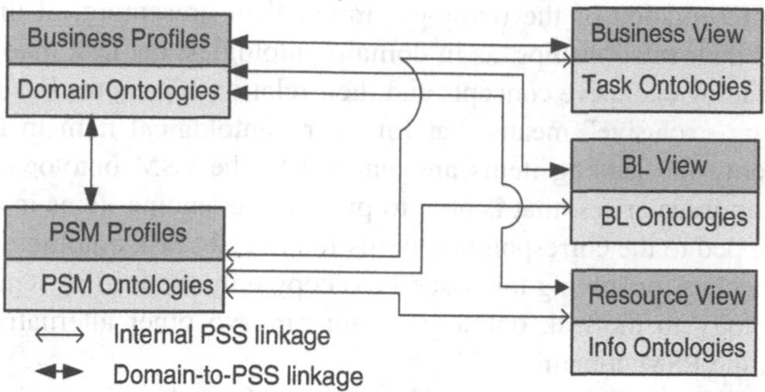

Fig. 12.4 Ontology-based system

with particular tasks (either domain independent or domain dependent). *Method ontologies* [22] describe the concepts used by the method on the reasoning process as well as the relationships between such concepts. For a given domain, reusable task ontologies can be built as *domain-task ontologies*; these ontologies are not across domains. *Application ontologies* [13] contain the necessary knowledge for modeling a particular application.

To our given problem—building infrastructure to support trading and mining—we classify the problem-solving ontology profile into the following ontological categories in terms of problem domains (business) and the interaction between problem-solver and business:

1. Business-oriented task ontologies
2. Problem-solving-oriented business logic ontologies
3. Information-oriented resource ontologies

Figure 12.4 shows the structure of ontological space in the problem domain and problem-solving domains. The domain ontologies in the figure have been discussed in the above section. In the following section, we will discuss the three types of PSM ontology individually.

12.2.3.1 Business-Oriented Task Ontologies

Business-oriented task ontologies are all mapped ontologies from problem domain ontologies to the problem-solving system. In our case, business-oriented task ontologies describe all the concepts, and the relationships between them, that we build for the problem-solving system; these ontologies are mapped from the domain ontologies to the problem-solving method ontologies.

The main difference between domain ontologies and business-oriented task ontologies is that the former are the real embodiment of the business domain, while the latter are the mirror of the domain ontologies to the problem-solving system. The mutual feature is that both are the embodiment of the problem domain.

In the representation of the ontology, rather than presenting all the potential concepts and their relationships, as in domain ontologies, the task ontologies only capture all leading exclusive concepts and their relationships in the business world. Here, "leading exclusive" means that for every ontological item in the domain ontologies, only the leading items are mapped to the PSM ontological domain. Thus, when an issue arises, that is how to present the leading items in the domain ontology mapped to the corresponding items in the PSM task ontology. One of the easiest approaches for solving this issue is to copy only the leading items from the domain ontology to the task ontological domain. No other alternatives will be embodied in the PSM domain.

Another problem is how to manage the mapping from domain ontologies to task ontologies. This requires strategies to manage the ontological space of task ontologies and the mapping between domain ontologies and task ontologies. The former will be discussed in Sect. 12.4; the latter will be described in Sect. 12.5.

Business-oriented task ontologies are embodied in the PSM ontologies. The visual model of task ontologies is similar to that in the domain ontologies, except that only the leading items will be positioned and linked for the task ontologies. The formal specifications will be discussed in Sect. 12.4.

12.2.3.2 Business Logic Ontologies

Business logic ontologies consist of ontological items for the management and execution of business logic in the problem-solving system. The following aspects are included in the ontological base:

1. Ontological elements consisting of the structure and organization of all relevant ontological items for the problem-solving system; this is related to system architecture and design patterns.
2. Ontological elements related to the workflow and processes; this may concern logical objects and processes of the system. Excerpts of the business logic ontologies are introduced below.

In the proposed F-TRADE, the basic design pattern follows the updated requester–mediator–provider (RMP), i.e., model–view–controller (MVC) computing model; this architecture requires the organization of the computing system in terms of M, V, and C. For each functional unit, therefore, it will be decomposed into three classes that are linked by logical relationships. On the level of knowledge management, some ontological elements must be specified for managing instantiated objects in terms of the computing model. For instance, the activity of executing an algorithm will be decomposed into four classes such as *AlgoExeAction*, *AlgoExeModel*, *AlgoExeController*, and *AlgoExeView* shown in Fig. 12.5a. The logical interactions are also shown in the figure.

The organization of the problem-solving system is reflected in the user interfaces. In Fig. 12.5b, the whole system is organized as five centers: Administration Center, Algorithm Center, Service Center, Control Center, and User Center. In each

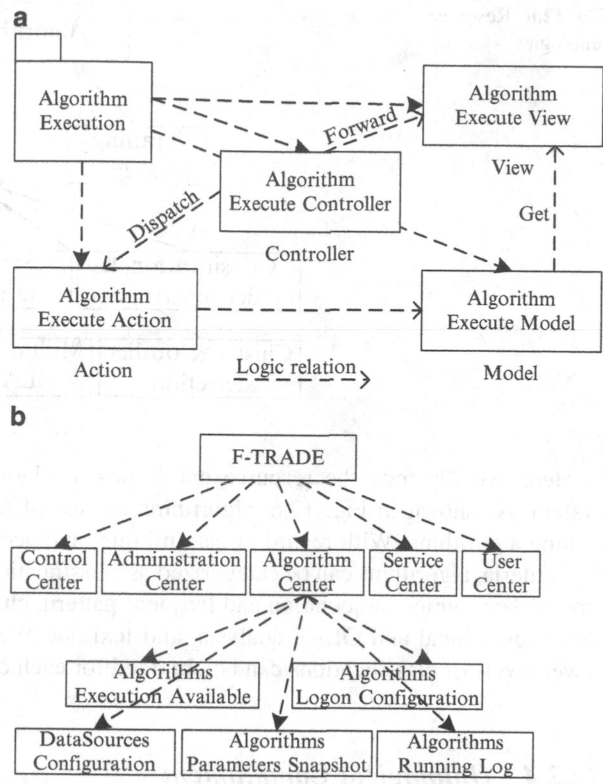

Fig. 12.5 (**a**) Business logic ontologies - Algorithm Execution Controller. (**b**) Business logic ontologies – F-Trade

center, specific entities and relationships are distributed at different levels from top down. For instance, with respect to the Algorithm Center, there are five lower levels: Algorithm Execution Available, Algorithms Logon Configuration, DataSources Configuration, Algorithms Parameters Snapshot, and Algorithms Running Log; the ontological elements for them are actually defined and arranged in the knowledge base.

12.2.3.3 Resource Ontologies

Resource ontologies capture all objects and relations between objects of all resources in the problem-solving system. Resources in an agent-based problem-solving system include the following:

1. Domain database storing the capital market data
2. Knowledge base storing all algorithms, models, rules, and the like
3. System base keeping all information for the configuration of the system and so forth

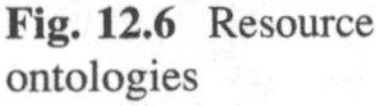

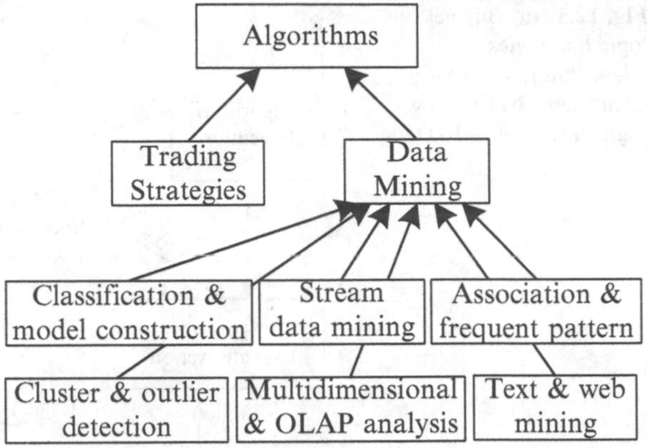

Fig. 12.6 Resource ontologies

Here, we illustrate the resource ontologies of algorithm management in the system. As shown in Fig. 12.6, algorithms consist of trading strategies and data mining algorithms. With regard to data mining, and according to some classification criteria, algorithms can be categorized as classification and model construction, stream data mining, association and frequent pattern, cluster and outlier detection, multidimensional and OLAP analysis, and text and Web mining. In some cases, lower levels of classifications can be obtained for each of these categories.

12.2.4 Ontological Commitment

Most of the time, a term or concept used in natural language has several meanings. The proper mapping between an ontology concept in natural language and its meanings in a given situation is quite important for the precise understanding and sharing of ontologies. This is investigated by ontological commitment. Ontological commitment defines agreement to use a shared ontology library in a coherent and consistent way. It determines precisely what meaning an ontology concept has in a given situation. It also permits ontologists to specialize in and instantiate the ontology as required with flexibility.

A formal specification (as shown in Formula (12.1)) is defined here for managing the abovementioned mapping, specialization, and instantiation in a given scenario.

Definition 12.1 The formal formula of an ontological commitment is a five-element tuple consisting of five attributes:

$$OC = (C, O, R, P, S). \tag{12.1}$$

where:

$C = \{c_i \mid i \in I\}$ is the set of domain-specific concepts in the given domain; it could also be the key business concept (word/phrase) entered by an end user for his/her preferences.

$O = \{o_j \mid j \in J\}$ includes a set of candidate ontologies (could also be meanings of the concept) in the problem-solving system for the given problem; it could also be the ontology element in the underlying mediator ontology base, which the user considers to be relevant and is mapped to a given c_i.

$R = \{r_i \mid i \in I\}$ is the set of semantic relationships between a domain-specific concept and its corresponding problem-solving ontology; it holds semantic relationship between c_i and o_i. $P = \{p_i \mid i \in I\}$ is set of user profile properties, which manages the type of association and correlation between business concept c_i and ontology element o_i.

$S = \{s_i \mid i \in I\}$ contains measures for ontological commitment, which manages the accuracy of coherence and consistency between business concept c_i and ontology element o_i.

In the above definition, I and J are two separate positive integer sets.

Definition 12.2 An atom item of ontological commitment can be explicitly and formally defined as follows:
<ontocommit> := (*<relation>* (*<concept>*, *<ontology>*) : *<property>*, *<similarity>*)*
<relation> ::= (Instantiation | Aggregation | Generalization | Substitution | Disjoin | Overlap | Association)
<concept> ::= ({*<name>* | *name* ∈ C})
<ontology> ::= ({*<name>* | *name* ∈ O})
<property> ::= (SingleUnrestricted | SingleRestricted | MultipleUnrestricted | MultipleRestricted)
<similarity> ::= ({ s | s ∈ [0, 1]})

A problem domain-oriented user-defined concept is an item in the business-oriented knowledge base. It is a reference ontology which allows the collection and retention of knowledge based on the meaning for the particular user of a keyword or key phraser. For each ontology element o_j in the problem-solving ontology base, there may be *several* related business concepts c_i in the user profile, since different users may have individual meanings for the same keyword/phrase for different interests in several subject domains.

For the administration of mapping and transformation between a business concept and its related ontology element(s), therefore, a *user profile property p* is introduced. The *user profile property* defines the type of association and correlation between the business concept and ontology element, namely, the number and constraint on a pair of $< o_j: c_i >$. Basically, the following four properties—single unrestricted, single restricted, multiple restricted, and multiple unrestricted—are used to describe the association between concepts and ontologies in two given domains:

- *Single Unrestricted* (SU): A single unrestricted property means that in the pair of $< o_i: c_i >$, an ontology element can have only one corresponding business concept, but any word/phrase can be entered.

- *Single Restricted* (SR): As a single restricted property, an ontology element can have only one partner of business concept, but the value is restricted to selection from a predefined list.
- *Multiple Unrestricted* (MU): For an ontology element in a multiple unrestricted property, the relevant business concepts can be multiple values, and any words/phrases can be entered.
- *Multiple Restricted* (MR): Unlike multiple unrestricted, users are restricted to selecting keywords/phrases from a predefined list.

In the real world of business operations, it may be difficult to precisely describe the extent to which the mapping between a business concept and an ontology element exists in terms of semantic meaning. As a personalized option, an end user (normally a domain expert or a knowledge engineer) can mark the *similarity* between an o_i and a c_i by assigning a heuristic measure s ($s \in [0,1]$), so that she/he can further rate the accuracy of the given semantic description. Here, $s = 0$ stands for a complete mismatch, whereas $s = 1$ stands for complete coherence.

The matching measure s in user profiles allows us to flexibly alter the precision of the search by setting the threshold factor. Only the business concepts and/or ontology elements from the user profile knowledge base whose similarity value is greater than the user-customized threshold will be used for keyword/phrase search and ontology discovery in the transformation; this has been discussed in Chap. 11 "Agent Services-Oriented Detailed Design". Ontological semantic relationships will be discussed in the next section.

12.3 Ontological Semantic Relationships

A classification benchmark is required to classify the relations among ontology concepts. In information retrieval systems, most classifications are based on taxonomies and semantic networks, and some make use of linguistic ontologies. The advantage of the latter is that the exploited semantic relationships among ontology concepts are richer than taxonomy relationships. As an observation from real requirements in industry, it would be more helpful for us to consider both lexical and semantic factors in the classification and organization of ontology elements.

In Sect. 12.2, we discuss ontology namespace, which focuses more on lexical taxonomies and the subtype/supertype relationship between ontology concepts and services. In this section, we first use ontological semantic relationships to specify semantic dependencies between different ontology concepts in the same namespace, the keywords/phrases in the user profiles, and the ontology concepts in domain ontology and problem-solving ontology.

As shown in the depiction of domain-specific ontology (Figs. 12.2 and 12.3), there are mainly three types of relationship, *subclass_of*, *part_of*, and *instance_of*, existing in the ontology elements for a specific domain. Other relationships like *similar_to*, *relate_to*, *disjoin_with*, and *overlap_to* may also exist in domain ontologies and problem-solving ontologies. They are categorized as seven classes of the semantic relationships: instantiation, aggregation, generalization, substitution, disjoin, overlap, and association, respectively. Their informal definitions are given below, in which the symbol O or o refers to an ontological element.

- *instance_of* (O, o): Two ontological elements O and o, where o is an instance of the ontological class O.
- *part_of* (O_1, O_2): Two ontological elements O_1 and O_2, where *O_2 is part/member of O_1 or O_1 is made of O_2.*
- *is_a*(O_1, O_2): The relationship between $O2$ and $O1$ is subtype/supertype or subsumption, i.e., O_2 is an O_1 or O_2 is a kind of O_1. The *is_a* is sometimes called *subclass_of*.
- *similar_to*(O_1, O_2): This indicates that O_1 and O_2 are identical or to a large degree similar; in this case, O_2 can be substituted by O_1.
- *disjoin_with*(O_1, O_2): This indicates that O_2 is independent of O_1.
- *overlap_to*(O_1, O_2): This represents that there is something shared by both O_1 and O_2, but the share percentage is not high enough for one to be substituted by the other.
- *relate_to*(O_1,O_2): This is the predicate over the ontologies O_1 and O_2; it represents a relationship between O_1 and O_2 which cannot be specified by any of the above six, but which shows that they are associated with each other by user-defined linkage.

Table 12.1 summarizes the definitions of ontology concepts and services for each of the ontology relationships.

Example 12.1 The following expression shows that ontology item billing consists of LocalBilling and RemoteBilling, while Localbilling and Remotebilling are in a Disjoin relationship:

disjoin_with(LocalBilling, RemoteBilling)
part_of(Billing, LocalBilling)
part_of(Billing, RemoteBilling)

Table 12.1 Ontology relationships

Relations	Description	Predicates
Instantiation	When ontology o is an instance of ontology O	*instance_of*
Aggregation	When ontology O_2 is part of ontology O_1	*part_of*
Generalization	When ontology O_2 is a kind of ontology O_1	*is_a* (*subclass_of*)
Substitution	When ontology O_2 is equal to ontology O_1	*similar_to*
Disjoin	When O_2 and O_1 have no share in common	*disjoin_with*
Overlap	When ontology O_2 and O_1 have partial share, but one cannot be replaced by the other	*overlap_to*
Association	When O_2 is related to O_1 in a relation except of the above six	*relate_to or user-defined associating predicates*

12.4 Ontological Representation

12.4.1 Ontology Modeling Techniques

The representation of ontologies has the following aspects:

1. How to make the representation: This includes the visual and formal representations.
2. How to organize the ontological space: This can be arranged in terms of concepts and their semantic relationships.

To this end, the modeling of ontologies can be highly informal if they are expressed in natural language, semi-informal if expressed in a restricted and structured form of natural language, semiformal if expressed in an artificial and formally defined language, and rigorously formal if they provide meticulously defined terms with formal semantics, theorems, and proofs of properties such as soundness and completeness [1].

In Sect. 12.2, ontologies are visually presented as a tree, where both the concepts and main semantic relationships are displayed on the figure. The main interlink relationships include meronymy (*part_of*), hypernymy (*subclass_of*), and instantiation (*instance_of*), which further capture four other types of semantic relations such as *disjoin_with*, *similar_to*, *overlap_to*, and *relate_to*. The opposite relation to hypernymy is hyponymy (*superclass_of*); similarly, holonymy (*has_a*) is contrary to meronymy. These two relations will not be discussed or reflected in this work.

Besides the visual representation, formal or semiformal specifications can present the ontological structure more precisely. The modeling of ontologies can be formally expressed through the following:

1. By AI techniques such as frames and first-order logics
2. By description logics theory such as KL-ONE [37] and LOOM [23]
3. By software engineering techniques such as UML and CASE tools
4. By database technique such as entity relationship diagram

In our work, formal specifications are presented in frames and first-order logics because:

1. These two mature in the area of artificial intelligence with reasoning and referring mechanisms available.
2. They are consistent with the formal specifications used in the OOA.

Based on these techniques, several ontological languages have been proposed for implementing ontologies. Traditional ontological languages include Ontolingua [24], KIF [25], LOOM [23], OKBC [26], OCML [27], and FLOGIC [28]. Based on existing Web markup languages like HTML and XML, new ontology markup languages have emerged; they are SHOE [29], XOL [30], RDF [31], RDF Schema (RDFS) [32], DAML + OIL [33], and OWL [17]. Table 12.2 summarizes the comparison of t XOL, OIL, DAML + OIL, and OWL [34]. In a sense, OWL seems to have a better chance of survival, while adaptations may be required for it to be easily used for future usage.

Table 12.2 Feature comparison of new ontology languages

	XOL	OIL	DAML+OIL	OWL
XML based	+	+		
RDF(S) based		+	+	+
Frame based	+	+		
DL based		+	+	+
Layered		+		"+"
Alive	No	No	No	Yes

Ontology languages are specified on the basis of underlying representation techniques. Five types of representation technique are commonly used to support the abovementioned ontology languages. They are

1. First-order predicate logics
2. Frame-based modeling primitives
3. Description logics (DL)
4. XML-based syntax
5. RDF(S)-based syntax

Two or more of these five are sometimes combined to express complex ontologies and ontological relations. For instance, OIL is represented by a combination of all five modeling techniques.

In our project, DL and RDF(S) are two alternatives for the representation of ontologies since some ontologies are stored and exchanged in the knowledge base, but others in XML. In the following section, we introduce their respective formal grammar.

12.4.1.1 DL-Based Ontological Grammar

Description logics [37], also known as terminological logics, stem from semantic networks and define formal and operational semantics for ontologies. They support intentional concepts and properties of concepts and allow the construction of composite descriptions. Systems based on DL include KL-ONE, CLASSIC, CRACK, BACK, FLEX, K-REP, LOOM, KRIS, and YAK. The grammar for expressing the ontological items is as follows:

```
/*DL-based grammar*/
<onto-item> ::= <item-name> |
(<logic-connective> <onto-item>+) |
(<bound-connective> <bound-number> <property>) |
(<cardinality-connective> (<property> <constraint>)) |
(<semantic-relationship> (<onto-item>+ | <property>+) (<onto-
item>+ | <property>+))
<item-name> ::= <symbol>
<logic-connective> ::= (AND | OR)
<bound-connective> ::= (AT-MOST | AT-LEAST)
<cardinality-connective> ::= (OO | OM | MO | MM)
<constraint> ::= (* | + | ? | " | ")
```

where cardinality connectives OO, OM, MO, and MM are as described in Sect. 7.4. Constraints *, +, and | are defined in Sect. 8.3 on first-order logic. The symbol "?" points to the situation as zero or one appearance of an item.

For the storage and management of ontologies, ontology code will be specified to distinguish every ontological item in the ontology structure. Coding systems such as NAICS (North American Industry Classification System) [35] and UNSPSC (United Nations Standard Products and Services Codes) [36] can be followed. Six digits are used for encoding; each level adds two digits to its previous level. To this end, a system of Capital Market Ontology Code needs to be built for the systematic classification of ontologies.

12.4.2 Representing Domain Ontologies

Domain ontologies capture general knowledge about concepts, terminology, and relationships from the viewpoint of business in the world. The objective of domain ontologies is to present users with domain-specific concepts, objects, and relationships between them (namely, their business relationships and rules). These ontologies are mainly used in the interactions between human and computer; they support the interaction and user interfaces in a user-friendly business-oriented profile.

In general, a name for an ontology in the business domain is not fixed. A specific element in capital market ontologies can take different names according to different organizations and businesspersons. For instance, a large order in one place is called block trade in another; opening price is also called open price; closing price is sometimes called close price; close date is also called end date or stop date. A synonym item can be added if necessary to manage the synonymous phenomenon in the knowledge representation of ontologies.

To describe the synonym phenomenon, one key term in the synonym list is nominated as the *leading item* (LI) to act as the representative of all items in a synonymous group. All other items are named *substitute items* (SI) and may be used by some staff in business life, but they have the same meaning as the leading item.

The output of the domain ontologies is a domain-specific conceptual ontology base, which includes a Concept Category Directory (CCD). The CCD, which is a hierarchical concept tree encompassing business namespace, lists and defines all terms and relationships abstracted in the daily business and generates a list of candidate concepts and expressions based on the business process and the activities that took place in the user's daily work.

Definition 12.3 Concept Category Directory Entry: A CCD entry can take form as "terms" or "relationships." A term or a relationship is expressed in the form of key–value tuples (KVT) as <*A*>k: v <\A>. A term or a relationship consists of a unique leading item (LI, an identifier) and optionally multiple substitute items (SI, recommended candidate concepts).

A synonymous group in the domain ontologies can be represented as follows:

```
/*CCD entry for domain ontologies */
<CCD entry> ::= (<term> | < relationships>)*
<term> ::= (<name>, <type> [, property] : <value>, <type>)*
<relationship> ::= (<name>, <type> [, property] : <value>, <type>)*
```

where the property is defined in Sect. 12.2 and the * donates one or more instances in the formula. Thus, a CCD may take the following formula:

```
{{{LI_name, LI, MO}:{LI_Value, Type}},
{{SI_name1, OM}:{SI_Value1} | {SI_name2, OM}:{ SI_Value2} | ...}}.
```

For instance, a term closing price can be informally expressed as follows:

```
{{{Closing_Price, LI, OO} : {13.23, Float}}.
{{Close_Price, SI, OM}:{13.23, Float} | {Daily_Price, SI, OM}:{13.23,
Float}}}.
```

or it can be formally expressed as follows in terms of the DL-based grammar. This means that there is at most one value of the closing price for a stock in a day; if a value exists, then its value is in float.

```
Closing Price ::=Closing_Price (OR Close_Price Daily_Price) (AT-MOST
  1 Stock_Code) (OO Float ?)
```

12.4.3 Representing Problem-Solving Ontologies

As discussed in Sect. 12.3, PSO consists of task ontologies and method ontologies from the problem-solving perspective. They can further be divided into ontologies for tasks, methods, business logics, and resources, respectively, from the interaction between business and the problem-solving perspective. The relationships between these four classes are as follows. A task is fulfilled by a method or methods. A method is instantiated into business logics and supported by resources. In most conditions, a task is divided into multiple subtasks; these subtasks are satisfied by alternative methods. Correspondingly, these alternative methods are implemented by alternative business logics and relevant resources. Figure 12.7 shows the relationships between these elements.

As shown in the figure, multiple (sub-) methods may be involved in a subtask. For instance, T_{1k} is fulfilled by methods M_2 and M_k. On the other hand, multiple business logics may be requested by one (sub-) model, while multiple resources may be relied on to execute a piece of business logic.

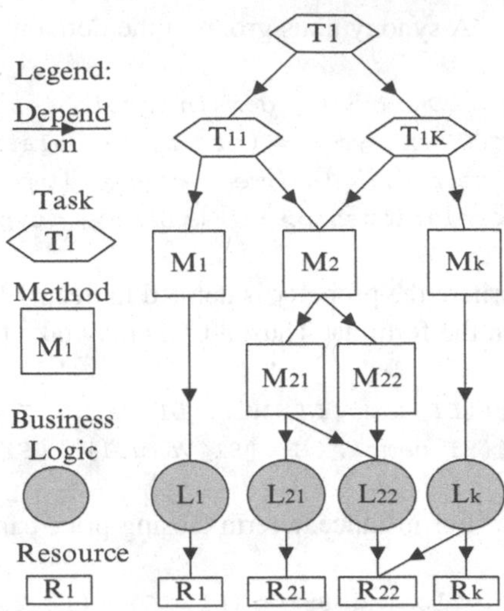

Fig. 12.7 Problem-solving ontological structure

For instance, the task Register Algorithm consists of three subtasks:

1. Check algorithm validity.
2. Fill in algorithm register ontologies.
3. Upload algorithm component.

The subtask Fill in algorithm register ontologies can be accomplished by human–computer interaction. Human–computer interaction is instantiated into the following business logic: algorithm registration ontologies are typed in through the user interface and r inserted into the ontological base. To support the typing and management of algorithm registration ontologies, a resource such as a database for storing the ontological base is required.

The above example encompasses set of ontological items for the task, the method, the business logic, and the relevant resource. A specification is built to represent these problem-solving ontologies, based on first-order logics as discussed in Chap. 8. The following shows the formula:

Definition 12.4 Problem-Solving Ontological Item: A problem-solving ontological item follows the specification of Concept Category Directory except that here the term name takes the form of one of the following four options: task, method, business logic, or resource.

A PSO item can be expressed as follows.

```
/*CCD entry for problem-solving ontologies*/
<CCD entry> ::= (<term> | < relationships>)*
<term> ::= (<name>, <type> [, property] : <value>, <type>)*
<name_type> ::= (task | method | business logic | resource)
<relationship> ::= (<name>, <type> [, property] : <value>, <type>)*
```

12.5 Ontological Semantic Aggregation and Transformation Cross Domains

The workflow of an ontology-based system (see Fig. 12.4) is as follows. Users from the business world interact with the system in user-defined business-oriented (or - domain-oriented) special terms. The ontology processing engine semantically aggregates and transforms these input terms to internal standard items used by the problem-solving system. Three aspects must be addressed to achieve the semantic aggregation and ontological transformation from user-defined keywords to ontological elements in the problem-solving ontological domain. They are:

1. Semantic aggregation between semantic relationships
2. Semantic aggregation of ontological items
3. Transformation from one ontology domain to another

All three types of transformation can occur in one ontological domain or in multiple domains. They are discussed in detail in the following three sections.

12.5.1 Semantic Aggregation of Semantic Relationships

Semantic aggregation of relationships studies whether there are rules for *transitivity*, *additivity*, and *antisymmetry* which can be performed between ontological semantic relationships. The aggregation of multiple semantic relationships simplifies the combination of semantic relationships and support to find the final reduced semantic relationships. This reduces the problem-solving sample space and speeds up the work of the engine.

Let $A(a)$, $B(b)$, and $C(c)$ be the arbitrary ontological items, where $A(a)$ means A or a. s, s_1, and s_2 are similarity values defined by users. Basic specifications are as follows:

Definition 12.5 "AND" and "OR" are logical connectives used to connect two ontological items which have the same grammatical function in a construction;

Definition 12.6 The resulting output of "(A AND B)" includes both A and B, while the output of "(A OR B)" is either A or B.

Definition 12.7 "(A AND B)" is equal to "(B AND A)"; similarly, "(A OR B)" is equal to "(B OR A)."

Definition 12.8 Boolean logic operators "$\wedge$" and "$\vee$" represent "and" and "or" relationships between semantic relationships or between logic formulae.

Definition 12.9 Similarity value s measures the degree to which two ontological items are related in a semantic relationship.

The metric s is usually used with relationships *similar_to*(), *overlap_to*(), and user-defined relationship *relate_to*(). For instance, *similar_to* (A, B, s_1) means that B is similar to A in a degree of s_1.

For all seven semantic relationships discussed in Sect. 12.3, rules will hold for semantic aggregation and combinations of the seven semantic relationships. The following shows an excerpt for some cases:

Rule 12.1 Let $A(a)$, $B(b)$, and $C(c)$ be associated by the instantiation relationship, then:

- ∀ (*instance_of* $(A, a) \wedge$ *is_a* $(B, A)) \Rightarrow$ *instance_of* (B, a)
- ∀ (*instance_of* $(A, a) \wedge$ *instance_of* $(B, a)) \Rightarrow$ *instance_of* $((A \text{ AND } B), a)$, which means a is an instance of both A and B

Here, a is an instance of class A, and A may be a subclass of class B.

Rule 12.2 Let A, B, and C be associated by the aggregation relationship, then:
- ∀ (*part_of* $(A, B) \wedge$ *part_of* $(B, C)) \Rightarrow$ *part_of* (A, C)
- ∀ (*part_of* $(A, B) \wedge$ *part_of* $(B, A)) \Rightarrow$ *similar_to*(A, B)
- ∀ (*part_of* $(A, B) \wedge$ *part_of* $(C, B)) \Rightarrow$ *overlap_to*$(A, C) \vee$ *similar_to*(A, C) depend on the intersection between A and C.

Rule 12.3 A, B, and C are associated by the generalization relationship, then:

- ∀ (*is_a*$(A, B) \wedge$ *is_a* $(B, C)) \Rightarrow$ *is_a*(A, C)
- ∀ (*is_a*$(A, B) \wedge$ *is_a*$(A, C)) \Rightarrow$ *is_a*$(A, (B \text{ AND } C))$
- ∀ (*is_a*$(A, B) \wedge$ *is_a*$(B, A)) \Rightarrow$ *similar_to*(A, B)

Rule 12.4 A, B, and C are associated by the substitution relationship, then:

- ∀ (*similar_to* $(A, B) \wedge$ *similar_to* $(B, C)) \Rightarrow$ *similar_to* (A, C)
- ∀ (*similar_to* $(A, B) \wedge$ *similar_to* $(A, C)) \Rightarrow$ *similar_to* (B, C)

Rule 12.5 A, B, and C are associated by the overlap relationship, then:

- ∀ (*overlap_to* $(A, B) \wedge$ *overlap_to* $(B, C)) \Rightarrow$ *overlap_to* $(B, (A \text{ AND } C))$

Accordingly, we can list many other aggregation rules for reducing combinations of the seven semantic relationships.

12.5.2 Semantic Aggregation of Ontological Items

Another semantic aggregation situation is to aggregate ontological items which are linked by logic connectors associated with semantic relationships. The objective of aggregating ontological items is to reduce the number of items and generate the resulting ontological items.

The rules for aggregating ontological items can be found in the above section. The following rules hold for semantic aggregation in some cases. These rules define the resulting logical output for each given input logical combination with a semantic relationship inside.

Rule 12.6

- $\forall$ (*A* AND *b*), $\exists$ *b* ::= *instance_of*(*B*, *b*)
  $\Rightarrow$ *A* AND *B*, the logical resulting output is *A* and *B*
- $\forall$ (*A* OR *b*), $\exists$ *b* ::= *instance_of*(*B*, *b*)
  $\Rightarrow$ *A* OR *B*, the logical resulting output is *A* or *B*

Rule 12.7

- $\forall$ (*A* AND *B*), $\exists$ *B* ::= *part_of*(*A*, *B*)
  $\Rightarrow$ *B*, the resulting output is *B*

Rule 12.8

- $\forall$ (*A* AND *B*), $\exists$ *B* ::= *is_a*(*A*, *B*)
  $\Rightarrow$ *B*, the resulting output is *B*

Rule 12.9

- $\forall$ (*A* AND *B*), $\exists$ *B* ::= *similar_to*(*A*, *B*)
  $\Rightarrow$ *A* OR *B*, the resulting output is *A* or *B*

Rule 12.10

- $\forall$ (*A* AND *B*), $\exists$ *B* ::= *disjoin_to*(*A*, *B*)
  $\Rightarrow$ *A* AND *B*, the resulting output is *A* and *B*

Rule 12.11

- $\forall$ (*A* AND *B*), $\exists$ *B* ::= *overlap_to*(*A*, *B*)
  $\Rightarrow$ *A* AND *B*, the resulting output is *A* and *B*

Rule 12.12

- $\forall$ (*A* AND *B*), $\exists$ *B* ::= *relate_to*(*A*, *B*)
  $\Rightarrow$ (*A* AND *B*) $\vee$ (*A* OR *B*), the resulting output is *A* and *B*, or either *A* or *B*; one of the three outputs will hold depending on the user-defined relationship

12.5.3 Transformation Between Ontological Items

This section discusses the transformation of ontological item to another in one or many domains. This could be a mapping from an arbitrary keyword in the business ontological domain to its relevant items in the problem-solving domain. The basic idea for the transformation of ontological items is as follows: given an input item, the candidate ontological items are checked by semantic relationship to find the most suitable candidate as the output item.

Rules for this transformation must be built so that the matched ontological item can be generated as output. The following lists some of the rules for the transformation, where C_i is an input item and O, O_1 and O_2 are candidate items in the target domain.

Rule 12.13

– $\forall C_i$, $\exists$: ($similar\_to(O, C_i) \vee is\_a(O, C_i) \vee instance\_of(O, C_i) \vee part\_of(O, C_i) \vee relate\_to(O, C_i)) \Rightarrow O$, the O is the output item

Rule 12.14

– $\forall C_i$, $\exists$: $(is\_a(O_1, C_i) \wedge is\_a(O_2, C_i)) \Rightarrow O_1$ AND O_2

Rule 12.15

– $\forall C_i$, $\exists$: $(part\_of(O_1, C_i) \wedge part\_of(O_2, C_i)) \Rightarrow O_1$ OR O_2

Rule 12.16

– $\forall C_i$, $\exists$: $(is\_a(O_1, C_i) \wedge part\_of(O_2, C_i)) \Rightarrow O_1$ AND O_2

Rule 12.17

– $\forall C_i$, $\exists$: $(similar\_to(O_1, C_i) \wedge similar\_to(O_2, C_i)) \Rightarrow O_1$ OR O_2

Rule 12.18

– $\forall C_i, C_j, (i \neq j)$, $\exists$: $(part\_of(O, C_i) \wedge part\_of(O, C_j)) \Rightarrow O$

Rule 12.19

– $\forall C_i, C_j, (i \neq j)$, $\exists$: $(is\_a(O, C_i) \wedge is\_a(O, C_j)) \Rightarrow O$

Rule 12.20

– $\forall C_i, C_j, (i \neq j)$, $\exists$: $(relate\_to(O, C_i) \wedge relate\_to(O, C_j)) \Rightarrow O$

12.6 Summary

Concepts, relationships, structures, and systems capture the main objects in a system, which are better managed in terms of ontological engineering. In this chapter, we showed how ontological engineering concepts and tools are used to conceptualize and organize objects and object relationships in a complex system.

Ontology profiles were built for business ontology, and problem-solving ontology, semantic relationships, and ontological representation were introduced to model concepts and relationships. Ontological semantic aggregation and transformation across domains were discussed to build a mapping relationship between concepts and relationships in different domains.

In the next chapter, case studies are introduced to apply computing and engineering techniques to a real-life system. This involves organization-oriented system analysis, organizational relationship model, organizational rationale model, formal analysis, and agent service-based design.

References

1. Gomez-Perez, A., Fernandez-Lopez, M., Corcho, O.: Ontological Engineering. Springer, London (2004)
2. Gruninger, M., Lee, J.: Ontology applications and design: introduction. Commun. ACM, ACM Press **45**(2), 39–41 (2002)
3. XC00086. FIPA ontology service specification. Accessed 2005
4. Storey, V.C.: Understanding Semantic Relationships. VLDBJ **2**(4), 455–488 (1993)
5. Cao, L.B., Luo, C., Luo, D., Liu, L.: Ontology services-based information integration in mining telecom business intelligence. In: Proceedings of PRICAI04, pp. 85-94. Springer, Berlin/Heidelberg (2004)
6. Kalfoglou, Y., Schorlemmer, M.: Ontology mapping: The state of the art. Knowl. Eng. Rev. ACM Press **18**(1), 1–31 (2003)
7. Neches, R., Fikes, R.E., Finin, T., Gruber, T.R., Senator, T., Swartout, W.R.: Enabling technology for knowledge sharing. AI Mag **12**(3), 36–56 (1991)
8. Gruber, T.R.: A translation approach to portable ontology specifications. Knowl. Acquis. **5**(2), 199–220 (1993)
9. Guarino, N., Giaretta, P.: Ontologies and knowledge bases: Towards a terminological clarification. In: Mars, N. (ed.) Towards Very Large Knowledge Bases: Knowledge Building and Knowledge Sharing (KBKS95), pp. 25–32. IOS Press, The Netherlands (1995)
10. Fridman-Noy, N., Hafner, C.D.: The state of the art in ontology design: A survey and comparative review. AI Mag **18**(3), 53–74 (1997)
11. Mizoguchi, R., Vanwelkenhuysen, J., Ikeda, M.: Task ontology for reuse of problem solving knowledge. In: Mars, N. (ed.) Towards Very Large Knowledge Bases: Knowledge Building and Knowledge Sharing (KBKS95), pp. 46–59ß. IOS Press, The Netherlands (1995)
12. http://suo.ieee.org
13. Van Heijst, G., Schreiber, A.T., Wielinga, B.J.: Using explicit ontologies in KBS development. Int. J. Hum. Comput. Stud. **45**, 183–292 (1997)
14. Tijerino, Y.A., Mizoguchi, R.: MULTIS II: enabling end-users to design problem-solving engines via two-level task ontologies. In: Aussenac, N., et al. (eds.) Proceedings of the 7th European Workshop on Knowledge Acquisition for Knowledge-Based Systems, pp. 340–359. Springer (1993)
15. Studer, R., Eriksson, H., Gennari, J.H., Tu, S.W., Fensel, D., Musen, M.: Ontologies and the configuration of problem-solving methods. In: Gaines, B.R., Musen, M.A. (eds.) Proceedings of the 10th Banff Knowledge Acquisition for Knowledge-Based Systems Workshop, Banff (1996)
16. Farquhar, A., Fikes, R., Rice, J.: The ontolingua server: a tool for collaborative ontology construction. Int. J. Hum. Comp. Stud. **46**(6), 707–727 (1997)

17. Dean, M., Schreiber, G.: OWL web ontology language reference, W3C working draft, http://www.w3.org/tr/owl-ref. Accessed 2005
18. Guarino, N.: Formal ontology in information systems. In: Guarino, N. (ed.) 1st International Conference on Formal Ontology in Information Systems (FOIS'98), pp. 3–15. IOS Press, The Netherlands (1998)
19. Stoll, H.R.: Market Microstructure. Working paper, financial markets research center. Accessed 2012
20. http://einstein.teknowledge.com:8080/FinancialOnt/, or http://ontology.teknowledge.com/
21. http://www.sirca.org.au/
22. Chandrasekaran, B., Josephson, J.R., Benjamins, V.R.: Ontologies: What are they? Why do we need them? IEEE Intell. Syst. Appl., **14**(1), 20–26 (1999), Special Issue on Ontologies
23. MacGregor, R.: Inside the LOOM classifier. ACM SIGART Bull. **2**(3), 88–92 (1991), Special Issue on Implemented Knowledge Representation and Reasoning Systems
24. http://ontolingua.stanford.edu
25. Genesereth, M.R., Fikes, R.E.: Knowledge interchange format. Version 3.0 reference manual. Technical report logic-92-1, Stanford University, 1992
26. Chaudhri, V.K., Farquhar, A., Fikes, R., Karp, P.D., Rice, J.P.: Open knowledge base connectivity 2.0.3. Technical report, SRI International/Stanford University, 1998
27. Motta, E.: Reusable Components for Knowledge Modeling: Case Studies in Parametric Design Problem Solving. IOS Press, The Netherlands (1999)
28. Kifer, M., Lausen, G., Wu, J.: Logical foundations of object-oriented and frame-based languages. J. ACM **42**(4), 741–843 (1995)
29. http://www.cs.umd.edu/projects/plus/SHOE/
30. Karp, P.D., Chaudhri, V., Thomere, J.: XOL: an XML-based ontology exchange language. Technical report, Version 0.3 (1999)
31. Lassila, O., Swick, R.: Resource Description Framework (RDF) Model and Syntax Specification. W3C Recommendation, World Wide,Web Consortium, Cambridge, MA (1999)
32. Brickley, D., Guha, R.V.: RDF vocabulary description language 1.0: RDF schema. W3C Recommendation (10 February 2004).
33. Harmelen, F., Patel-Schneider, P., Horrocks, I.: Reference Description of the DAML + OIL (March 2001) Ontology Markup Langauge, March 2001. http://www.daml.org/2001/03/reference.html
34. Fensel, D.: Ontologies: A Silver Bullet for Knowledge Management and Electronic Commerce, 2nd edn. Springer, Berlin/Heidelberg (2004)
35. http://www.naics.com
36. http://www.unspsc.org
37. Brachman, R.J., Schmolze, J.G.: An overview of the KL-ONE knowledge representation system. Cogn. Sci. **9**(2), 171–216 (1985)

Chapter 13
OSOAD Case Study

13.1 Organization-Oriented System Analysis

In Chap. 9, an integrative modeling framework was proposed for modeling open agent systems. In Chap. 6, we have advocated an ORGANISED framework to undertake the organizational abstraction of an agent organization in terms of organizational metaphor. Collections of visual model-building blocks were subsequently specified and defined in Chap. 7 for the concrete, graphic analysis of an open agent organization. Chap. 8 presented a system of formal representation of the ORGANISED framework through temporal logics. This body of work constitutes organization-oriented analysis system of open agent systems in terms of organizational metaphor.

This chapter will focus on a case study that considers how to utilize the above integrative model-building blocks to model the F-TRADE infrastructure both diagrammatically and formally. We will not build detailed and individual models for every member and relationship in F-TRADE to generate a full and concrete picture in terms of the ORGANISED framework. Partly this is because most elements have been dealt with in the preceding chapters; the remaining issue concerns time rather than the technical aspect. More significantly, we believe there are important problems that might emerge in the process of linking those building blocks to generate a high-level view of the system. For instance:

- Problems could arise from the process of instantiation and combination of the model-building blocks in constructing the complicated high-level structure of the system.
- They might reflect on theoretical completeness and integrity; for instance, whether mechanisms for refinement and model checking are available or how to support them.

L. Cao, *Metasynthetic Computing and Engineering of Complex Systems*,
Advanced Information and Knowledge Processing,
DOI 10.1007/978-1-4471-6551-4_13

- They might also result from issues of theoretical soundness; for instance, the incompatibility of the current OOA system with scenarios in real open agent systems.

We will therefore pay more attention to the above problems in the sections that follow. The main content of this chapter concerns:

1. Building a high-level visual *organizational relationship model*
2. Building a high-level visual *organizational rationale model*
3. Formally analyzing elements in the above visual models
4. Refinement of the analysis with scenarios

In building the high-level organizational relationship and rationale models, we adopt the framework of the SD model and SR model in *i** and TROPOS. However, we update and extend them on the basis of our ORGANISED framework and the model-building blocks. In our system, the SD model is a subset of the organizational relationship model. In fact, the SD model corresponds to our organizational dependency model, which encompasses members linked mainly by the dependency relationship. The SR model can correspondingly be part of our organizational rationale model.

13.2 Organizational Relationship Model

At this stage, the modeling support for organizational relationships is still under investigation. For instance, the four structural rules can be instantiated, but more precise model-building blocks are still under development, with the exception of dependency. For this situation, the dependency relationship has been the main analytical object in building our organizational relationship model. We introduce the main elements and the procedure we follow to build the model with the F-TRADE case study.

The organizational relationship model embodies and presents both functional and nonfunctional requirements in terms of organizational relationships. The main model-building blocks utilized to develop the organizational relationship model include actors, goals, and structural rules. The actors can be agent services and their roles, and the workspace, entities and resources on demand. The goals can be those for nonfunctional requirements (softgoal in TROPOS), and their subgoals or tasks and subtasks. All structural rules will be used on demand. Detailed dependencies such as single dependency, bilateral dependency, and/or composition dependency will be the main forms in the analysis of relationships. The system boundary marked by broken lines is also shown on the model, which clearly separates the organization from the system environment.

For simplicity, we analyze and present only the functional dependencies in the following modeling. Figure 13.1 shows an excerpt of the functional organizational relationship model for the proposed F-TRADE. The main human actors include

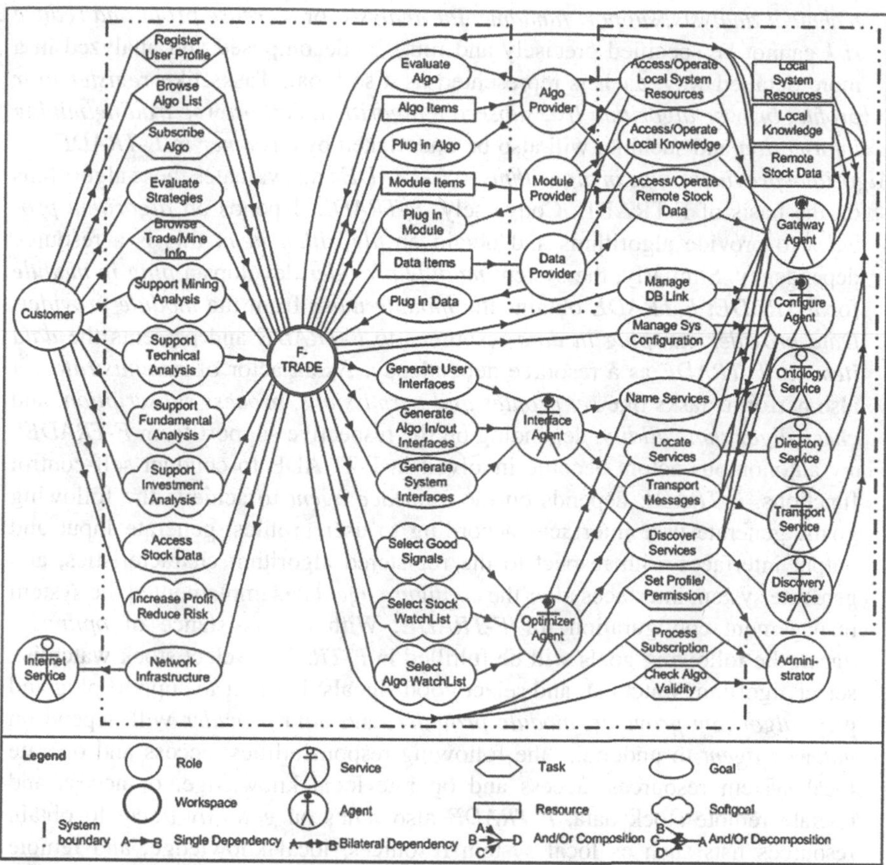

Fig. 13.1 An excerpt of the functional organizational relationship model

customer, *evaluator*, *provider*, and *administrator*. The agents include interface agents, plug-in agents, configure agents, gateway agents, and optimizer agents. Services are composed of ontology services, directory services, transport services, discovery services, and Internet services. *F-TRADE* is the workspace actor in this system. The workspace is the infrastructure supporting all goals, tasks, and softgoals to implement the proposed objectives. It is the drive center of both the target system and the requirement engineering; we therefore mark it as a special actor, different from the generic human roles or other actors. All the above-mentioned actors are represented by different diagrammatic symbols.

The principles and processes of F-TRADE are as follows:

Customer depends on *F-TRADE* to fulfill its goals including: *evaluate strategies* of trading and mining algorithms, *access stock data*, *support mining analysis*, *support technical analysis*, *support fundamental analysis*, *increase profit and reduce risk*, and so forth. Since the dependum *support mining analysis*, *support technical analysis*, *support fundamental analysis*, or *increase profit and reduce risk* cannot be specified precisely and must be decomposed and analyzed in a more detailed way, each is represented as a softgoal. Tasks like *register user profile*, *browse algorithm list*, *subscribe algorithm*, and *browse trading/mining information* and suchlike will also be undertaken by *customer* in *F-TRADE*.

Algorithm provider plugs in algorithms to *F-TRADE* and evaluates these algorithms on the basis of F-TRADE. Conversely, *F-TRADE* depends on *algorithm provider* to provide algorithms and obtain an *algorithm item*; this is a resource dependency. Similarly, the system *module provider* develops a *plug-in module* for *F-TRADE*; *F-TRADE* obtains the *module items* from the *module provider*. *Data provider* can *plug in data* resources to *F-TRADE* and presents the *data items* to *F-TRADE* as a resource dependency. As an actor of *administrator*, it also performs tasks like *set profiles and permissions*, *process subscription*, and *check algorithm validity*, depending on the respective support from *F-TRADE*.

Many autonomous actors become involved in F-TRADE to conduct self-control functions. *F-TRADE* depends on the *interface agent* to achieve the following goals: generate user interfaces according to user profiles, generate input and output interfaces with respect to the registered algorithm characteristics, and generate system interfaces after the *configure agent* has implemented the system management configuration for *F-TRADE*. With the assistance of *optimizer agent*, the following goals will be fulfilled in *F-TRADE*: select stock watchlist, select algorithm watchlist, and select good signals. In order to support plug and play, *algorithm provider*, *module provider*, and *data provider* will depend on *gateway agent* to undertake the following responsibilities: access and operate local system resources, access and operate local knowledge, or access and operate remote stock data. *F-TRADE* also relies on *gateway agent* to obtain resources lists such as local system resources, local knowledge, and remote stock data.

Service actors are taking on increasingly important levels of responsibility in open enterprise application integration with the emergence and widespread use of service-oriented architecture (SOA). The introduction of *service* actor brings the following probability into effect: an open enterprise system is integrated on top of multiple heterogeneous operational systems based on the system architecture of SOA. The *service* actor further complements the *agent* actor with explicit activity support. All agent actors, such as the *interface agent*, *optimizer agent*, *gateway agent*, and *configure agent*, depend on the agent *ontology service* to conduct name services, *directory service* to locate services, *transport service* to transport messages, and *discovery service* to discover services. In addition, service actors like *ontology service*, *directory service*, *transport service*, and *discovery service* depend on *F-TRADE* to obtain resources such as local system resources, local knowledge, and remote stock data. *F-TRADE* depends on the service actor *Internet service* to provide the Internet infrastructure.

13.3 Organizational Rationale Model

In the organizational relationship model, the relevant stakeholders and their goals in the proposed system have been identified and specified. We also get to understand why the processes in the system are structured in a certain way. However, the system hasn't told us how these goals can actually be fulfilled through the interactions and contributions of related actors. This can be achieved by the organizational rationale model.

As discussed in the beginning of this section, the *organizational rationale model* is the updated version of the traditional strategy rationale model in the *i** framework and TROPOS. The new model introduces (new) organizational actors, relationships, and rules into itself by following the organizational metaphor. In the organizational rationale model, goals, tasks, interactions, principles, and processes in high granularity in organizational relationship model are decomposed and refined into grain subgoals and subtasks in terms of business logic process and business information flow, goal decomposition, means–ends analysis, structural rules, and other problem-solving rules. Correspondingly, more concrete and detailed responsibilities, specific relationships, and causalities will be discovered and exhibited in the model. The main model-building blocks which may be utilized include those for actors, environment, rules, interaction, goals, structures, and dynamics.

As an instance for studying the process of building the organizational rationale model, we illustrate and discuss an excerpt of the organizational rationale model for the actor *algorithm provider*. The details are given below.

Figure 13.2 shows an excerpt of the organizational rationale model for *algorithm provider* and its environment. The main work for an algorithm provider is to code his/her algorithm, register it into the F-TRADE, and evaluate it as he/she likes. In the center of the figure, one basic task *Algo Coding* is refined into *Implement Algo API*, *Implement Algo Logic*, *Stock Fields Accessing*, *Algo Ontology Define*, and *Set Data Source* concurrently. After the programming of an algorithm has been completed, the provider can log onto the F-TRADE and *Call Algo Plugin Interfaces* to *Register Algo*, which will be further decomposed into *FillinAlgoRegisterOntologies* and *UploadAlgoComponent* in parallel. After *Check Algo Validity* by *administrator*, *Algo Provider* moves to *Evaluate Algo*. This includes such subtasks as *Call Algo*, *Parameter Value Setting*, and *Execute Algo*. The *Algo Output Interface* will be generated automatically and will show outputs and reports in *Visual Outputs*, *Detailed Outputs*, *Simulated Transactions*, and *Summary Report*, respectively.

Partially nonfunctional organizational rationale is also shown in the figure. Softgoal contributions to model sufficient/partial positive (++ and +, respectively) or negative (−− or −, respectively) support to make softgoals *Flexible*, *Available*, *Secure*, *Easy*, and *Adaptable*. For instance, to provide sufficient support to the *Flexible* softgoal, a number of functional tasks are specified to deal with registering the algorithm, calling the *algorithm plug-in interfaces*, and iteratively setting parameter values. Other tasks in the figure, such as *setting data source*,

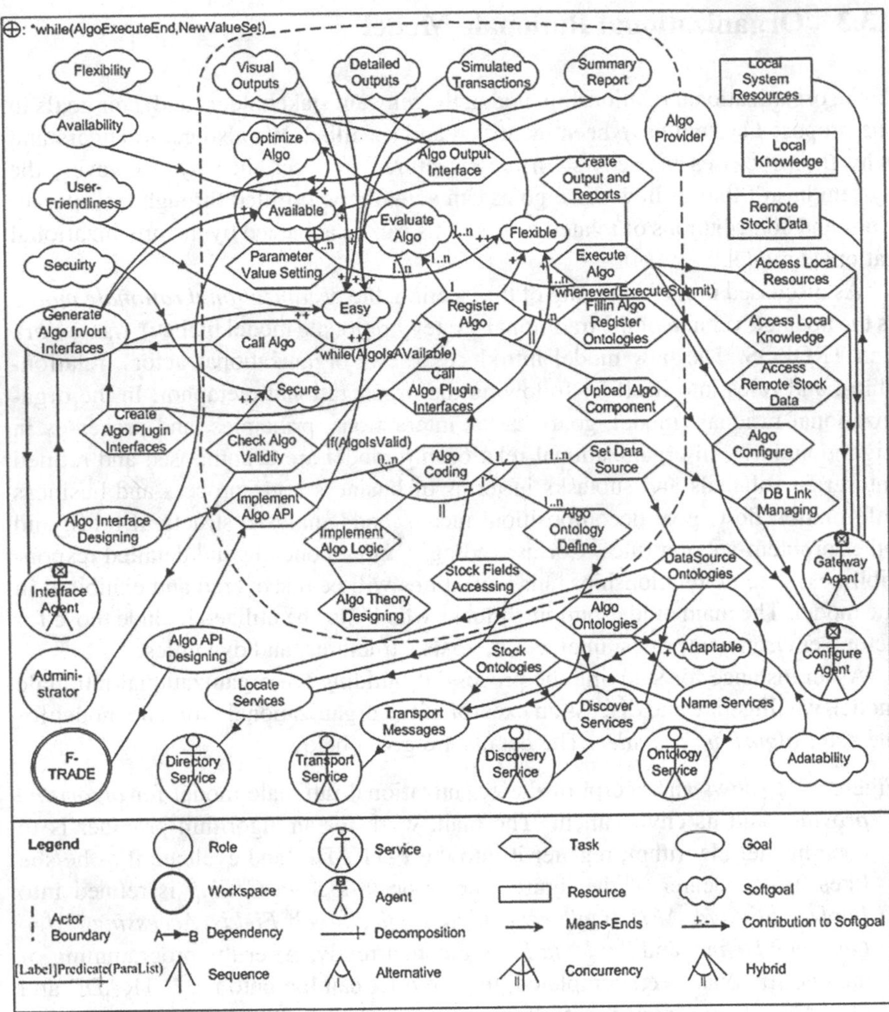

Fig. 13.2 An excerpt of the organizational rationale model for algorithm provider

defining algorithm ontologies, *designing algorithm API*, and so forth, can also contribute to the flexibility of the F-TRADE. These are not fully shown.

To more precisely and completely support the representation of complex relationships, the problem-solving rules such as rules for goal composition and decomposition—sister relationships and parent–child relationships—are explicitly shown as being associated with certain subtasks or subgoals. For instance, the algorithm can only be evaluated while the algorithm is available and after the execution request has been submitted. Iterative evaluation can be conducted when the previous execution of the algorithm has been completed and new parameter combinations have been set.

The organizational rationale model assists us to extract the rationale of the system more explicitly and precisely. Apart from goal composition and decomposition, and the refinement of responsibilities, other links exist between (sub)tasks or (sub)goals. This will further explore the relationships between ontology concepts, which are semantic annotations of dependums in the interdependent organization. We can abstract these links and express them as ontology relationships [1] between ontology concepts (dependums). Basic ontology relationships which may exist in the system take the form of instantiation, aggregation, generalization, substitution, disjuncture, overlap, and association, respectively. The above agent ontologies are not only significant for system analysis but are also closely involved in system design and implementation.

13.4 Formal Analysis

With the first-order linear-time temporal logics and formal grammars, we can conduct the *formal analysis* for refining the requirements specification in visual models. In this section, we take the algorithm plug-in and registration by *Algorithm Provider* as an example to introduce formal analysis. Formal analysis is an iterative process. The initial work is to generate a formal starting skeleton. This can be obtained from the organizational relationship model and organizational rationale model by mapping actors and intentional elements into corresponding formal elements and by adding non-intentional entities of the domain, if any [2].

Figure 13.3 presents an example of the formal analysis of algorithm plug-in and registration. In this activity, the design pattern for Algorithm, and Component *AlgorithmComponent* and *PluginInterfaces* will be involved. *Algorithm Provider* is the owner of the goal *RegisterAlgo*, so *Provider* is its actor. In registration, *Algorithm* will be referred to; it is the attribute of the goal. This goal is handled by multiple tasks. A *Provider* fulfills the task *CallPluginInterfaces* by reference to classes of *PluginInterfaces* and *Algorithm*. Whether or not *PluginInterfaces* is called as the relevant state is defined by the Boolean attribute *called*. Similarly, the task *UploadAlgoComponent* is defined by the attributes *pi* and *ac* and *uploaded*; *FillinAlgoRegisterOntologies* is specified by the attributes *pi*, *ac*, *algo*, *aro*, and *filled*; *ConfigureAlgo* is achieved by the actor *AlgoConfigureAgent*, and its attributes include *algo* and *aro*. For all the intentional elements, the modality of the fulfillment of the goal or task is the mode attribute *achieve*. For the dependency *Ontology*, the mode attribute *achieve* refers to the *Algorithm* (attribute *algo*) that is registered and to the *AlgoRegisterOntologies* which are plugged (attribute *aro*). The *AlgoConfigureAgent* depends on the *OntologyService* to perform the configuration of the registered algorithm; this configuration and registration support is based on *Ontologies*. The mode of the dependency includes *achieve* and *maintain*; this means the *AlgoConfigureAgent* wants to reach a state in which it can configure the ontologies and keep the condition continuously maintained.

```
Pattern Algorithm
Component AlgorithmComponent
Component PluginInterfaces
Actor Provider
Actor Administrator
Actor AlgoConfigureAgent
Actor OntologyService
Goal CodeAlgo
Actor Provider
Mode achieve
Attribute constant algo: Algorithm
Goal RegisterAlgo
Actor Provider
Mode achieve
Attribute constant ca: CodeAlgo
Attribute constant algo: Algorithm
Task CallPluginInterfaces
Actor Provider
Mode achieve
Attribute constant pi: PluginInterfaces
Attribute constant algo: Algorithm
        called: boolean
Task UploadAlgoComponent
Actor Provider
Mode achieve
Attribute constant pi: PluginInterfaces
Attribute constant ac: AlgorithmComponent
        uploaded: boolean
Task FillinAlgoRegisterOntologies
Actor Provider
Mode achieve
Attribute constant pi: PluginInterfaces
Attribute constant algo: Algorithm
Attribute constant ac: AlgorithmComponent
Attribute constant aro: AlgoRegisterOntologies
        filled: boolean
Task ConfigureAlgo
Actor AlgoConfigureAgent
Mode achieve
Attribute constant ac: AlgorithmComponent
Attribute constant aro: AlgoRegisterOntologies
        configured: boolean
Resource Dependency Ontology
Depender AlgoConfigureAgent
Dependee OntologyService
Mode achieve & maintain
Attribute constant algo: Algorithm
Attribute constant aro: AlgoRegisterOntologies
defined: boolean
```

Fig. 13.3 An excerpt of the initial formal declaration of algorithm registration

13.5 Formal Refinement Using Scenario-Based Analysis

All goals and dependencies in the above initial formal specification can be further improved, refined, and even extended by a step-by-step refinement and scenario-based analysis [2–4]. Scenarios [2] present possible ways to use a system to accomplish desired functions or implicit purposes. Typically, a scenario is a temporal sequence of interaction events between the intended software and its environment (composed of other systems or humans). The content of a scenario can describe either system environment interactions or events inside a system.

A scenario can be expressed in such forms as narrative text, structured text, images, animation or simulations, charts, maps, etc. In this work, scenarios are represented by scenario diagrams [2], where the different time instances of a scenario are distinguished by temporal symbols like t_0, t_1, … t_n. The creation of an object is indicated by a symbol ●. The states of creation and fulfillment of relevant objects are shown on the sequential scenario diagram.

The excerpt of the refined formal specification for the goal *RegisterAlgo* is shown in Fig. 13.4, where both informal and formal definitions are given. We first take a close look at the refinement of the goal RegisterAlgo with the goal- and scenario-based modeling techniques. As shown in Fig. 13.4, the creation condition for an instance of goal *RegisterAlgo* is that its predecessor goal *CodeAlgo* has already been fulfilled. The invariant shows constraints on the lifetime of class instances. For the goal *RegisterAlgo*, the invariant binds a *RegisterAlgo* object to its predecessor object. To fulfill the goal *RegisterAlgo*, only one *Algorithm-Component* instance exists for each algorithm that has been coded and will be uploaded at a time t_2 in the future by calling *CallPluginInterfaces* at future time t_1 ($t_1 \leq t_2$); one legal instance of *FillinAlgoRegisterOntologies* is filled at t_2. The *FillinAlgoRegisterOntologies* are allowed for multiple instances by following the rules in the class *Ontology* for each *AlgorithmComponent*. One scenario for refinement is shown in Fig. 13.5.

The relevant tasks *CallPluginInterfaces*, *FillinAlgoRegisterOntologies*, and *UploadAlgoComponent* can be further refined. Here, we take another task *ConfigureAlgo* as an instance for refinement (as shown in Fig. 13.6). The task *ConfigureAlgo* is created after the fulfillment of the parent tasks *UploadAlgo-Component* and *FillinAlgoRegisterOntologies*. The dependee object of task *UploadAlgoComponent* is bound to the *ConfigureAlgo* object; additionally, the algorithm and ontologies attributes of the above two objects are the same. The task will be fulfilled after the ontologies of the registered algorithm have been configured and the state will always hold. Figure 13.7 shows one scenario for configuring algorithms. The refined specification of the *AlgoConfigureAgent* configuring algorithms follows below.

Domain ontology is a key resource dependency for supporting the registration and configuration of algorithms. Given a scenario (as shown in Fig. 13.8), the creation of the ontologies must allow *ConfigureAgent* to configure the algorithm component provided by the algorithm *Provider*, and the filled ontologies of the

Goal RegisterAlgo
InformalDef *When an algorithm component has been coded and the algorithm isn't available from the system at the moment, this algorithm component can be registered into the system by calling plug-in interfaces, filling in algorithm registration ontologies, and upload the algorithm module.*
FormalDef
Actor Provider
Mode achieve
Attribute constant ca: CodeAlgo
Attribute constant algo: Algorithm
registered: boolean
Creation condition
• Fulfilled(ca) ∧ ¬ Existed(algo)
Invariant ca.actor = actor
Fulfillment condition
∀ ac: AlgorithmComponent (ac.algo = algo →
$\Diamond_{\le t1}$ ∃ cpi: CallPluginInterfaces (cpi.actor = actor ∧ Fulfilled(cpi) ∧ pi.Called) ∧ $\Diamond_{\le t2}$(∃ faro: FillinAlgoRegisterOntologies (faro.depender = actor ∧ Fulfilled(faro) ∧ aro.Filled) ∧ ∃ uac: UploadAlgoComponent (uac.depender = actor ∧ Fulfilled(uac) ∧ ac.uploaded)))

Fig. 13.4 An excerpt of formal specifications for provider's goal algorithm registration

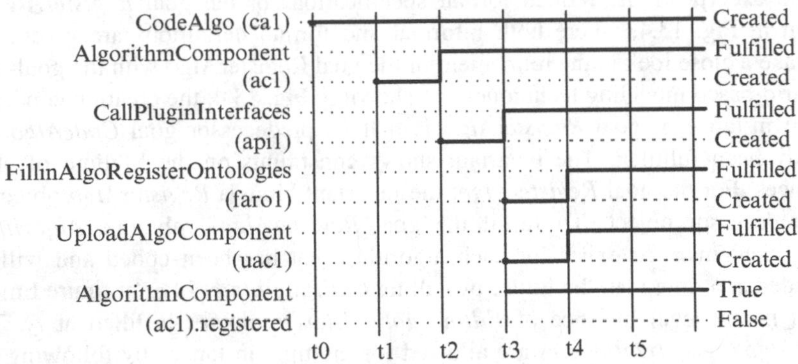

Fig. 13.5 A sequential scenario for *provider* to register an algorithm

Goal ConfigureAlgo
InformalDef *The algorithm registered will be configured in the algorithm knowledge base in terms of domain ontology and configuration management*
FormalDef
Actor AlgoConfigureAgent
Mode achieve
Attribute constant ac: AlgorithmComponent
Attribute constant aro: AlgoRegisterOntologies
Attribute constant ra: RegisterAlgo
configured: boolean
Creation condition
• (∃ uac: UploadAlgoComponent (uac.depender = actor ∧ uploaded(ac)) ∧ ∃ faro: FillinAlgoRegisterOntologies (faro.depender = actor ∧ Filled(aro)) ∧ ¬ Fulfilled(ra))
Invariant
uac.dependee = actor ∧ faro.algo = uac.algo ∧ faro.onto = uac.onto
Fulfillment condition
□ ∃ onto: Ontology (onto.depender = actor ∧ Fulfilled(onto) ∧ onto.configured)

Fig. 13.6 An excerpt of specifications for configuring algorithm by *ConfigureAgent*

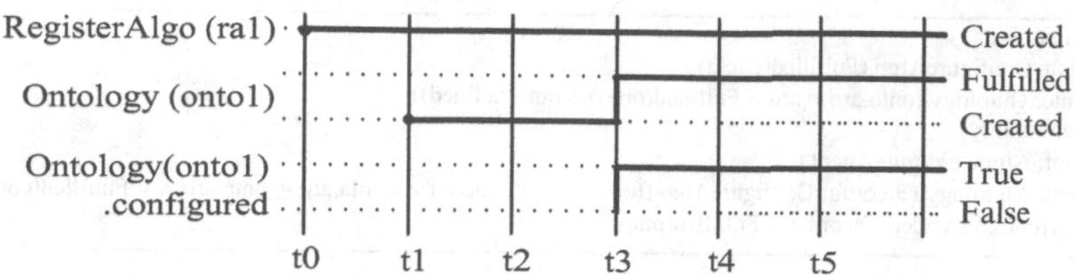

Fig. 13.7 A sequential scenario for configuring algorithm by *ConfigureAgent*

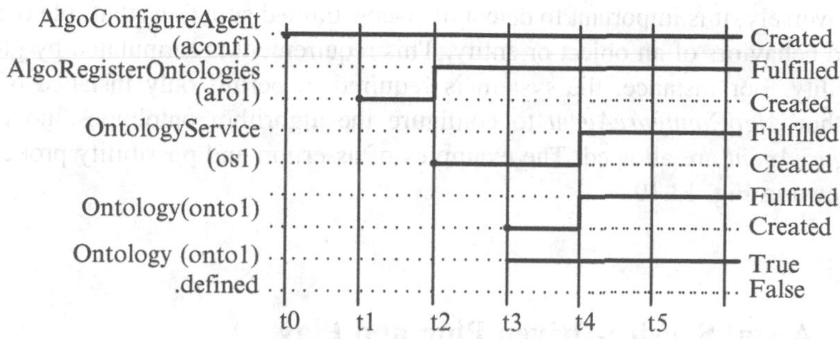

Fig. 13.8 Ontology service defines algorithm registration ontologies

---

InformalDef *After fulfillment of configuring algorithm and filling in the algorithm registration ontologies, algorithm ontologies will be uniquely defined and cannot be changed.*
FormalDef
Depender AlgoConfigureAgent
Dependee OntologyService
Mode achieve & maintain
Attribute constant aro: AlgoRegisterOntologies
Attribute constant confa: ConfigureAlgo
defined: boolean
Creation condition for domain
• (∃confa: ConfigureAlgo (confa.dependee = depender ∧ confa.depender = dependee ∧ confa.aro = aro ∧ Fulfilled(confa)) ∧ ∃ aro: AlgoRegisterOntologies (aro.dependee = depender ∧ aro.depender = dependee ∧ aro.filled)))
Invariant
□ (Fulfilled(confa) → configured ∨ Fulfilled(self) → defined) ∧ • ∃ aro: AlgoRegisterOntologies (aro.actor = depender)

Fig. 13.9 An excerpt of specifications for dependency *Ontology*

algorithm must conform to the domain algorithm ontology in the knowledge base. Algorithm ontologies cannot be changed once defined; this leads to the invariant constraint on the dependency *Ontology*. An excerpt of both informal and formal specifications are illustrated in Figs. 13.8 and 13.9.

The configuration and definition of algorithm ontologies must be present once the ontology configuration and service definition has been fulfilled, so that the value of the attribute *configured* of the task *ConfigureAlgo* and the value of attribute *defined* of the dependency *Ontology* will remain stable.

Global assertion
 $\forall$ confa: ConfigureAlgo (Fulfilled(confa) $\rightarrow$
  $\exists$ onto: Ontology (onto.aro = aro $\wedge$ Fulfilled(onto) $\wedge$ onto.defined))
Global possibility
 $\forall$ aconf: AlgoConfigureAgent (
  $\exists$ onto: Ontology ($\exists$ confa: ConfigureAlgo ((confa.actor = aconf $\wedge$ confa.aro = onto.aro $\wedge$ $\Diamond$ Fulfilled(confa)) $\wedge$
   (onto.depender = aconf $\wedge$ $\Diamond$ Fulfilled(onto))))

Fig. 13.10 Example of assertion and possibility

Conversely, it is important to detect over-constrained situations that rule out the desired behaviors of an object or entity. This requirement is formulated by global possibility. For instance, the system is required to permit only instance of the algorithm *AlgoConfigureAgent* to configure the algorithm ontologies; no other *ConfigureAgent* are allowed. The examples of assertion and possibility properties are shown in Fig. 13.10.

13.6 Agent Service-Driven Plug and Play

13.6.1 Plug and Play Modeling

Plug and play is involved in many functions and the evolutionary life cycle of F-TRADE, as discussed above. In this section, we shall take the plug-in of an algorithm as an instance and introduce the conceptual model, role model [7], agent services [5, 6, 8], and generation of the user interface to plug in an algorithm.

We use the Agent Unified Modeling Language (AUML) technology [9] to model the agent service-based plug and play. In Agent UML, package is one of the two techniques recommended for expressing agent interaction protocols. We use packages to describe the agent interaction protocols in plug-in support.

There are four embedded packages to support the plug-in process: To Implement, To Input, To Register, and To Generate, as shown in Fig. 13.11. They represent the process of agent interactions in the plug-in activities. The package To Implement programs an algorithm in AlgoEditor by implementing the AlgoAPIAgent and ResourceAPIAgent. The To Input package types in the agent ontologies of the programmed algorithm and makes the request to plug in the algorithm. The real registration of the algorithm is done by the package To Register; the naming, directory, and class of the algorithm are stored in an algorithm base, and linkage to the data resources is set at this moment. Lastly, the input and output user interfaces for the algorithm are managed by the package To Generate.

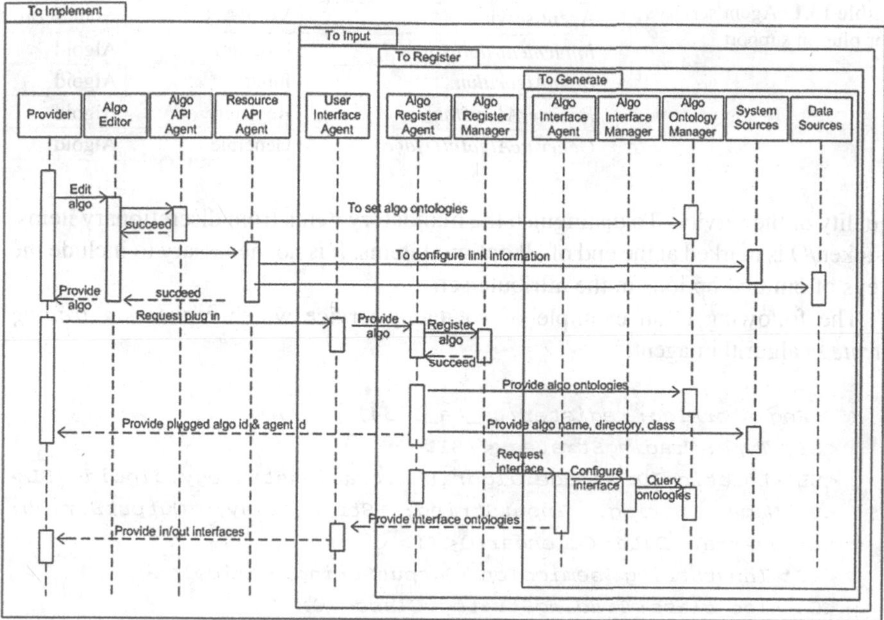

Fig. 13.11 Package diagram of the plug and play

13.6.2 Agent Service-Driven Plug and Play

In this section, we focus on discussing the analysis and design of the agent service-driven plug and play. We first introduce the role model for plug and play, after which the agent service for plug and play is presented in detail. Lastly, the user interface for the algorithm and system module plug-in is presented.

13.6.2.1 Agent Service Description

We use multiple-attribute items to describe an agent service. They are as follows: <ST>k : v </ST> is a service type item atom, <SL>k : v</SL> is a locator item atom, <I>k : v</I> is an input item atom, <IC>k : v</IC> is a precondition atom which defines constraints on service, <O>k : v</O> is an output item atom, the item <IO>k : v</IO> defines constraints across inputs and outputs, and <OC>k : v</OC> is a postcondition item of service. Furthermore, each item atom may have a list of items, called item list. For instance, an item list of *j* inputs is a list of j ($j \geq 1$) items as <I>$k_1{:}v_1, k_2{:}v_2, \ldots, k_j{:}v_j$</I>.

For each item atom in a service, there may be zero, one, or many items as required. Some of these items are mandatory (e.g., some of the service inputs must be specified), while others are not required (optional) for the service to be provided, but their provision may increase either the efficiency of the provided service or the

Table 13.1 Agent services for plug-in support

Agent service	Verb	Noun term
ImplementAlgorithm	Program	Algoid
InputAlgorithm	Input	Algoid
RegisterAlgorithm	Register	Algoid
GenerateAlgoInterface	Generate	Algoid

quality of the service. To distinguish the mandatory items from discretionary items, a token/O is marked at the end of all optional items. It is not necessary to include the keys of omitted options in the attribute set.

The following is an example of an agent service which registers a *trading strategy* algorithm agent.

```
<S>RegisterAlgo: register(ts_ta_1234)
  <ST>Type: TradingStrategy</ST>
  <SL>LocationID: ftrade.algorithm.tradingstrategy.algoid</SL>
  <I>Name: String, InputString: StringArray, OutputString:
  StringArray, Date: Calendar/O</I>
  <IC>InputString: semicolon, OutputString: semicolon</IC>
  <O>Algo_class: TradingStrategyClass</O>
  <OC>Status: successful</OC>
  <IO>Relation: one-one</IO>
</S>
```

This service registers a trading strategy algorithm for technical analyses with a globally unique algoid *ts_ta_1234*. It has three mandatory inputs and one optional item as the algorithm creation date. All inputs and outputs of the trading strategy are separated by semicolons and stored into an array in Java, generating one new trading strategy class if registration is successful.

We define four agent services (as shown in Table 13.1) to support the plug-in function.

13.6.2.2 Role Model for Plug and Play

In the agent-based F-TRADE, there is a role PLUGINPERSON which is in charge of the function of plug and play. A role model can be built for the PLUGINPERSON which describes the attributes of permissions, responsibilities, protocols, and activities of the role. Figure 8.1 shows the role schema of PLUGINPERSON.

This role plugs an algorithm, which is non-existent in the algorithm base, into F-TRADE. The agent playing this role will execute the protocol ReadAlgorithm, followed by the activities *ApplyRegistration* and *FillinAttributeItems*, and then will then execute the protocol SubmitAlgoPluginRequest. The role has rights to read the algorithm from the non-plug-in directory and to change the application content for

AgentService
RegisterAlgorithm(algoname;inputlist;inputconstraint;outputlist; outputconstraint;)
Description:
This agent service involves accepting the registration application submitted by role PluginPerson, checking the validity of attribute items, creating the name and directory of the algorithm, and generating a universal agent identifier and a unique algorithm id.
Role: PluginPerson
Pre-conditions:
- A request of registering an algorithm has been activated by protocol SubmitAlgoPluginRequest
- A knowledge base storing rules for agent and service naming and directory

Type: algorithm.[datamining/tradingsignal]
Location: algo.[algorithmname]
Inputs: inputlist
InputConstraints: inputconstraint[;]
Outputs: outputlist
OutputConstraints: outputconstraint[;]
Activities: Register the algorithm
Permissions:
- Read supplied knowledge base storing algorithm agent ontologies
- Read supplied algorithm base storing algorithm information

Post-conditions:
- Generate a unique agent identifier, naming, and a locator for the algorithm agent
- Generate a unique algorithm id

Exceptions:
- Cannot find target algorithm
- There are invalid format existing in the input attributes

Fig. 13.12 Agent service RegisterAlgorithm

the registration and attributes of the algorithm. As a precondition, the agent is required to ensure that two constraints in safety responsibility are satisfied.

13.6.2.3 Agent Service Specification

In reality, many agents and services are involved in the plug and play to make it successful. There are three directly related agent services which handle the plug and play. They are the *InputAlgorithm*, *RegisterAlgorithm*, and *GenerateAlgoInterface* services, respectively. Here, we take just one service, *RegisterAlgorithm,* as an example and introduce it in detail in Fig. 13.12. More information about agent service-oriented analysis and design and about the plug and play can be found in Chaps. 10 and 11.

13.6.3 Implementation

User interfaces must be implemented in association with the plug and play. Figure 13.13 shows the user interface for plugging an algorithm into F-TRADE. The ontologies include all the parameters of the algorithm, and the specifications for, and constraints on, every ontology element must be defined and typed here. After submitting the registration request, the input and output interfaces for this algorithm

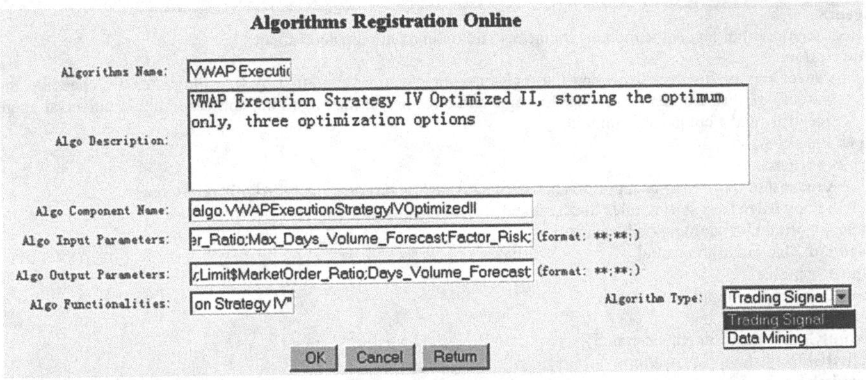

Algorithms Registration Online

Algorithms Name: VWAP Executi

Algo Description: VWAP Execution Strategy IV Optimized II, storing the optimum only, three optimization options

Algo Component Name: algo.VWAPExecutionStrategyIVOptimizedII

Algo Input Parameters: er_Ratio;Max_Days_Volume_ForecastFactor_Risk; (format: **;**;)

Algo Output Parameters: ;Limit$MarketOrder_Ratio;Days_Volume_Forecast (format: **;**;)

Algo Functionalities: on Strategy IV"

Algorithm Type: Trading Signal | Trading Signal | Data Mining

OK Cancel Return

Fig. 13.13 User interface for an algorithm plug-in

will be generated automatically. As previously discussed, plug and play can not only be used for algorithms but also for data sources and functional agents and services.

13.7 M-Space for Macroeconomic Decision Support

Complex economic systems [16] and economy-related decision support [13] belong to the OCGS family. In 1999, a major program grant was allocated to investigate macroeconomic decision support [13], which involved multidisciplinary researchers from more than 10 organizations including institutes in the Chinese Academy of Sciences and such universities as Tsinghua University.

The project studied the methodological, technological, and engineering support of metasynthesis-based macroeconomic decision support. An M-space prototype was built for this purpose [8, 10, 14, 15] which fused knowledge and techniques from multiple areas such as artificial intelligence, machine learning, system modeling and simulation, quantitative economic modeling, group cognition and consensus building, knowledge management and discovery, software engineering and networking, and systems science and engineering. Figure 13.14 illustrates the system structure of the M-space for macroeconomic decision support.

The system consists of major subsystems including M-space access points, M-space infrastructure support, M-space applications, M-space services, and M-space resources. M-space access points take over user access locally and remotely. M-space infrastructure support consists of system modules of single sign-on, registration, M-interaction modes, templates, workflow, life cycle, mediation and log management, reporting, and sub-M-space gateway. M-space applications are composed of model, case, method and script builder, knowledge acquisition and discovery, information retrieval and Web information processing, information cooperation, case-based reasoning, text processor, and reporting.

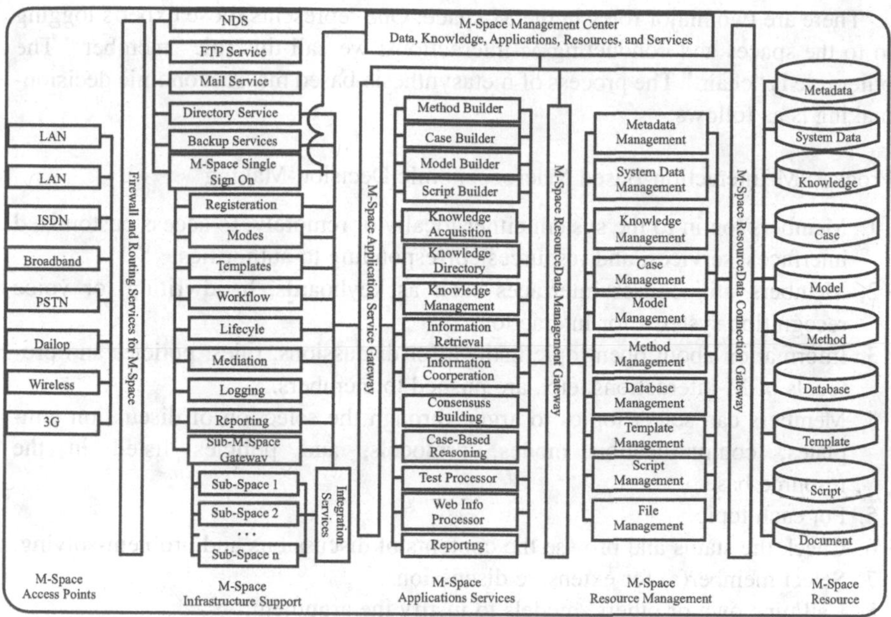

Fig. 13.14 An M-space for macroeconomic decision support

M-space applications and services consist of services catering for metadata, system data, knowledge, cases, models, methods, databases, templates, scripts, and files. M-space resources consist of bases for storing knowledge, models, methods, data, templates, files, and scripts. All these subsystems are linked and managed by corresponding gateways; for instance, the M-space applications and services are managed by the M-space application/service gateway.

The macroeconomic decisions are made through the following mechanism. Issues/topics to be discussed are identified and distributed to relevant sub-M-spaces. All relevant experts from multiple domains log on to the M-space and are then distributed into respective sub-M-spaces. An expert can attend one or multiple sub-M-spaces to discuss the corresponding topics with group members. An expert can also call his/her own models and methods to support an argument or exchange ideas with other participants to form a new argument. He/she can also retrieve information from the M-space repositories or Internet to learn about the outputs of others and to search for evidence. For each sub-M-space, a chair is in charge of topic management, and schedule and resources management. The chair issues M-space policies, rules, norms, and argument protocols to all attendees to maintain fairness, flexibility, and free interaction and to balance individual benefits with group goals by invoking mechanisms for risk management, conflict resolution, and agreement/consensus building. Through the M-interactions, individual views merge into group opinions and collective intelligence as a result of consensus-building mechanisms.

There are two major roles in the M-space. One represents those experts logging in to the spaces and conducting M-interactions; we call this role "member." The other role is "chair." The process of metasynthesis-based macroeconomic decision-making is as follows:

Process: M-Interaction-Based Macroeconomic Decision-Making

1. Members log in to the system either locally or remotely, to access customized interfaces, services, and resources corresponding to authorities.
2. Members utilize user interfaces such as keyboards, handwriting, or voice recognition systems for interaction.
3. Information about open topic listings for discussions, rules, policies and protocols of M-interactions, etc., are pushed to members.
4. Members can select topics to argue through the selection of discussion templates, communication modes, protocols, and policies listed in the resource base.
5. For each topic.
6. Check the status and browse the contents of discussions and problem-solving.
7. Select member(s) for extensive discussions.
8. Call/tune own or others' models to justify the argument.
9. Receive the summary and recommendation from the sub-M-space chair.
10. Join N-sets of intensive discussions called by the chair.
11. Chair announces the rule, policy, protocol, and norm to be followed.
12. Chair broadcasts models, methods, modes, and resources available for the argument.
13. Members prompt a specific macroeconomic topic to the chair for decision.
14. Members vote on the macroeconomic decisions recommended by the chair for discussion.
15. Chair aggregates and recommends the topics for public discussion.
16. Chair opens the extensive discussion phase.
17. Members conduct bilateral or multilateral extensive discussion and debate on the topic against public and/or private models, methods, and knowledge base.
18. Chair scrutinizes, filters, and aggregates the opinions and arguments progressively and shares the summary with members for feedback.
19. Chair summarizes the discussions to consolidate initial understanding of the topic and possible/potential directions for the problem-solving.
20. Chair shifts the discussion to the intensive discussion phase.
21. For the selected topic.
22. Members conduct intensive discussions by following discussion scripts on the selected topic.
23. Members communicate, argue, and debate opinions and conclusions by calling models, methods, algorithms, and evidence.
24. Members adjust themselves to others' arguments or form new arguments through negotiation.

25. Chair tries to fuse the outputs from every member and draw preliminary conclusions if any for the current set of discussions.
26. Chair may issue new models/methods/modes for the next-set discussion.
27. Chair determines when to terminate the intensive discussions.
28. Chair initiates a voting and consensus-building phase.
29. Chair nominates the models/methods for consensus building for feedback from members.
30. Chair summarizes the arguments and conclusions emerging from the N-sets of discussions.
31. Members vote on the arguments and conclusions.
32. Chair calls consensus-building models to consolidate the arguments for final recommendation.
33. Output the finally agreed findings.
34. Close the M-interactions.

Experiments show that metasynthesis-based macroeconomic decision support can lead to solutions for issues such as Chinese macroeconomic trends that could not possibly be achieved by the sole use of traditional theories of economics or single lines of economists. The system prototype has been positively assessed by the expert group organized by the National Natural Science Foundation of China.

13.8 Summary

Engineering complex systems [11, 12] is not an easy task. Building on the computing and engineering techniques for complex systems introduced in the previous chapters, this chapter has illustrated the application of some such systems. Taking an online financial trading and data mining system and a macroeconomic decision-support system as examples, this chapter has demonstrated the use of some of the tools discussed. In particular, examples of the analysis of a system based on organization-oriented abstraction and analysis in both visual and formal manners have been presented, as well as details of how to refine formal models through scenario-based analysis. Agent service-based implementation is discussed to support the plug and play of agent services.

Two of the three remaining chapters, Chaps. 14 and 15, will discuss general issues and challenges in mining complex data and problems and learning complex behavioral and social problems from a broader perspective. These issues are becoming increasingly important for business, government, and academia and fundamentally challenge existing theoretical and practical systems. The main challenges arise from ubiquitous intelligence and critical system complexities, which have not been effectively or comprehensively addressed in the current theories and tools.

References

1. Storey, V.C.: Understanding semantic relationships. VLDBJ **2**(4), 455–488 (1993)
2. Fuxman, A., Liu, L., Mylopoulos, J., Pistore, M., Roveri, M., Traverso, P.: Specifying and analyzing early requirements in TROPOS. **9**(2), 132–150 (2004)
3. Rolland, C., Grosz, G., Kla, R.: Experience with goal-scenario coupling in requirements engineering. In: Proceedings of the 4th IEEE International Symposium on Requirements Engineering, Limerick, Ireland (1999)
4. Kazman, R.: Using scenarios in architecture evaluations. SEI Interactive, On-line at http://interactive.sei.cmu.edu/Columns/The_Architect/1999/June/Architect.jun99.htm. Accessed 30 May 2013
5. Baer, J.: Creativity and Divergent Thinking: A Task-Specific Approach. Lawrence Erlbaum, Hillsdale (1993)
6. Boumen, R., de Jong, I.S.M., Mestrom, J.M.G., van de Mortel-Fronczak, J.M., Rooda, J.E.: Integration and test sequencing for complex systems. IEEE Trans. Syst. Man. Cybern. A. Syst. Hum. **39**(1), 177–187 (2009)
7. Cao, L. B.: Some critical issues in agent-based open giant intelligent systems. Ph.D. dissertation, Chinese Academy of Science, Beijing (2002)
8. Cao, L.B.: Metasynthetic computing for solving open complex problems. In: Cao, L.B., Dai, R. W., Gorodetsky, V. (eds.) Proceedings of 1st IEEE EOCS-MCP/32nd Annual International Computer Software Applications Conference, COMPSAC 2008, Turku, Finland, pp. 896–901 (2008)
9. Cao, L.B., Dai, R.W.: Software architecture of the hall for workshop of metasynthetic engineering. J. Softw. **13**(8), 1430–1435 (2002)
10. Cao, L.B., Luo, D., Luo, C., Zhang, C.: Systematic engineering in designing architecture of telecommunications business intelligence system. In: Abraham, A., Koppen, M., Franke, K. (eds.) Design Application Hybrid Intelligent. Systems, pp. 1084–1093. IOS Press, Amsterdam (2003)
11. Cao, L.B., Dai, R.W.: Open Complex Intelligent Systems. Posts Telecom, Beijing (2008)
12. Cao, L.B., Zhang, C.Q., Zhou, M.C.: Engineering open complex agent systems: a case study. IEEE Trans. Syst. Man Cybern. C. Appl. Rev. **38**(4), 483–496 (2008)
13. Dai, R.W., Wang, J., Tian, J.: Metasynthesis of Intelligent Systems. Zhejiang Science and Technology Publishing House, Hangzhou (1995)
14. Du, C.Z.: The application of metasynthesis in intelligent building systems. Intell. Build. **6**, 22–26 (2003)
15. Lei, Y.J., Li, X.Y.: Metasynthesis for military supply chain management based on knowledge management. Commer. Res. **17**, 146–149 (2006)
16. Li, X., Dai, R.W.: Conceptual system structure and metasynthesis. Technical Report, Chinese Academy of Science, Beijing (1998)

Chapter 14
Actionable Knowledge Discovery and Delivery

14.1 Introduction

This chapter illustrates the issues in mining complex data and problems for knowledge that will support decision-making actions and shows how complex problems are analyzed through consideration of the concepts and thinking in metasynthetic computing. Typically, mining complex data to deliver actionable knowledge is becoming increasingly challenging. These challenges arise from the following issues:

1. The limitations of existing KDD methodologies and systems, such as purely data-driven techniques or the poor involvement of business and domain coupling with data.
2. The characteristics and challenges of complex data in the real world, which involve many different kinds of complexities as discussed in Chaps. 1, 2, 3 and 4.
3. Arguably, methodologies and systems available in the current KDD literature rarely present a systematic and comprehensive guide from system sciences and multidisciplinary aspects. These instead play an important role in the study of open complex systems.
4. The majority of the existing work focuses on mining simple and manipulated data and problems, which are abstracted from the complexities of real-life problems, and we therefore face critical challenges in addressing real problems and their complexities. Real-life problems present as complex systems, and taking a systematic and comprehensive view is thus very important.

Treating a data mining problem as a complex system is not an easy thing to do. It requires training and knowledge in complex systems. In recent years, an increasing number of researchers have recognized the importance and value of thinking of KDD in a way that considers subjective factors, domain-specific characteristics,

L. Cao, *Metasynthetic Computing and Engineering of Complex Systems*, Advanced Information and Knowledge Processing,
DOI 10.1007/978-1-4471-6551-4_14

and decision-support expectations. In [1–10], we propose a KDD methodology, domain-driven data mining, to highlight the need to:

1. Evaluate the gaps between manipulated data and the nature of the real-life problem
2. Evaluate the gaps between the theoretical findings and real-life decision-support actions
3. Understand the gaps between existing theoretical systems and complexities and challenges appearing in real-life data and problems
4. Explore effective next-generation KDD methodology to promote the paradigm shift from data-driven pattern mining to actionable knowledge-oriented discovery and delivery

Accordingly, we discuss the following topics in this chapter:

- Issues with existing KDD theories, systems, and tools
- Gaps in the delivered outcomes discovered by existing KDD systems and what is required to address real problems
- A new framework to support actionable knowledge discovery
- The means to apply such actionable knowledge discovery methodology
- Case studies to show the need for and effectiveness of this methodology

14.2 Issues with Existing KDD

Actionable knowledge "is not only relevant to the world of practice, it is the knowledge that people use to create that world" [11]. Actionable knowledge is not a new concept in the domains of social science and business. It has been discussed very intensively in areas such as business management [12, 13], organization science [14], management science [15, 16], and suchlike. However, the engagement of actionable knowledge with data mining has only taken place in recent years [17–19], especially, in retrospection, on deliverable effectiveness in supporting decision-making action-taking.

Data mining seeks to extract interesting patterns from data. In recent decades, data mining has boomed as an emerging discipline and has been featured by an increasingly large number of publications. In applications, however, we see only a small number of commercially available data mining products (some of them including statistical aspects) on the market. These products often lead to patterns, or so-called knowledge discovered in data, which either evidence existing domain observations or common sense, or cannot be used at all for taking decision-making actions. This reflects the emergence of an extreme and overwhelming imbalance between a massive number of research publications and rare workable products/systems.

In recent years, more and more data mining researchers with strong practices or first-hand industrial experience have recognized the critical problems and challenges associated with data mining research [1–4, 20–25]. This realization has triggered the first round of a paradigm shift, namely, from "data mining" to "knowledge discovery" [24], to discover hidden and interesting knowledge from data. Typical research includes research on subjective measures [26–28], unexpectedness [29, 30], novelty [31], actionable rules [32], action rules [33, 34], and interpretability [21]. However, it is argued that the "knowledge" discovered from data is not powerful enough for "direct" and "decisive" [35] problem-solving. Here, "direct" means there is no need to further manipulate the knowledge discovered and "decisive" indicates the decision-making actions needed to achieve the desired results when it is used for problem-solving.

In industry and business, an even more obvious trend is that more and more practitioners are urging the transformation from data to actionable knowledge, to make data mining useful for real-world applications [36] in health [37], retail [35], intrusion detection [38], and Web logs [39], and to generally upgrade organizational competitive advantage [15].

The reality facing knowledge discovery in the database (KDD) community seems to be unpleasant. The imbalanced situation has never been thoroughly addressed; rather, it seems to become more and more serious [1]. From the system science perspective, KDD-based problem-solving is a system which involves not only data itself but also the environment of knowledge discovery and the delivery of decisions to application needs. However, in the traditional KDD framework, data are the focus, and the environment is often simplified, if not ignored. Decisions are often too simple, straightforward, or far removed from the expectations of business problem-solving. A typical and increasingly common scenario is that a very limited proportion of the growing enormous number of annual publications in the community is able to support operable/workable decision-making in the real world.

In KDD projects, we often face the following scenarios:

- Data miner: "I have found something interesting!" "Many patterns have been found!" "They satisfy my technical metric thresholds very well!"
- Business people: "So what?" "They are just commonsense." "I don't care about them." "I don't understand them." "How can I use them?"

From the technical and engineering perspective, many issues seem to be overlooked in classic data mining. Let us take financial data mining as an example:

- Problem dynamics and interaction in a system: market dynamics such as coupling between two stock prices are often overlooked in modeling.
- Problem environment: a time-series model is built on closing prices which are applied to represent the market dynamics.
- Business processes, organizational factors, and constraints: a trading pattern is discovered without differentiating between types of orders, market, and investors.
- Human involvement: a trading rule captures historical trading patterns but not the investor's intention.

- Knowledge discovered: if frequent pattern mining is used to identify frequent trading patterns, we may find very many "interesting" patterns. However, almost all of them may be unworkable, because they may only reflect majority trading behaviors which are not interesting to investors.
- Evaluation: a frequent trading pattern with high confidence but low *Sharpe ratio* when applied to a market.

The above problems arise from the gap between academic objectives and business goals, and between academic outputs and business expectations. To utilize the unique power of KDD to enable smart businesses and transform business and industry by providing smart decisions, it is worthwhile to ask ourselves: What is "wrong" with "discovering knowledge from data"? What is "inconsistent" about the underlying KDD methodologies, research intentions, and the focus on/needs of real-life problem-solving? Why and where does the imbalance occur? How do these gaps arise? Most importantly, how should the existing KDD paradigm be transformed into one that can produce actionable knowledge for decision-making?

14.3 Gap Analysis

To gain a better understanding of knowledge actionability [40], we explore the gaps that appear in data mining. Although it is complicated to scrutinize the true reasons and to discover effective solutions for the supposed "wrongness" or "inconsistency" identified above, deep gap analysis and thorough retrospection about traditional KDD, and therefore innovative thinking and interdisciplinary interaction, are helpful for determining possible actions.

14.3.1 Gaps Between Delivered and Desired

There may be gaps between knowledge, power, and action [41, 42] in existing data mining methodologies. Let us first try to understand where the gaps are located. We observe this from the macrolevel by focusing on the methodological issues surrounding traditional KDD research [4].

On the one hand, from the research culture perspective, we often:

- Concentrate on innovative algorithms and patterns
- Only check the interestingness of identified patterns from the technical significance perspective
- Do not really perceive or care about the needs of business people
- Do not take the business environment into account
- Oversimplify data, surroundings, and problem definition

On the other hand, practitioners and business analysts value something different, for example:

- Can it solve my business problem, or will it lead to what I expect?
- Has the surrounding social, environmental, and organizational factors been considered?
- Can I interpret it in my business language, experience, and knowledge?
- Can I easily adjust it as I need by following my business rules and processes?
- Can it make my job more efficient rather than creating new issues?
- Can it be easily integrated into my business rules, operational systems, and workflow?
- What could be the impact on business if I use it? Is that manageable?

Consequently, we see the gaps between academia and business in goals, factors involved, outputs, deliverable presentation modes, evaluation, and impact:

- Gap between a converted research issue and its actual business nature
- Gap between academic objectives and business goals
- Gap between technical significance and business interest
- Gap between identified patterns and the deliverables expected by business
- Gap between the deliverables from data miners and the eventual entities deployed into problem-solvers

The above gaps result in the imbalance emerging in the KDD community and this imbalance is embodied in many aspects, for instance:

- Algorithm imbalance: many published algorithms vs. few that are actually workable in the business environment
- Pattern imbalance: many patterns mined vs. a very small portion, if any, eventually used
- Evaluation imbalance: technical performance validated vs. no check for business interest or business impact
- Decision power imbalance: impressive (technical) performance claimed vs. very little that can either be used directly or be converted to support decision-making actions and achieve business expectations

The knowledge we often see is "passive," presenting information at a surface level with little context or background. Such passive knowledge tells us very little about how to act and on what to act. What decision-makers need is "active" knowledge with the power to work; knowledge which is compelling and powerful for action-taking and decision-making. To narrow these gaps, we need to convert passive data (knowledge) into active knowledge, or directly produce active knowledge.

14.3.2 Aspects for Narrowing Gaps

The above gap analysis shows that it is not easy to discover actionable knowledge [43], and we therefore discuss which aspects can be explored further to narrow the gaps. We observe this from both macrolevel and microlevel perspectives [1, 4, 44].

On the macrolevel, aspects are related to methodological and fundamental issues, including the key elements of environment, human role, process, infrastructure, dynamics, evaluation, risk policy, and deliverability.

- Environment: This refers to any factor surrounding data mining models and systems; for instance, domain factors, constraints, expert groups, organizational factors, social factors, business processes, and workflows. Some factors, such as constraints ,have been considered in the current data mining research, but many others have not. It is essential to represent, model, and involve all these factors in KDD systems and processes.
- Human role: To handle many complex problems, human-centered and human–mining-cooperated KDD is necessary. Challenging problems related to this include how to involve domain experts and expert groups in the mining process and how to allocate the roles between human and mining systems.
- Process: Real-world problem-solving has to cater for the dynamic and iterative involvement of environmental elements and domain experts along the way.
- Infrastructure: The engagement of environmental elements and humans at run time in a dynamic and interactive way requires an open system with closed-loop interaction and feedback [45]. KDD infrastructures need to provide facilities to support such scenarios.
- Dynamics: To deal with the dynamics in data distribution from training to testing and from one domain to another is essential in relation to domain and organizational factors, human cognition and knowledge, the expectation of deliverables, and business processes and systems.
- Evaluation: Interestingness needs to be balanced between technical and business perspectives from both subjective [28] and objective [46, 47] aspects; special attention must be paid to deliverable formats, their actionability, and generalizable capability, as well as to securing the support of domain experts.
- Risk: Risk needs to be measured in terms of its presence and magnitude, if any, in conducting a KDD project and system.
- Policy: Data mining tasks often involve policy issues such as security, privacy, and trust which exist not only in the data and the environment but also in the use and management of data mining findings in an organization's environment.
- Delivery: This includes determining the right form of delivery and presentation of KDD models and findings so that end users can easily interpret, execute, utilize, and manage the resulting models and findings and integrate them into business processes and production systems.

On the microlevel, aspects related to technical and engineering issues that support KDD need to be addressed. Listed below are a few dimensions that address these concerns: architecture, procedure, interaction, adaptation, actionability, and deliverability.

- Architecture: KDD system architectures need to be effective and flexible to incorporate and consolidate specific environmental elements, KDD processes, evaluation systems, and final deliverables.
- Procedure: Tools and facilities supporting the KDD process and workflow are necessary, from business understanding, data understanding, and human–system interaction to the assessment, delivery, and execution of deliverables.
- Interaction: To cater for interaction with business people throughout the KDD process, appropriate user interfaces, user modeling, and servicing are required to support individuals and group interactions.
- Adaptation: Data, environmental elements, and business expectations change all the time.

It is necessary for KDD systems, models, and evaluation metrics to be adaptive to handling differences and changes in dynamic data distribution, cross domain situations, changing business situations, and user needs and expectations.

- Actionability: What do we mean by "actionability"? How can we measure it? What is the trade-off between the technical and business sides? Do subjective and objective perspectives matter? These issues require essential metrics and integration mechanisms to be developed.
- Deliverable: End users certainly feel more comfortable if the models and patterns delivered can be presented in a business-friendly way and be compatible with business operational systems and rules. In this sense, it is necessary for KDD deliverables to be easily interpretable and convertible into presentation in a business-oriented way, such as business rules, and to be linked to decision-making systems.

The above discussions contribute to this insight: KDD-based problem-solving is expected to be a process and system for discovering and delivering actionable knowledge [48]. Such *actionable knowledge discovery and delivery* (AKD) needs to systematically consider/involve problems, data, environment, model, and decisions, as well as optimization [49] in KDD. This brings us to the methodology of *domain-driven data mining* [1].

14.4 An AKD Framework

To enable the discovery of actionable knowledge, AKD is proposed to narrow the gaps in KDD. Domain-driven data mining (*D3M*) [1] has been proposed as a framework for AKD, to analyze the underlying problems and challenges facing traditional KDD methodology and systems. The purpose is to develop appropriate methodology and techniques to tackle the problems and changes that will enable AKD as well as the real-life problem-solving and decision support yielded by KDD deliverables.

14.4.1 AKD Problem Statement

Rather than focusing on what happens in the current KDD, we prefer to observe the nature of KDD-based problem-solving from an interdisciplinary perspective, by integrating the methodologies from other disciplines including system sciences, cybernetics, and complex systems [50]. This perspective produces a multidimensional view of AKD-based problem-solving. AKD is a six-dimension-based optimization process o [1–6]:

$$AKD ::= optimization(problem, data, environment, model, decision). \quad (14.1)$$

1. *Problem*: a problem is the KDD target, composed of data, business, environment, and needs, contributing to corresponding KDD objectives and business analytical goals and to the eventual evaluation and validation of KDD findings.
2. *Data*: extracted in a business problem, reflecting a mapping between a business problem-based space and an extracted/converted object-based space for decisions; risk often starts from conducting the mapping.
3. *Environment*: a problem and its data are enclosed in a certain environment, embodied through organizational or social factors, which need to be considered for a complete and genuine understanding of the problem; however, they are easily neglected, filtered, destroyed, or simplified in data extraction and analysis.
4. *Model*: a model is an appropriate tool to connect the data to proper decisions by addressing the underlying problem within an environment; a model is biased if it is not fully reflective of the problem, data, environment, or decision.
5. *Decision*: a decision is presented in terms of identified patterns or knowledge through KDD, which is believed to be an overarching solution addressing the underlying problem.
6. *Optimization*: optimization seeks a perfect match between the model and problem, data, environment, and decision against expectations.

Let us further discuss how the above AKD framework can be executed in the real world. First, AKD is a problem-solving process that progresses business problems (Ψ, with problem status τ) to problem-solving solutions (Φ) [2, 7, 8]:

$$\Psi(\cdot|\tau) \rightarrow \Phi(). \quad (14.2)$$

The process of finding problem-solving solutions ($\Phi()$) is a procedure to find the *actionable pattern set* $\widetilde{P}$ by employing all valid models M.

$$AKD^{mi \in M} \rightarrow Op \in PAct(p), \quad (14.3)$$

where $P = P^{m_1} UP^{m_2}, \ldots, UP^{m_n}$, $\text{Act}(\cdot)$ is the evaluation function, and $O(.)$ is the optimization function to extract actionable pattern $\widetilde{p}(\widetilde{p} \in \widetilde{p} \subset p)$, where $\text{Act}(\widetilde{p})$ beats a given benchmark.

For a pattern p, $\text{Act}(p)$ can be further measured in terms of *technical actionability* (*ta*(p)) and *business actionability* (*ba*(p)).

$$\text{Act}(p) = I(t_i(p), b_i(p)), \tag{14.4}$$

where $I(.)$ is an objective function for aggregating the contributions of all particular aspects of actionability from problem, data, environment, model, and decision.

Further, $\text{Act}(p)$ can be described in terms of *objective* (o) and *subjective* (s) factors from both *technical*

(t) and *business* (b) perspectives:

$$\text{Act}(p) = t_o(\mathbf{x}, \_p) \wedge t_s(\mathbf{x}, \_p) \wedge b_o(\mathbf{x}, \_p) \wedge b_s(\mathbf{x}, \_p), \tag{14.5}$$

c"$\wedge$" indicates the "aggregation" of the specific aspect of actionability. We say that p is truly *actionable* (i.e., $\widetilde{p}$) both to academia and business if it satisfies the following condition:

If:

$$\forall p \in \widetilde{P}, \exists \mathbf{x} : t_o(\mathbf{x}, p) \wedge t_s(\mathbf{x}, p) \wedge b_o(\mathbf{x}, p) \wedge b_s(\mathbf{x}, p) \to Act(p) \tag{14.6}$$

Then:

$$p \to \widetilde{p}. \tag{14.7}$$

Further, let $\widetilde{P} = (\widetilde{p}_1, \widetilde{p}_2, \ldots, \widetilde{p}_z)$ be an *actionable pattern set* mined by method m_n for the given problem Ψ (its data set is *DB*), in which each pattern $\widetilde{p}_z$ is *actionable* for the problem-solving if it satisfies the following conditions:

- 1.a. $t_i(\widetilde{p}_z) \geq t_{i,0}$; indicating the pattern $\widetilde{p}_z$ satisfying technical actionability t_i with threshold $t_{i,0}$.
- 1.b. $b_i(\widetilde{p}_z) \geq b_{i,0}$; indicating the pattern $\widetilde{p}_z$ satisfying business actionability *bi* with threshold $b_{i,0}$.
- 1.c. $R : \tau_1 \overset{A, m_n(\widetilde{p}_z)}{\longrightarrow} \tau_2$; the pattern can support business problem-solving (R) by taking action A and can correspondingly transform the problem status from initially nonoptimal state τ_1 to greatly improved state τ_2.

Actionable knowledge (patterns) can lead to effective actions for better results (decision, answer, conclusion, etc.). The process of discovering actionable knowledge, or AKD, forms a framework that engages knowledge discovery on data with problem and environment toward optimal evaluation and decisions to satisfy both technical and business expectation from objective and subjective perspectives. This differentiates AKD from the normal KDD process.

Due to the inconsistency that often exists in different aspects, we frequently find that the identified patterns only fit into one of the following subsets:

$$\text{Act}(p) \rightarrow \{\{t_i^{act}, b_i^{act}\}, \{\neg t_i^{act}, b_i^{act}\}, \{t_i^{act}, \neg b_i^{act}\}, \{\neg t_i^{act}, \neg b_i^{act}\}\} \quad (14.8)$$

where "¬" indicates that the corresponding element is not satisfactory. In real-world data mining, it is often very challenging to find the actionable patterns that are most associated with both "optimal" a_i^{act} and "optimal" b_i^{act}. Clearly, AKD favors patterns that confirm the relationship $\{a_i^{act}, b_i^{act}\}$.

14.4.2 Actionability Computing

Actionability means the power to work, which is an optimal outcome and objective of AKD through the best integration of six core dimensions. Consequently, actionability is embodied through each dimension and its integration:

1. *Actionability on problem* reflects the depth and width of our understanding of the underlying problem, its surroundings, constraints, and expected outcomes from AKD.
2. *Actionability on data* reflects the depth and width of our understanding of the underlying data complexity, structure, volume, dimensionality, type, speed, and dynamics.
3. *Actionability on environment* reflects the depth and width of our understanding of human, domain, organizational, and social aspects, as well as the interactions and dynamics surrounding the problem and the data.
4. *Actionability on model* reflects the quality of the models selected to understand the problem, data, and environment.
5. *Actionability on decision* reflects the operational power of the AKD deliverables for direct and effective problem-solving.
6. *Actionability on optimization* reflects the best mapping from the underlying problem to the expected decisions made by AKD models and the best combination of all dimensions.

In essence, actionability is the quality and power of AKD outcomes for effective decision-making and problem-solving. Actionability may be interpreted in varying terms for different purposes, for instance:

- *Autonomy* of the deliverables for direct use in an unattended problem-solving process or system
- *Deliverability* and *transferability* of the identified patterns and knowledge from data miners to business people and from one domain to another
- *Dependability* of the identified patterns and knowledge
- *Explainability* and *interpretability* [51] of the identified patterns and knowledge
- *Impact* of the deliverables leading to what is expected by business

- *Repeatability* of the proposed algorithms and methods
- *Semantics* and *understandability* of deliverables for seamless integration into business ontology and machine-based understanding and use
- *Trust* of the proposed algorithms and methods, as well as identified patterns and knowledge, without security and privacy offense or risk to the underlying problem and environment

Actionability computing therefore emerges as an interesting research issue in AKD. Formula (14.5) reflects the overall interpretation of actionability computing from the quantitative perspective, but there are nevertheless many open issues to be further explored; for instance, how to represent and quantify the trust, autonomy, and semantic quality of fraud detection models for online banking fraud control. This leads to many new opportunities, as suggested by the following topics, for further exploration in creating actionable knowledge [16]:

- What are the key attributes for actionable knowledge? How can knowledge be both scientifically rigorous and practically useful?
- One aspect of actionable knowledge involves the dissemination of research to practitioners so that they understand it and are willing to act on it. What forms of diffusion are most effective for this purpose? What kinds of communication and messages attract practitioners' attention and understanding? What are the mechanisms for translating research into practice?
- What research methods are likely to contribute to actionable knowledge? How open are we to different research methods? How can research questions be formulated and examined so that subsequent findings are likely to be implemented?
- A good deal of actionable knowledge is tacit and exists only in practice. How do we capture and make sense of such knowledge? How do we study it scientifically?
- Generating actionable knowledge involves an inherent tension between two radically different cultures: science that seeks knowledge that is internally valid and generalizable, and practice that asks for useful answers to situation-specific problems. How might these competing demands be managed so that there is greater appreciation and dialogue between the two cultures? What does each culture stand to gain and lose from interacting with the other? What should be the relationship between practitioners and researchers?
- How can practitioners help researchers formulate, conduct, and disseminate their research in more actionable ways? How can they inform researchers about the tacit dimensions of their practice? What valuable lessons can practitioners teach researchers, and how can this be done so that researchers will listen?
- How can we help practitioners become better consumers of knowledge about management? Can they be inoculated against fads?
- What can our scholarly journals do to close the gap between research and practice? Should authors be held accountable for reflecting on the action

possibilities of their findings? Should the implications for practice be more than an afterthought?

14.4.3 AKD Concept Map

A high-level concept map for developing AKD methodology and techniques consists of the following layers: domain problem, ubiquitous intelligence, theoretical foundation, supporting technique, and actionability computing:

- Domain problem: This, in general, targets complex knowledge from complex data in domain-specific applications and problems that cannot be well handled by existing data mining and knowledge discovery techniques. Such problems may include domain problems from retail to government to social networking, from either a sector or a specific business problem perspective.
- Ubiquitous intelligence: This refers to the intelligence surrounding AKD problem-solving, from data to domain, organizational, social, and human aspects, and the representation, synthesis, and consolidation of respective intelligence for AKD-based problem-solving.
- Theoretical foundation: This refers to the fundamental theories to enable AKD, either borrowed from many relevant disciplines from the information sciences to social sciences or invented for data sciences and analytics sciences, targeting the establishment of a family of scientific foundations for dealing with increasingly emergent complexities and challenges in data and analytics.
- Supporting technique: This refers to AKD techniques and tools to engage and consolidate ubiquitous intelligence, support knowledge representation and deliverables, cater for project and process management, and implement decision-making pursuant to the findings.
- Actionability computing: This refers to the quantification of the decision-making power of identified knowledge and deliverables by AKD and the presentation, delivery, and impact of AKD findings for direct decision-making.

Seriously addressing the above key components in AKD demands the necessary engagement and support to cater for problem, data, environment, model, decision, and optimization in KDD and the reshaping of KDD processes, modeling, and outcomes from technical, procedural, and business perspectives.

14.4.4 Ubiquitous Intelligence

The success of AKD relies on involving and integrating ubiquitous intelligence [1, 9] in a domain-specific application. This involves data intelligence, domain intelligence, network intelligence, human intelligence, and social intelligence, as well as the synthesis of ubiquitous intelligence.

14.4.4.1 Data Intelligence

Data intelligence indicates interesting information and stories about the formation of a business problem or driving forces. Typical efforts focus on handling data complexity such as on large-scale, multidimensional/high-dimensional, online/real-time, social media, multimedia, dynamic, highly frequent, uncertain, noisy, mixed structure aspects. Apart from the usual focus on exploring the complexity of data structure, quantity, speed, and characteristics from the individual data object perspective, coupling and interaction between data objects have not been seriously considered, yet they present challenges to AKD, such as what may happen if dependency between objects is considered in a similarity-based clustering process.

14.4.4.2 Domain Intelligence

Domain intelligence emerges from domain factors and resources that not only wrap a problem and its target data but also assist in problem understanding and problem-solving. Domain intelligence involves qualitative and quantitative aspects. These are instantiated in terms of such aspects as domain knowledge, background information, prior knowledge, expert knowledge, constraints, organizational factors, business processes, and workflow, as well as environment intelligence, business expectations, and interestingness.

14.4.4.3 Human Intelligence

Human intelligence refers to the explicit or direct involvement of human empirical knowledge, belief, intention, expectation, run-time supervision, evaluation, and expert groups in AKD. It also concerns the implicit or indirect involvement of human intelligence such as imaginary thinking, emotional intelligence, inspiration, brainstorming, reasoning inputs, and embodied cognition such as convergent thinking through interaction with other members in dynamic data mining and assessing identified patterns.

14.4.4.4 Network Intelligence

Network intelligence emerges from both Web intelligence and broad-based network intelligence such as information and resource distribution, linkages between distributed objects, hidden communities and groups, information and resources from networks, and, in particular, the Web, information retrieval, searching, and structuralization from distributed and textual data. The information and facilities from the networks surrounding the target business problem either consist of the problem constituents or contribute to useful information for actionable knowledge discovery and should therefore be catered for in AKD.

14.4.4.5 Social Intelligence

Social intelligence refers to the intelligence that lies behind group interactions, behaviors [10], and corresponding regulation. Social intelligence covers both human social intelligence and animated/agent-based social intelligence. Human social intelligence is related to aspects such as social interaction, group goals and intention, social cognition, emotional intelligence, consensus construction, and group decision. Social intelligence is often associated with social network intelligence and collective interaction, as well as business rules, law, trust, and reputation for governing the emergence and use of social intelligence.

The use of ubiquitous intelligence may take one of the following two paths: *single intelligence engagement* and *multiaspect intelligence engagement*. Examples of single intelligence engagement are the involvement of domain knowledge in data mining and the consideration of user preferences in data mining.

Multiaspect intelligence engagement aims to integrate ubiquitous intelligence as needed. It is very challenging but inevitable in mining complex enterprise applications.

It is often very difficult to integrate every type of intelligence into one data mining system, in addition to the challenges of modeling and involving a specific type of intelligence. New data mining methodologies and techniques need to be developed to involve ubiquitous intelligence in AKD. The theory of metasynthetic engineering [50, 52–54], agent mining [55–58], and integration of ubiquitous intelligence [59] may provide useful clues for synthesizing ubiquitous intelligence in the AKD process.

14.5 Deployment

14.5.1 Opportunities

To learn and analyze complex data in complex problems and systems, we have to address the issues facing existing KDD and apply the thinking outlined above. The gaps between the aims of AKD and the existing situation of KDD research and development reveal great opportunities for AKD research and development through incorporating the cognition and recognition of system complexities and methodologies in metasynthetic computing and engineering, if the problem more or less fits the nature of complex systems as discussed earlier.

Below, we list a number of observations from which we can further develop KDD by taking AKD as a complex system and problem-solving process:

- Complex applications: Although any application can be linked to AKD, we are particularly interested in complex applications (represented in the complexity of a problem, data, or environment). Complex enterprise applications will present major functional and nonfunctional requirements that cannot be handled by existing KDD approaches and will drive the development of AKD toward novel and effective methodologies, algorithms, and tools [54].

- Complex data: Data is becoming more complex in many aspects from type, volume, structure, speed, and dimensionality to dynamics. More powerful tools are needed to tackle such data complexity, as well as to consider similarity, dependency, and interaction between data points.
- Complex behaviors: Behavior can be seen everywhere and is an essential object in analyzing applications and data. Limited techniques are available for the effective general analysis and mining of complex behaviors, especially the representation and reasoning of complex behaviors, and the learning and mining of coupled behaviors, behavior networking, group behaviors, behavior convergence and divergence, impact analysis of group behaviors, and detection of complex behavior interaction patterns in group behaviors.
- Complex environments: Next-generation knowledge discovery will have to discover knowledge in complex environments, mixing elements of human, domain, organizational, and societal factors. Environment complexities present characteristics such as dynamics that need to be catered for in the architecture, model, and process.
- Actionability measure: New performance metrics will be developed to quantify the actionability [40] of KDD deliverables and services which are dependable, user-friendly and business-friendly, explainable, actionable, reliable, safe, trustworthy, repeatable, and transferable.
- Deliverable semantics: Besides patterns, the delivery format and semantics will be important issues to study, so that not only are the deliverables and properties of deliverables specified, they can also be transparently and seamlessly integrated into the operational environment for decision-making. Semantic interfaces and services [60] may be developed to embed AKD deliverables into operational systems.
- Decision power: The decision power of AKD deliverables is not determined by performance such as accuracy; rather, it is the utility and usability that can be directly taken over by business people and plugged into the operational environment, leading to better decisions or expected impact.
- AKD as service: Next-generation knowledge discovery may work in a cloud environment, which will trigger the development of knowledge discovery as a service for problem-solving and decision-making in an organization.
- In a standard situation, knowledge discovery services will be provided by AKD specialists in a highly advanced knowledge discovery center rather than fostering a knowledge discovery team in every organization,.
- This requires the development of corresponding infrastructure, networking and privacy-processing facilities, and protocols for defining, communicating, subscribing, and monitoring services.

14.5.2 AKD Architectures

To support the involvement of ubiquitous intelligence and the delivery of actionable knowledge, it is essential to develop effective system architectures for constructing AKD systems and effective techniques for supporting AKD. We briefly introduce a few flexible frameworks [1, 4, 7, 8] below:

Postanalysis-based AKD (PA-AKD) [61, 62] is a two-step pattern extraction and refinement exercise. First, generally interesting patterns (which we call "general patterns") are mined from data sets by technical interestingness (*to*(), *ts*()) associated with the algorithms used. The mined general patterns are then executed, distilled, and summarized into operable business rules (or embedding actions, which we call "deliverables") in terms of domain-specific business interestingness (*bo*(), *bs*()) and involving domain and meta-knowledge.

Unified interestingness-based AKD (UI-AKD) develops unified interestingness metrics which are defined for capturing and describing both business and technical concerns. The mined patterns are converted into deliverables based on domain knowledge and semantics. UI-AKD looks just the same as the normal data mining process except for three inherent characteristics. One is the interestingness system, which combines technical significance (t_i()) with business expectations (b_i()) into a unified interestingness-based AKD system (i()). This unified interestingness system is then used to extract truly interesting patterns. The second is that the domain knowledge and environment must be considered in the data mining process. Finally, the outputs are actionable patterns and operable business rules.

Combined interestingness-based AKD (CMAKD) comprises multiple steps of pattern extraction and refinement on the whole data set. First, J steps of mining are conducted based on business understanding, data understanding, exploratory analysis, and goal definition. Second, generally interesting patterns are extracted based on technical significance (t_i()) (or unified interestingness (i())) into a pattern subset (P_j) in step j. Third, knowledge obtained in step j is further fed into step $j+1$ or relevant remaining steps to guide the corresponding feature construction and pattern mining (P_{j+1}). Fourth, after the completion of all individual mining procedures, all identified pattern subsets are merged into a final pattern set based on environment, domain knowledge, and business expectations (b_i). Lastly, the merged patterns are converted into business rules as final deliverables (patterns and business rules) that reflect business preferences and needs.

14.5.3 AKD Implementation

Corresponding AKD techniques are necessary to fulfill the methodology of AKD.

14.5.3.1 Constrained Knowledge Delivery Environment

Actionable knowledge is discovered in a constraint-based context that mixes environmental reality, expectations, and constraints in the knowledge discovery and delivery process. Several types of constraints play a significant role in the AKD process: *domain constraints*, *data constraints*, *interestingness constraints*, and *deliverable constraints*. Efforts are needed to develop both generic and domain-specific tools and systems to cater for these constraints.

14.5.3.2 Cooperation Between Human and KDD Systems

The real-life requirements for discovering actionable knowledge in a constraint-based environment determine that real-world data mining is more likely to follow a man–machine-cooperation mode—in other words, a human–mining cooperation rather than an automated process and system. Human involvement is embodied through the cooperation between humans (including users and business analysts, mainly domain experts) and a data mining system. This is because of the complementation between human qualitative intelligence, such as domain knowledge and field supervision, and the quantitative intelligence of KDD systems, such as computational capabilities. Therefore, real-world complex data mining presents as a human–mining-cooperative interactive knowledge discovery and delivery process. The tasks are to develop theories and tools to support the involvement of humans and human intelligence into an AKD system.

14.5.3.3 Interactive and Parallel KDD Support

To support AKD, it is important to develop interactive mining support that involves domain experts and human–mining interaction. Interactive facilities are also useful for evaluating data mining findings by involving domain experts in a closed-loop manner.

On the other hand, parallel mining support is often necessary for dealing with concurrent applications of distributed and multiple data sources. In cases with intensive computation requests, distributed and parallel mining [63] can greatly upgrade real-world data mining performance. There are huge areas to explore in terms of developing interactive, visual, parallel, and distributed systems and tools for AKD in a complex environment, such as those involving multiple sources of data, information, resources, and humans.

14.5.3.4 Closed-Loop and Iterative Refinement

Actionable knowledge discovery in a constraint-based context is more likely to be a closed-loop process rather than an open-loop one. A closed-loop process indicates that the outputs of data mining are fed back to adjust/inform relevant parameter or factor tuning in particular stages. It is worthwhile to study what should be in the loop, how to engage and optimize the components in the loop, and when to terminate the iteration.

14.5.3.5 Mining In-Depth Patterns

Greater effort to uncover in-depth patterns in data is essential. "In-depth patterns" (or "deep patterns") are not straightforward and can only be discovered through more powerful models following thorough data and business understanding, and effectively involving domain intelligence or expert guidance. An example is to mine for insider trading patterns in capital markets.

Without deep understanding of the business and data, a naive approach is to analyze the price movement change by partitioning data in terms of pre-event, on the event, and post-event. A deeper pattern analysis on such price difference analysis may involve domain factors such as considering market or limit orders, market impact, fusion of price, index and announcement information, and checking the performance of *potential abnormal return*, *liquidity*, *volatility*, and *correlation*.

14.5.3.6 Post Mining

Post mining [62] handles the following problems: How to read and understand discovered patterns, which are often in thousands or more. Which are the most interesting? Is the model accurate and what does the model tell us? How should we use the rules, patterns, and models? To answer these questions and present useful knowledge to users, it is necessary to conduct post mining to further analyze the learned patterns, evaluate the built models, refine and polish the built models and discovered rules, summarize them, and use visualization techniques to make them easy to read and understand.

14.5.3.7 Combined Mining

Combined mining [8] is a general method of analyzing complex data for identifying complex knowledge. The deliverables of combined mining are *combined patterns*. For a given business problem (ψ), we suppose the following key entities are associated with it in discovering interesting knowledge for business decision support: data set D, feature set F, method set R, interestingness set I, impact set T, and pattern set P. On the basis of the above variables, a general pattern discovery process can be described as follows: patterns $P_{n,m,l}$ are identified through data mining method R_l deployed on features F_k from a dataset D_k in terms of interestingness $I_{m,l}$ [8]:

$$P_{n,m,1} : Rl(Fk) \rightarrow Im, l, \tag{14.9}$$

where $n = 1, \ldots, N$; $m = 1, \ldots, M$; $l = 1, \ldots, L$. From a high-level perspective, combined mining represents a generic framework for mining complex patterns in complex data.

14.5.3.8 Agent-Driven Actionable Knowledge Discovery

Agent-driven actionable knowledge discovery [53, 59, 64–70] can contribute to the problem-solving of many data mining issues; for example, multiagent data mining infrastructure and architecture, multiagent interactive mining, multiagent-based user interaction, automated pattern mining, multiagent-distributed data mining, multiagent dynamic mining, multiagent mobility mining, multiagent multiple data source mining, multiagent peer-to-peer data mining, and multiagent Web mining. Agent technology can help with these challenges by incorporating autonomy, interaction, dynamic selection and gathering, scalability, multistrategy, and collaboration. Other challenges include privacy, mobility, time constraints (e.g., with streaming data there is no time to extract and then mine), and computational costs and performance requests.

14.5.3.9 Knowledge Discovery as Service

This is the era of efficiently mining for actionable knowledge to support specific, ad hoc, and intention-driven needs for organizational decision-making by involving distributed, highly heterogeneous, dynamic, and ubiquitous intelligence in complex environments, such as intranets, extranets, cloud, and grid. This will require breakthroughs in managing and integrating different resources from different channels, implementing knowledge discovery as services, and making services available to satisfy diverse and changing requirements.

14.5.4 Knowledge Delivery

Well-experienced data mining professionals attribute the weak executable capability of existing data mining findings to the lack of proper tools and mechanisms for implementing the ideal deployment of the resulting models and algorithms by business users rather than analysts. In fact, the barrier and gap arise from the weak, if not nonexistent, capability of existing data mining deployment systems and services, found in the presentation, deliverable, and execution aspects. They form the AKD delivery system, which is much beyond the identified patterns and models themselves:

- Deliverable: studies how to deliver data mining findings and systems to business users so that the findings can be readily reformatted, transformed, or cut and pasted into their own business systems to be presented on demand, and ensures that the systems can be understood and taken over by end users.
- Presentation: studies how to present data mining findings that can be easily recognized, interpreted, and taken over as needed.
- Execution: studies how to integrate data mining findings and systems into production systems and how the findings can be executed easily and seamlessly in an operational environment. Supporting techniques need to be developed for AKD presentation, deliverability, and execution. Some such techniques are as follows:
- Deliverable: business rules are widely used in business organizations, and one method for delivering patterns is to convert them into business rules; for this, we can develop a tool with underlying ontologies and semantics to support the transfer from pattern to business rules.
- Presentation: typical tools such as visualization techniques are essentially helpful; visual mining could support the whole data mining process in a visual manner.
- Execution: tools to make deliverables executable in an organization's environment need to be developed; one such effort is to generate Predictive Model Markup Language (PMML) to convert models to executables so that the models can be integrated into production systems and run on a regular basis to provide cases for business management.
- Communication plan: a document to craft the right information, sell the right stories, communicate the value of the deliverables for targeted clients, set up possible goals to achieve, and identify services, products, processes, and tools to disseminate or share [35].

14.6 An Example

Basket analysis is a typical application for illustrating the power of association mining for business applications. In practice, frequent association rules such as *{diaper, beer}* are arguably not profitable for a department store because putting two such items as these together would significantly reduce the potential for purchasing other items on the way from the bottle shop to the baby product section. In many countries, it is even not permitted to put beer and diapers in the same section.

Let us take a fruit retail shop as an example and explain how to make rules more actionable. We extract shopping transactions from 100 customers and identify the following association rules by using the Apriori algorithm:

$R1$: {apple, banana, 95}
$R2$: {apple, apple–mango juice, 80}
$R3$: {apple, Harry Potter, 20}

Which rule would you choose if you were the shop owner? According to the association rule theory, R_1 is recommended because it has the highest *support*.

In practice, no shop owner would follow this rule to put apples and bananas on the same shelf, since they are normally in the same section, and owners may want to promote other fruit during a customer's search path from apple to banana. Thus, R_1 should be filtered. Let us further add the item price: apple, \$2/kg; apple–mango juice, \$4/bottle; and Harry Potter, \$39.99/copy. By adding the prices, we have R_2 and R_3:

R_2: {(apple, \$2), (apple–mango juice, \$4), 80}
R_3: {(apple, \$2), (Harry Potter, \$39.99), 20}

Suppose each customer definitely only buys 1 unit per shopping trip, which rule would you choose if you were the shop owner? Technically, R_2 is recommended because:

$\text{supp}(R_2) > \text{supp}(R_3)$.

From the revenue perspective, R_3 is more profitable:

$\text{profit}(R_2) = \$48 > \text{profit}(R_3) = \$84$.

This shows the importance of involving additional data and evaluating the business impact of the findings during pattern mining.

The above findings arise from the traditional association rule framework, frequency-based similarity analysis, which treats transactions independently. In reality, we know that different family members may visit the same shopping center to purchase consumables for the family's use. This makes the transactions from the same family dependent on each other. Let us look at a retail database as shown in Table 14.1. If *min supp* $= 2$, based on traditional association rule mining, we have three shopping patterns for individual customers:

R_4: {Beer, iPhone}
R_5: {Cherry}
R_6: {Scooter}

However, if we consider the dependency between customers in terms of family role and address:

C_1-Father, $C2$-Mum, $C3$-Son;
C_1-Mum, $C2$-Father, $C3$-Son;
$C_1.\text{Address} = C2.\text{Address} = C3.\text{Address}$;
$C_4.\text{Address} = C5.\text{Address} = C6.\text{Address}$;

and the constraint that a family member does not purchase the same item that other members have purchased for the same week:

Table 14.1 Basket analysis database

Sequence ID	Date	Customer role	Address	Shopping cart
C_1	2011-10-1	Father	A	Beer, diaper, banana, Harry Potter, iPhone
C_2	2011-10-2	Mum	A	Apple, cherry, blackberry, plum
C_3	2011-10-3	Son	A	Pencil case, rubber, Lego, scooter
C_4	2011-10-4	Mum	B	Pear, cherry, peach, plum, melon, apple
C_5	2011-10-4	Father	B	Beer, iPhone, fish, meat
C_6	2011-10-4	Son	B	Scooter, pen, notebooks

C_1.Week = C_2.Week = C_3.Week;
C_4.Week = C_5.Week = C_6.Week;

we have a single family shopping pattern:

R_7: {Father: Beer, iPhone; Mum: Cherry; Son: Scooter}
R_4, R_5, and R_6 have the same concerns as R_1.

By involving object relationships, this family pattern is clearly more informative for a shopping center's marketing campaign. For instance, the following marketing strategy may be designed to promote iPhone:

Definition 1 ***(Marketing Strategy 1)*** *If: Father buys Beer, Mum buys Cherry, and Son buys Scooter*

Then: 30 % off for another iPhone or a promotion pricing package for family shopping

Definition 2 ***(Marketing Strategy 2)*** *Individual: Beer, \$30/box; iPhone, \$900; Cherry, \$15/kg; Scooter, \$100*

Package: Beer, \$30/box; iPhone, \$900; Cherry, \$15/kg; Scooter, \$0

The above sample marketing strategies illustrate the delivery of AKD-based association mining for improving retail business.

14.7 Summary

AKD is a challenging and fundamental task in data mining and the knowledge discovery revolution. It involves the discovery and delivery of knowledge from increasingly complex data, problems, and systems. It is necessary to tackle system complexities in designing new KDD, as well as machine learning methodologies, theories, and tools to address the complexities. System methodologies such as metasynthetic computing are helpful for deepening our understanding of complex data and problems and discovering actionable knowledge that can be transparently and seamlessly incorporated into an operational environment for smart information use and decision-making. It is distinguished from traditional data mining by the extensive involvement of ubiquitous intelligence in the KDD process and the elevation of the decision power of the findings.

AKD complements other data mining methodologies and techniques by explicitly involving human, domain, organizational, and social intelligence and their synthesis rather than simplifying or overlooking these factors in the modeling and evaluation process. This empowers the end user and decision-maker to easily understand the outcomes and take actions on the findings to enable the transparent and seamless integration of data mining outputs into the operational environment.

The direct engagement of ubiquitous intelligence and the delivery of actionable knowledge into research and development have not been fully investigated, however. As a result, there are increasing gaps and imbalance between business expectation and problem-solving needs, and research focus and deliverables. We argue that this is because much of the research has been driven by innovating algorithms and exploring new data complexities, rather than focusing on the intrinsic complexities and challenges of the underlying problems and the expectations of enterprise applications.

With the already substantial development of algorithms and techniques, next-generation data mining will arguably benefit from practice-based research innovation and development, concentration on the underlying business problem itself, the reduction of oversimplification and abstraction of the problem in constructing solutions, and in-depth observations of core challenges. This may lead to breakthrough methodology and techniques for data mining and render the outputs workable and actionable for real-life problem-solving.

Many opportunities may materialize during the deep investigation of fundamental problems. These may lie in those areas widely explored, such as the involvement of domain knowledge in the dynamic data mining process, as well as emerging issues such as engaging organizational and social intelligence in the KDD modeling process. This makes furthering the paradigm shift from knowledge discovery toward AKD a very promising proposition.

References

1. Cao, L.: Domain-driven data mining: challenges and prospects. IEEE Trans. Knowl. Data Eng. **22**, 755–769 (2010)
2. Cao, L., Zhang, C.: The evolution of KDD: towards domain-driven data mining. Int. J. Pattern Recogn. Artif. Intell. **21**, 677–692 (2007)
3. Cao, L., Zhang, C.: Domain-driven actionable knowledge discovery in the real world. In: Proceedings of PAKDD2006. LNAI, vol. 3918, pp. 821–830. Springer, Heidelberg (2006)
4. Cao, L., Yu, P., Zhang, C., Zhao, Y.: Domain Driven Data Mining. Springer, New York (2010)
5. Cao, L.: Domain-driven actionable knowledge discovery. IEEE Intell. Syst. **22**, 78–89 (2007)
6. Cao, L., Zhang, C.: Domain-driven data mining: a practical methodology. Int. J. Data Warehousing Min. **2**, 49–65 (2006)
7. Cao, L., Zhao, Y., Zhang, H., Luo, D., Zhang, C.: Flexible frameworks for actionable knowledge discovery. IEEE Trans. Knowl. Data Eng. **22**, 1299–1312 (2010)

8. Cao, L., Zhang, H., Zhao, Y., Luo, D., Zhang, C.: Combined mining: discovering informative knowledge in complex data. IEEE Trans. SMC Part B **41**, 699–712 (2011)
9. Cao, L., Luo, D., Zhang, C.: Ubiquitous intelligence in agent mining. In: Proceedings of ADMI 2009. LNCS, vol. 5680, pp. 23–35. Springer, Heidelberg (2009)
10. Cao, L.: In-depth behavior understanding and use: the behavior informatics approach. Inf. Sci. **180**, 3067–3085 (2010)
11. Argyris, C.: Knowledge for Action: A Guide to Overcoming Barriers to Organizational Change. Josssey-Bass, San Francisco (1993)
12. Argyris, C.: Actionable knowledge: intent versus actuality. J. Appl. Behav. Sci. **32**, 441–444 (1996)
13. Argyris, C.: Actionable knowledge: design causality in the service of consequential theory. J. Appl. Behav. Sci. **32**, 390–406 (1996)
14. Cross, R., Sproull, L.: More than an answer: information relationships for actionable knowledge. Organ. Sci. **15**, 446–462 (2004)
15. Morgan, K., Morabito, J., Merino, D.: Creating actionable knowledge within the organization to achieve a competitive advantage. www.coalescentknowledge.com/Slides/Presentation-2.pdf (2012). Accessed 13 Jan 2012
16. Cummings, T., Jones, Y.: Creating actionable knowledge. http://meetings.aomonline.org/2004/theme.htm (2012). Accessed 13 Jan 2012
17. He, Z., Xu, X., Deng, S.: Data mining for actionable knowledge: a survey. http://arxiv.org/abs/cs/0501079 (2012). Accessed 13 Jan 2012
18. Cras, Y.: Turning raw data into actionable knowledge: known challenges and new complexities. In: Proceedings of CSDM. Springer, Heidelberg (2010)
19. Barrett, J., Caruthers, E., German, K., Hamby, E., Lofthus, R., Srinivas, S., Ells, E.: From data to actionable knowledge: a collaborative effort with educators. In: Proceedings of KDD. ACM Press (2011)
20. Adomavicius, G., Tuzhilin, A.: Discovery of actionable patterns in databases: the action hierarchy approach. In: Proceedings of KDD1997, pp. 111–114. ACM Press (1997)
21. Aggarwal, C.: Towards effective and interpretable data mining by visual interaction. ACM SIGKDD Explor. Newslett. **3**, 11–22 (2002)
22. Ankerst, M.: Report on the SIGKDD-2002 panel—the perfect data mining tool: interactive or automated? ACM SIGKDD Explor. Newslett. **4**, 110–111 (2002)
23. Boulicaut, J., Jeudy, B.: Constraint-based data mining. In: The Data Mining and Knowledge Discovery Handbook, pp. 399–416. Springer, New York (2005)
24. Fayyad, U., Smyth, P.: From data mining to knowledge discovery: an overview. In: Fayyad, U., Piatetsky-Shapiro, G., Smyth, P., Uthurusamy, R. (eds.) Advances in Knowledge Discovery and Data Mining, pp. 1–34. AAAI Press, Menlo Park (1996)
25. Fayyad, U., Shapiro, G., Uthurusamy, R.: Summary from the KDD-03 panel—data mining: the next 10 years. ACM SIGKDD Explor. Newslett. **5**, 191–196 (2003)
26. Silberschatz, A., Tuzhilin, A.: On subjective measures of interestingness in knowledge discovery. In: Proceedings of KDD, pp. 275–281. ACM Press (1995)
27. Silberschatz, A., Tuzhilin, A.: What makes patterns interesting in knowledge discovery systems. IEEE Trans. Knowl. Data Eng. **8**, 970–974 (1996)
28. Liu, B., Hsu, W., Chen, S., Ma, Y.: Analyzing subjective interestingness of association rules. IEEE Intell. Syst. **15**, 47–55 (2000)
29. Padmanabhan, B., Tuzhilin, A.: A belief-driven method for discovering unexpected patterns. In: Proceedings of KDD-98. ACM Press (1998)
30. Suzuki, E.: Autonomous discovery of reliable exception rules. In: Proceedings of KDD. ACM Press (1997)
31. Al Hegami, A., Bhatnagar, V., Kumar, N.: Novelty framework for knowledge discovery in databases. In: Proceedings of DaWaK. Springer, Heidelberg (2004)
32. Kaur, H.: Actionable rules: issues and new directions. Trans. Eng. Comput. Technol. World Info. Soc. **5**, 61–64 (2005)

33. Ras, Z.W., Wieczorkowska, A.: Action rules: how to increase profit of a company. In: Proceedings of PKDD00. LNCS/LNAI **1910**, 587–592 (2000)
34. Tsay, L.S., Ras, Z.W.: Action rules discovery system DEAR2, method and experiments. J. Exp. Theor. Artif. Intell. **17**, 119–128 (2005)
35. Dunoff, A.: Turning data into actionable knowledge. mcmorrowreport.com/WPs/WPActionableKnowledge.pdf (2007). Accessed 12 June 2013
36. Dynamia2004: Generate actionable knowledge to improve business performance. www.statoo.com/en/PDF/Dynamia2004.pdf (2012). Accessed 13 Jan 2012
37. Kumar, S.: Transforming data for actionable knowledge. http://www.youtube.com/watch?v=DV8-Hk7HnAQ&noredirect=1 (2012). Accessed 13 Jan 2012
38. Julisch, D.: Intrusion detection alarms for actionable knowledge. In: ACM SIGKDD International Conference on Knowledge Discovery and DataMining, pp. 366–375. ACM Press (2002)
39. Yang, Q., Ling, C., Gao, J.: Mining web logs for actionable knowledge. In: Zhong, N., Liu, J. (eds.) Intelligent Technologies for Information Analysis. Springer, Heidelberg (1998)
40. Cao, L., Zhang, C.: Knowledge actionability: satisfying technical and business interestingness. Int. J. Bus. Intell. Data Min. **2**, 496–514 (2007)
41. Saha, B., Kakani, R.: Knowledge, power and action: towards an understanding of implementation failures in a government scheme. AI Soc. **21**, 72–92 (2007)
42. Montecel, R.: Knowledge and action from dropping out to holding on. IDRA Newsletter, Nov–Dec (2006)
43. Sexton, M., Lu, S.: The challenges of creating actionable knowledge: an action research perspective. Constr. Manag. Econ. **27**, 683–694 (2009)
44. Kleinberg, J., Papadimitriou, C., Raghavan, P.: A microeconomic view of data mining. Data Min. Knowl. Discov. **2**, 311–324 (1998)
45. Xin, D., Shen, X., Mei, Q., Han, J.: Discovering interesting patterns through user's interactive feedback. In: Proceedings of the 2006 ACM SIGKDD International Conference on Knowledge Discovery and Data Mining (KDD'06), pp. 773–778. ACM Press (2006)
46. Freitas, A.: On objective measures of rule surprisingness. In: Proceedings of PKDD98, pp. 1–9. Springer, Heidelberg (1998)
47. Hilderman, R., Hamilton, H.: Applying objective interestingness measures in data mining systems. In: Proceedings of PKDD00, pp. 432–439. Springer, Heidelberg (2000)
48. Yang, Q., Yin, J., Ling, C., Pan, R.: Extracting actionable knowledge from decision trees. IEEE Trans. Knowl. Data Eng. **19**, 43–56 (2007)
49. Freitas, A.: A critical review of multi–objective optimization in data mining—a position paper. SIGKDD Explor. **6**, 77–86 (2004)
50. Cao, L., Dai, R., Zhou, M.: Metasynthesis: M–space. M-interaction and M-computing for open complex giant systems. IEEE Trans. Syst. Man. Cybern. A **39**, 1007–1021 (2009)
51. Yao, Y., Zhao, Y.: Explanation-oriented data mining. In: Wang, J. (ed.) Encyclopedia of Data Warehousing and Mining, pp. 492–497. Idea Group Reference, Hershey (2005)
52. Qian, X., Yu, J., Dai, R.: A new discipline of science—the study of open complex giant system and its methodology. China J. Syst. Eng. Electron. **4**, 2–12 (1993)
53. Cao, L., Dai, R.: Agent-oriented metasynthetic engineering for decision making. Int. J. Inf. Technol. Decis. Mak. **2**, 197–215 (2003)
54. Cao, L., Dai, R.: Open Complex Intelligent Systems. Posts and Telecom Press, Beijing (2008)
55. Cao, L.: Integrating agent, service and organizational computing. Int. J. Softw. Eng. Knowl. Eng. **18**, 573–596 (2008)
56. Cao, L., Weiss, G., Yu, P.: A brief introduction to agent mining. J. Auton. Agents Multiagent Syst. **25**, 419–424 (2012)
57. Gorodetsky, V., Karsaev, O., Samoilov, V.: Multi-agent technology for distributed data mining and classification. In: Proceedings of IAT 2003, pp. 438–441. IEEE Computer Society Press (2003)
58. Symeonidis, A., Mitkas, P.: Agent Intelligence Through Data Mining. Springer, Heidelberg (2006)

59. Cao, L.: Data Mining and Multiagent Integration. Springer, Heidelberg/New York (2009)
60. Singh, M., Huhns, M.: Service Oriented Computing: Semantics. Processes and Agents, Wiley, Chichester (2005)
61. Liu, B., Hsu, W.: Post-analysis of learned rules. In: Proceedings of AAAI/IAAI **1**, 828–834 (1996)
62. Zhao, Y., Zhang, C., Cao, L.: Post-Mining of Association Rules: Techniques for Effective Knowledge Extraction. Information Science Reference, Hershey (2009)
63. Kargupta, H., Chan, P., Kumar, V.: Advances in Distributed and Parallel Knowledge Discovery. AAAI Press, Menlo Park (2000)
64. Cao, L., Gorodetsky, V., Mitkas, P.: Agent mining: the synergy of agents and data mining. IEEE Intell. Syst. **24**, 64–72 (2009)
65. Cao, L., Luo, C., Zhang, C.: Agent-mining interaction: an emerging area. In: Proceedings of AIS-ADM, pp. 60–73. Springer, Heidelberg (2007)
66. Cao, L., Gorodetsky, V., Liu, J., Weiss, G.: Agents and Data Mining Interaction, Proceedings of ADMI 2009. LNCS, vol. 5680, pp. 23–35. Springer, Heidelberg (2009)
67. Baik, S., Cho, J., Bala, J.: Performance evaluation of an agent based distributed data mining system. Adv. Artif. Intell. **3501**, 91–99 (2005)
68. Dasilva, J., Giannella, C., Bhargava, R., Kargupta, H., Klusch, M.: Distributed data mining and agents. Eng. Appl. Artif. Intell. **18**, 791–807 (2005)
69. Davies, W.: Agent-based data-mining. http://www.agent.ai/doc/upload/200403/davi941.pdf (2012). Accessed 13 Jan 2012
70. Davies, W., Edwards, P.: Distributed learning: an agent based approach to data-mining. In: Proceedings of Machine Learning 95 Workshop on Agents that Learn from Other Agents (1995)

Chapter 15
Learning Complex Behavioral and Social Data

15.1 Introduction

Behavioral and social applications are ubiquitous, ranging from business and online applications to social and organizational applications and domains. With the increasing and continuous development of such applications, an emerging need is to develop an in-depth understanding of the underlying working mechanism, driving force, dynamics and evolution of a behavioral and/or social system, as well as the impact on business and context. To this end, building on the classic theories and tools available in behavioral science, social science, behavior informatics [1, 2], and social informatics [3][1] have recently been studied to "formalize," "quantify," and "compute" complex behavioral and social applications.

As an emerging area of research, behavior and social informatics is in its earliest stage and features many challenges and opportunities. A canonical trend is to develop theories, tools, and algorithms based on the classic outcomes available in extant disciplines including statistics, data mining, and machine learning. Typically, frequent pattern mining, clustering, and classification of behavioral and social applications are conducted by expanding the corresponding existing theories and algorithms. In this chapter, we discuss the potential issues and risk in pursuing this path for complex behavioral and social applications by explicitly or implicitly taking the IIDness assumption, and thus reveal the need for developing non-IIDness learning for behavior and social informatics.

Arguably, most of the existing theories, tools, and systems in statistics, data mining, and machine learning are built on the IIDness assumption, which assumes the independence and identical distribution of the underlying objects, attributes, and/or values. Based on a high-level abstraction, it is assumed that objects,

[1] See more from the IEEE Task Force on Behavioral, Economic, and Socio-cultural Computing: www.bsic.info.

L. Cao, *Metasynthetic Computing and Engineering of Complex Systems*,
Advanced Information and Knowledge Processing,
DOI 10.1007/978-1-4471-6551-4_15

attributes, and values are independent and identically distributed, with most existing learning theories, models, and algorithms being proposed on the basis of this assumption. This works well in simple business applications and abstract problems with weakened and avoidable relations and heterogeneity, and serves as the foundation of classic analytics, mining and learning theories, algorithms, systems, and tools.

Complex behavioral and social applications often exhibit strong coupling relations (which are beyond the usual dependency relation) and heterogeneity between objects, object attributes, and attribute values, which cannot be abstracted or weakened to the extent of satisfying the IIDness assumption. Couplings may be presented in a variety of forms and levels on different objects, attributes, and/or values. Heterogeneity is reflected through multiple or mixed structures or distributions within objects, attributes, and/or values. This makes it necessary and unavoidable to consider coupling and heterogeneity in behavior and social informatics. Accordingly, non-IIDness learning emerges as a crucial issue, even though it has not been extensively studied or recognized in the statistics, data mining, and machine learning communities [4–19].

Motivated by the above challenges and prospects, this chapter focuses on a high-level discovery of the IIDness nature of the classic analytics and learning systems and the intrinsic need and fundamental principles of non-IIDness learning for tackling complex analytics and learning problems. As the underlying problem is so novel but widespread and challenging, it is not our intention to provide a unified solution or framework here as it is beyond our existing capability. However, we intend to share some of the preliminary efforts made toward considering non-IIDness in complex analytics and learning tasks.

In particular, we discuss the characteristics of complex behavioral and social problems in Sect. 15.2. Extended discussions are given in Sect. 15.3 about the non-IIDness feature of behavioral and social data. Section 15.4 presents the issues associated with the IIDness-based algorithms in classic behavior analysis, social media and recommendation systems, and in social network analysis. Section 15.5 introduces high-level concepts and principles of non-IIDness learning. Preliminary explorations for non-IIDness learning and case studies are given in Sect. 15.6, followed by the conclusions drawn in Sect. 15.7.

15.2 Complex Behavioral and Social Problems

15.2.1 Behavioral and Social System and Intelligence

From an abstract perspective, a behavior and a social event can be described in terms of a four-element tuple [1, 20], consisting of *actor* (subject and/or object), *operation* (activity and activity properties), *relation* (interactions), and *context* (including environment). For an in-depth understanding of behaviors [1], interaction, coupling relationships, semantics, dynamics, change, and impact and utility are important factors to consider.

In [1], an abstract behavior model is presented. A behavior is a vector with properties that describe key aspects including subject, object, action, status, time, place, goal, plan, belief, constraint, context, associate, and impact. The couplings between behaviors from one or more actors may take place on a behavior property or across different properties, such as the temporal relationship between the behaviors of a blogger, or a causal relationship between cars involved in an accident.

A social system may be interpreted in terms of the theory of open complex systems [21–24], such as a complex multiagent system. One way to explore a complex social system if this is applicable is to adopt the OSOAD methodology [23, 24]. The OSOAD methodology argues that a complex system consists of the following key working components and mechanisms: organization, goal, actor, role, rule, relationship, interaction, and environment, and also involves ubiquitous intelligence including data intelligence, domain intelligence, human intelligence, behavior intelligence, social intelligence, organizational intelligence, and network intelligence.

Accordingly, the emerging field of behavior informatics [1, 2] and social informatics [3] aims to reveal deep behavior intelligence and social intelligence in behavioral and social systems.

Behavior intelligence refers to the intelligence generated through analyzing the process, impact, and utility of a collection of activities conducted by a range of actors in a certain context. From the scale perspective, we may be interested in *individual behavior intelligence*, *group behavior intelligence*, or *collective behavior intelligence* [25].

An example of implementing individual behavior intelligence can be seen when an investor purchases a stock that accrues an expected profit. Pool manipulation reflects negative group behavior intelligence. Financial crisis is a negative presentation of collective behavior intelligence.

Social intelligence refers to the intelligence that emerges from group interactions, behaviors, and the corresponding regulation during a process within a context. We are concerned about *human social intelligence* and *animated/agent-based social intelligence* [25].

Human social intelligence is embodied in aspects such as social cognition, emotional intelligence, culture, consensus construction, and group decision. Animated/agent-based social intelligence involves swarm intelligence, action selection, and the foraging procedure. Both sides also engage social network intelligence and collective interaction as well as social regulation rules, law, trust, and reputation for governing the emergence and use of social intelligence.

Our goal here is to analyze, mine, and learn deep behavior and social intelligence from behavioral and social systems by developing corresponding theories and techniques. Before we specify our task in non-IIDness learning for deep behavior and social intelligence, we discuss the complexity embedded in complex behavioral and social systems.

15.2.2 Complexity of Behavioral and Social Systems

Complex behavioral and social problems exhibit intricacies that greatly challenge existing theories and techniques. The discussions about open complex intelligent systems [21, 22] and ubiquitous intelligence [25] provide high-level hints for an in-depth understanding of a complex system.

According to the theory of open complex intelligent systems, system complexity is embodied in human engagement, openness, interaction, environment, hierarchy, and evolution. These aspects are embodied in behavioral and social systems in terms of specific corresponding entities and attributes. For instance, from the hierarchical perspective, a social system may consist of multiple levels of components, forming subsystems, subsystem constituents, and constituent properties. Interaction may take place on a variety of levels and in various forms, such as on global and local levels and in terms of following and followed roles between or within a social group, subgroup, or nodes.

The complexities and characteristics of complex behavioral and social problems may therefore be discussed from the following aspects: *openness*, *large scale*, *heterogeneity*, *hierarchy*, *networking*, *coupling relationships*, *societal characteristics*, and *dynamic characteristics* [21–24, 26, 27].

Openness reflects the exchange of energy, information, and materials between a behavioral/social system and its external environment. A behavioral/social system often involves or is composed of hundreds or even millions of actors and/or operations [22], forming a very *large scale*. There may be many types or forms of behaviors, behavioral actors, data sources, relationships, and even impact making up the system components. This results in strong *heterogeneous characteristics*. Such system components in a behavioral/social system are likely to be organized in a *hierarchical structure* in a network. The *networking* that exists between system components is the intrinsic driving force of behavior/social intelligence emergence. Networking is further driven by different *coupling relationships* between actors, behaviors, and context from temporal, inferential, combinational, and party-based aspects [20]. Couplings existing in behavioral/social systems cause the underlying objects to be dependent on each other.

The *societal characteristics* of a behavioral/social system may be embodied in many social factors such as the laws of business, politics, organizational factors, and business processes. In addition, behavioral/social systems are *dynamic* in the sense that they may change states, working mechanisms, constituents, and internal and external interaction mechanisms at any time, often beyond imagination.

The discussions about ubiquitous intelligence [25, 27] offer additional aspects from which to explore the complexity in a complex behavioral and social system. The notion of ubiquitous intelligence argues the need to consider the following types of intelligence embedded explicitly or implicitly in a complex system: human intelligence, domain intelligence, behavior intelligence, data intelligence, organizational intelligence, social intelligence, and networking intelligence.

In analyzing complex behavioral and social systems, we may specifically explore system complexity from the perspective of *data*, *domain*, *context*, and *impact* respectively.

Data represents the information generated directly by behavioral and social systems and by the management systems that govern behavioral and social problems. *Domain* refers to the broad area in which the underlying behavioral and social problems exist or reside. *Context* is the particular environment which surrounds a specific behavioral and social problem. *Impact* is indicated by the outcomes produced by behavioral and social systems.

This analysis of the underlying characteristics and complexities in behavioral/social systems discloses that behavior/social systems are strongly dependent and heterogeneous. This is inconsistent with the assumption of *IIDness*, i.e., that they are independent and identically distributed.

15.3 Non-IID Behavioral and Social Problems

The above analysis shows that non-IID characteristics are intrinsic to complex behavioral and social systems. Here, we further specify the coupling and heterogeneity aspects in behavioral and social systems.

15.3.1 Coupling

Following the abstract behavior model in [1], coupling may take place within and between behavior attributes, on different levels in a system. As previously discussed, interactions in open complex intelligent systems may take place within and between system elements, subsystems, and system and environment.

Couplings in complex behavioral and social systems may have different forms and structures which are often mixed with each other. Such couplings may need to be explored from structural, semantic, probabilistic and mathematical, dynamic, and/or graphical perspectives.

Different types of coupling relationships exist in behavioral and social systems. As discussed in [20, 28], the following couplings may appear between users, between items, and between users and items in a social media system, or between the elements and components of a behavioral system.

1. *Serial coupling*: One behavior takes place after another, or one item is purchased after another; for example, one comment in a blog triggers another comment on the same topic.
2. *Causal coupling*: One behavior causes the occurrence of another behavior, or one social state is caused by another; for instance, a breaking news item causes a significant increase in concern in social media.

3. *Synchronous coupling*: All behaviors or social events occur at the same time; for instance, two bloggers comment on the same issue on different social media at the same time.
4. *Exclusive coupling*: Different events happen on a mutually exclusive basis; for instance, two opposing groups express different views on the same social event in a blog.
5. *Dependent coupling*: Some behaviors or social events require dependents such as prefix or postfix components; for instance, the occurrence of a behavior is associated with the pre-occurrence of a series of other behaviors.

By targeting different types of behavior or social events, couplings may present in different forms. Couplings in numeric data are very different from those in categorical data. From the number of involved behavioral and social attributes, couplings include single attribute-based couplings, such as temporal coupling, and compound couplings, such as hierarchical coupling. From the knowledge representation aspect, syntactic coupling, semantic coupling, and inferential coupling can be explored. In [20, 28], different temporal couplings and inferential couplings are discussed for coupled behaviors.

In addition, there are couplings on different levels, from value, attribute, object, method, and measure to pattern. Such couplings, which are more comprehensive and complex than correlation and association, refer to the relations that exist explicitly or implicitly between source and destination entities. A source or destination entity can be a value, attribute, object, method, or pattern in a behavioral or social system.

For example, there are user–user couplings, item–item couplings, and user–item couplings in a recommendation system. Item attributes such as item price and quantity are often associated with each other. The price of one item may affect the price of another. An item may influence the sale market of another. In recommendation modeling, different methods may focus on specific aspects, and there may be a need to integrate multiple methods to cater for comprehensive couplings between item attributes, users, and items, and between users and items.

The above comprehensive couplings are often ignored in related work. Only certain relation or correlation is considered. For example, in recommendation systems, only user–user influence or item–item co-occurrence is considered.

15.3.2 Heterogeneity

In behavioral and social systems, heterogeneity may appear in different aspects, input data sources, value types, object types, etc. Often a behavioral or social system involves multiple heterogeneous (multi-structured or mixed-structured) data sources. They may be composed of divided value distributions, heterogeneous attributes, nonidentical distributions of data subsets, and thereafter heterogeneous objects.

1. Values: Often different types of values present in a system, such as categorical, numerical, audio and/or video, textual, and graphical data. Accordingly, there are different value characteristics such as value distributions.
2. Attribute: Similar to value types, various types of attribute are often engaged in a system. Different attributes may generate separated value ranges, distributions, frequencies, etc.
3. Object: Represented by attributes and values, object heterogeneity presents objects in different ways. The same object may be presented in different forms in respective systems or at respective times.
4. Source: A behavioral or social system may involve multiple sources of information, presenting in heterogeneous values, attributes, and/or objects to form multiple heterogeneous information, media, or channel sources.
5. Subset: A subset of a value set, attribute set, object set, or source set may be selected for analysis or is the only practically available option. There is still heterogeneity in the subset, as discussed above.

Heterogeneity plays an essential role in understanding the difference embedded in a behavioral or social system. For instance, learning algorithms have to consider the significant difference incorporated in the attribute value range distribution and/or the value frequency distribution and further difference existing between attributes, objects, and sources.

Further, depending on coupling forms, heterogeneity in a complex behavioral and social system may present in different forms, such as structural heterogeneity, semantic heterogeneity, probabilistic/mathematical heterogeneity, dynamic heterogeneity, and/or graphical heterogeneity.

1. Structural heterogeneity: there may be different structural forms between behavioral and social components on one level or across multiple levels.
2. Semantic heterogeneity: various semantic relations may exist in a behavioral and social system.
3. Probabilistic heterogeneity: behaviors and social events may follow different probabilistic distributions.
4. Mathematical heterogeneity: behaviors and social events may be captured by different mathematical mechanisms and tools.
5. Dynamic heterogeneity: various types of interactions and evolutionary mechanisms may exist in one or many behavioral and social systems.
6. Graphical heterogeneity: behavioral and social systems may be best represented by different graphical models.

The heterogeneity discussed above needs to be aligned with couplings in behavioral and social study. If heterogeneity can be converted into homogeneous cases, then the classic approaches are sufficient. Unfortunately, in complex behavioral and social systems, it is difficult or sometimes even not possible to transform a heterogeneous system into a homogeneous one. This is because any partition and transformation would seriously cut off the intrinsic and sophisticated couplings between heterogeneous components. A transformed system would behave very differently from the original system if such couplings were destroyed.

Another issue is that heterogeneity is very much related to personalization. However, very limited outcomes are available on truly personalized learning, such as personalized information retrieval and personalized recommendation. In existing research, one tries to simplify the heterogeneity and personal characteristics in behavioral and social systems. For instance, most of the existing work in social media community learning and information retrieval treats all target objects similarly, and a model is built on a population of equally treated nodes or queries. Although the profiles of individuals are involved, the outcomes reflect population-oriented features rather than personal profiles. The resultant recommendations or retrieval outcomes are based on the behaviors of many objects rather than on an individual entity. This may be the key reason that existing search algorithms often bring about many irrelevant or uninteresting results.

15.4 Issues in Classic Behavioral and Social Learning

In this section, we analyze the IIDness nature of several classic learning algorithms, including classic sequence analysis and recommendation algorithms.

15.4.1 Classic Behavior Analysis

Behavior analysis is widely seen in areas such as Web mining, data mining, machine learning, social network analysis, and business intelligence. The so-called behavior in classic research usually refers to a weak and specific, and sometime very rough or virtual, concept. Its definition is not as clear and comprehensive as the abstract behavior model discussed in [1]. Correspondingly, behavior analysis has been very widely used in behavioral science and social science and more recently in computing science, referring to anything that present activities or movements, without a formal definition of what behavior is, we call such an analysis *soft behavior analysis*. In [1, 2] about behavior informatics and computing, we target a *hard behavior analysis* approach, which is based on a concrete behavior model defining what a behavior is, such as the one proposed in [1]. This hard behavior analysis aims to invent and develop computing methodologies, techniques, and tools for modeling, representing, reasoning about and checking behavior-oriented systems; for modeling, analysis, discovery, and learning of dynamics, networking, group/community formation and deformation, divergence and convergence, pattern and exception, impact, risk, and utility; and for the management and emergence of behaviors and behavioral systems.

The above aims and objectives are far beyond the classic efforts made in behavioral science [29, 30], social science, and specifically behavioral finance and economics [31]. In these fields, behavior is usually not solidly presented; couplings and heterogeneity are overlooked or weakly addressed.

Let us take sequence analysis, a very recent focus in data mining, as an example to explore the issues in classic behavior analysis. Sequence analysis is a typical approach for analyzing behavioral sequences. Classic sequence analysis research and algorithms focus on positive (occurring) sequences which are composed of actions only, without the involvement of action properties and interactions. They consider only the ordering relationship between sequential elements. The comprehensive couplings discussed in Sect. 15.3.1 are ignored.

Similarly, although typical algorithms in negative sequence analysis, including e-NSP and GA-NSP [32–35], incorporate one more relation, namely, the negation of a sequential element, other couplings are overlooked, and there is no differentiation between sequences and/or between sequential elements. In Sect. 15.6.1, we introduce the problem of coupled behavior analysis and a model for capturing coupled sequences. In [28], we discuss many different relationships between patterns, most of which are applicable for sequences. The consideration of complex couplings in behavioral sequences will create new types of sequential patterns, namely, various relational sequences.

15.4.2 Classic Social Media and Recommendation Systems

Recommendation systems are widely used in areas such as social media, Web service, and online business. Typical recommendation algorithms include *collaborative filtering* and *matrix factorization*. Here, we analyze the underlying assumption behind these two algorithms and their relation to IIDness.

Collaborative filtering (CF) is the process of filtering that involves collaboration between objects. Depending on the main entity of focus, CF takes the form of user-based CF or item-based CF. User-based CF assumes that if user A shares the same opinion as B, then A is likely to accept B's opinion on another issue. Item-based CF makes the assumption that users who buy X also buy Y.

Several variants of CF are proposed in the literature to handle filtering by addressing respective issues. Let us take the original memory-based CF algorithm [36] as an example to explore its underlying issues. Equation (15.1) represents its predicted vote for user *a* based on user *i*'s vote $v_{i,j}$ on item *j* and mean vote $\overline{v}_i$ (where *w(a,i)* is the weight of similar user *i* on *a*):

$$Pa, j = \overline{v}_a + k\sum\nolimits_{i=1}^{n} w(a,i)\left(v_{i,\,j} - \overline{v}_i\right) \tag{15.1}$$

Pa,j assumes there is only a weak correlation between users and does not substantially consider:

1. The coupling between the votes of user *i* on all items, namely, between $v_{i,\,j_1}$ and $v_{i,\,j_2}$
2. The influence between votes on different items for user *i*
3. The coupling between different users, namely, between $v_{i_1,\,j}$ and $v_{i_2,\,j}$
4. The aggregation of both couplings
5. The couplings between item attributes and between attribute values

Such couplings, if involved, could disclose intrinsic complexities of social networking and contribute to more informative and meaningful findings for determining the collaboration between i and a. If the properties of users and items can be incorporated into the above couplings, as in coupled similarity [37], the prediction could be based on much more solid support.

The matrix factorization (MF) approach supports a matrix R with users and items as two dimensions, with the values of users' ratings on items. MF predicts the missing ratings of some users on future items based on the approximate factorization of the matrix R into two matrices P and Q: $R \approx PXQ^T = \hat{R}$, where P represents the association between a user and the latent features and Q captures the association between an item and the latent features. The prediction of a rating $\hat{r}_{ij}$ of an item j by a user I is the dot product of the two vectors corresponding to i and j:

$$\hat{r}_{ij} = p_i^T = \sum_{i=1}^{K} p_{ik} q_{kj} \tag{15.2}$$

As we can see in the matrix and the predictive function (Eq. 15.2), MF is based on a direct business linkage between users and items, but it does not take the coupling between items and between users into consideration, much less the couplings between properties describing users and items.

The above analysis of the fundamental approaches used in CF and MF shows that IIDness has been taken into account in the basic CF and MF working processes.

15.4.3 Classic Social Network Analysis

Social network analysis (SNA) [38, 39] has been a hot topic in many fields recently. The basic ideas of analyzing social networks involve key concepts such as graphs and matrices formed by nodes in a network to build graphical models or adjacency matrices. The similarity or dissimilarity between nodes (or any objects including actors in a network) is measured by the relation of some nodes (objects) to others. The contribution strengths can then be represented by measuring and weighting the communication, connections, information flow, similarities/affiliations, and/or social interactions between nodes (objects).

Based on the above basic concepts, SNA then studies typical issues such as how to represent various social networks, how to identify linkages, how to measure the strength of linkages, how to identify key/central nodes (actors) in a network, how the influence is transferred in a network, how some nodes are connected to form a community, and how to measure the overall network structure.

In the above SNA tasks, relation, linkage, interaction, and influence are some of the core aspects for analyzing the working mechanisms and opportunities and problems in social networks. Similar to social media analysis and recommendation systems, there is usually a very weak focus in SNA on uncovering the node–node coupling, actor–actor coupling, node–actor coupling, and couplings between and/or within objects, subgroups, subgraphs, and communities. Heterogeneity between entities (including the above aspects) is usually ignored by simply treating all entities equally. The existing SNA approaches and algorithms usually focus on the explicit linkage between objects but ignore the couplings between object properties and between property value sets.

In summary, the above discussions about classic behavior analysis, social media recommendation systems, and social network analysis show that IIDness-based learning and analysis has been widely accepted as a fundamental assumption in complex behavioral and social applications.

15.5 Non-IIDness Learning

Here, we illustrate the assumptions of IIDness and non-IIDness, respectively, and compare their different settings. Given a learning problem consisting of three heterogeneous objects from different data sets or varied feature sets (as shown in the three different symbols), as shown in Fig. 15.1a, our goal is to determine the position of $O3$; for instance, whether it belongs to the same cluster of $O1$ and $O2$ or which label it can be classified as. Figures15.1.1b and 1c illustrate the main working mechanisms of IIDness learning and non-IIDness learning, respectively, and their differences. Rather than probing a specific learning task, our discussion here focuses on the assumptions and working mechanisms that are adopted for IIDness and non-IIDness learning, which can be any specific learning objectives.

Figure 15.1b illustrates the approach of IIDness learning, which treats all objects as homogeneous (identically distributed, as shown in the circle) and independent (no connection between objects). The similarity or distance d is calculated between O_3 and its baseline O, say $d = \|O - O_3\|$, for which we ignore the relations between O_3 and other objects.

Fig. 15.1 IIDness learning vs. non-IIDness learning

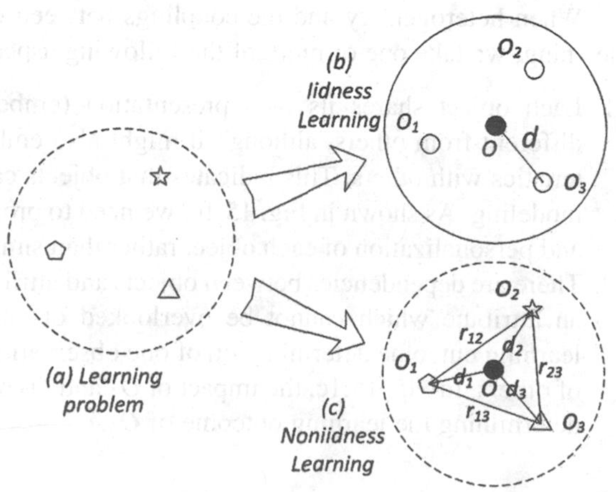

In conclusion, IIDness learning relies on the assumption that all objects are independent and identically distributed, which is applied to objects, object attributes, attribute values, learning objective function determination, evaluation criteria, etc. Correspondingly,

1. We treat all observations (as well as their elements, attributes, and attribute values) equally and are only concerned about the similarity (or dissimilarity) between an observation and the reference (for instance, central point or mean).
2. The reference to determine the belongingness of any observation is either a global benchmark (say, the minimum support) or is obtained in the same way (the same mean value or the selection of a central point).
3. The interactions between objects and the influence of one object on others are ignored in the belongingness determination.
4. The influence of objects on the reference is usually weakened. In Fig. 15.1a, all objects including O_1, O_2, and O_3 are treated independently; only the similarity between O_3 and the global reference O is concerned in determining O_3's belongingness.
5. Objects are also treated as being identically distributed; thus, the same objective function is applied to all objects during the learning process.

Figure 15.1c illustrates the concept of non-IIDness learning for solving the learning problem in Fig. 15.1a. The determination of O_3 considers:

1. Its direct relation with O_1, i.e., r_{13}, and its relation with O_2, i.e., r_{23}, as well as the indirect relation r_{12} between O_1 and O_2.
2. The calculation of similarity d_3 between O_3 and the baseline needs to involve r_{13} and r_{23}, probably even r_{12}, say $d_3 = \|O - O_3(r_{12}, r_{13}, r_{23})\|$.
3. The baseline O for each object may be different when the heterogeneity between the three objects is considered.
4. The functions for calculating d_1 and d_2 may also be different since O_1 and O_2 may follow different distributions, and the interactions between O_1 and O may differ from those between O_2 and O.

When heterogeneity and the couplings between objects feature in non-IIDness learning, we take one or more of the following aspects into consideration:

1. Each object shares its own presentation (embodied in properties) which is different from others, although it might also embrace certain common characteristics with others. This indicates that objects cannot be treated equally in the modeling. As shown in Fig. 15.1c, we need to protect the original characteristics and personalization of each object rather than simply convert them to be similar.
2. There are dependencies between objects and attributes and between the values of an attribute which cannot be overlooked or simplified. This means that the learning outcome determination of one observation has to consider the influence of others. In Fig. 15.1c, the impact of O_1 and O_2 on O_3 needs to be considered in determining the learning outcome of O_3.

3. In determining the learning objective function, the benchmark (for instance, mean or central point) determination has to consider its position in the local or global space to reflect the intrinsic characteristics shared by those observations with similar distributions. In Fig. 15.1c, we may need to develop a different baseline O for the three objects, and the similarity (distance) to O_1, O_2, and O_3 may have diverse functions.

These aspects will be reflected in building the corresponding learning objective functions. In Fig. 15.1c, data characteristics analysis is conducted on the objects, making the observation that they share different distributions and should be treated in three subspaces, in which objects belonging to the same subspace share more similarities with each other than they do with objects in other spaces. Accordingly, three "local" benchmarks rather than one are determined for each subspace. In determining each object's belongingness, for instance, O_3, the coupling relationships of the object with other objects within the subspace, such as with O_1 and O_2, are considered in the objective function. In this case, we overlook the influence of couplings with objects in the other two subspaces, since for the sake of simplicity, they are weak enough to be ignored.

15.6 Non-IIDness Learning Case Studies

The assumptions and abstraction made in IIDness learning techniques seriously mismatch the reality and complexities in complex behavioral/social systems such as the coupled behavior analysis problem [40]. As we see in social media, users are interrelated and influenced by one another in various aspects and for various reasons. Each user and his/her behaviors present specific characteristics and preferences which are usually different from those of others.

Such strong couplings and heterogeneity are particularly embodied in complex behavioral/social systems, forming the major driving forces of behavioral/social networking and evolution, which are slightly, and sometimes greatly, different from traditional applications which can be highly abstracted into an IIDness-based problem. This determines that typical approaches that expand the classic IIDness-based algorithms and frameworks often lead to limited or incremental improvement and cannot fundamentally solve the problem. This is challenging when handling large scale behavioral/social data and general big business data, in which heterogeneity and couplings, existing in objects, interactions, behaviors, and context, are two intrinsic working mechanisms and driving forces of system dynamics and their evolution.

We here briefly introduce the main principles of several preliminary attempts with case studies, to show that they can handle the coupling aspects of non-IIDness in the respective learning and analytics of behavioral and social problems. They consist of:

1. *Coupled behavior analysis* for analyzing intra-couplings between an actor's behaviors and the inter-couplings between behaviors of different actors for group behavior understanding

2. *Coupled item recommendation* by applying coupled object similarity [37] to analyze the coupling between items and convert the item-based collaborative filtering to coupled item-based item recommendation
3. *Term–term relation-based document analysis* to analyze the semantic relations between terms appearing within and between documents

Rather than giving a detailed introduction to each technique, our intention is to introduce a few exemplar approaches to show that non-IIDness issues are manageable, and that the involvement of non-IIDness can lead to improved or substantial outcomes. These techniques are general and can be widely used and expanded for analyzing complex behavioral and social problems. Interested readers can find a detailed introduction in the cited references as well as some of our other efforts on quantifying similarity in categorical [37] and numerical objects [41], coupled clustering by incorporating coupled object similarity [37], analyzing non-IIDness at the method level to explore couplings between clusterings for coupled ensemble clustering [42], and considering the relations between patterns [28] and rules [43] for pattern relation analysis.

15.6.1 Coupled Behavior Analysis

In this section, we discuss the couplings between the behaviors of an individual and between the behaviors of different actors. We present a formal statement of couplings in group behaviors and the problem of analyzing such coupled group behaviors. This case study instantiates the concept of object couplings to complex behavior relations, in which behavior is a ubiquitous entity in social and business applications. Here, behavior presents heterogeneous attributes and is undertaken by many actors. The couplings between behaviors are captured in terms of probabilistic relations and the Markov assumption.

In the discussion that follows, behaviors refer to actions, operations, events, and activity sequences conducted within certain contexts and environments in either virtual or physical organizations [1]. In practice, behaviors from the same or different actors are often associated with each other, and we call them *coupled behaviors* [40]. Coupled behaviors play a more fundamental role than individuals in the cause, dynamics, and effect of business problems [1, 40]. The fundamental characteristic and challenge of understanding coupled behaviors is reflected through the intra-couplings embedded in behaviors from the same actor, the inter-couplings between those from different actors, and the aggregative couplings of both intra- and inter-couplings.

Suppose there are I actors (customers) $\{E_1, E_2, \ldots, E_I\}$, an actor Ei undertakes J behaviors $\{B_{i_1}, B_{i_2}, \ldots B_{i_J}\}$, actor E_i's jth behavior B_{ij} is a K-variable vector, and its variable $p_{ij\ k}$ reflects the kth behavior property. For the set of behaviors $\{B_{ij} \mid 1 \leq i \leq I, 1 \leq j \leq J\}$, each element B_{ij} can be expressed as the vector

$-\!\rightarrow B_{ij} = ([p_{ij}]_1, [p_{ij}]_2, \ldots, [p_{ij}]_K)$, where $[p_{ij}]_k$ $(1 \le k \le K)$ is the kth property of the behavior B_{ij}. Then, coupled behaviors are defined as follows:

Definition 15.1 (Coupled Behaviors) *Coupled behaviors Bc refers to behaviors $B_{i_1 j_1}$ and $B_{i_2 j_2}$ that are coupled in terms of relationships $f(\theta(\cdot), \eta(\cdot))$, where $(i_1 \neq i_2) \vee (j_1 \neq j_2) \wedge (1 \le i_1, i_2 \le I) \wedge (1 \le j_1 \le J_1) \wedge (1 \le j_2 \le J_2)$,*

$$\begin{aligned} B_c &= \left(B_{i_1}^{\theta} * B_{i_2}^{\theta}\right)^{\eta} :: \\ &= B(E, O, C, R) \\ &\quad \times \sum\nolimits_{i1,i2=1}^{I} \sum\nolimits_{j1=1}^{J1} \sum\nolimits_{j2=1}^{J2} f(\theta(\cdot) j1, j2, \eta(\cdot) i1i2) \odot (\mathrm{B}i1j1, \mathrm{B}i2j2), \end{aligned} \tag{15.3}$$

where $f(\theta(\cdot)i_1, i_2, \eta(\cdot)i_1 i_2)$ is the coupling function denoting the corresponding relationships between $\mathrm{B}i_1 j_1$ and $\mathrm{B}i2j2$, $\sum_{i1,i2=1}^{I} \sum_{j1=1}^{J1} \sum_{j2=1}^{J2} \odot$ means the subsequent behaviors of B *are* $\mathrm{B}i1j1$ *coupled with $f(\theta(\cdot)j_1, \eta(\cdot)i_1 i_2)$, $\mathrm{B}i_2 j_2$ with $f(\theta(\cdot)j_2, \eta(\cdot)i_1 i_2)$, and so on, with non-determinism.*

Corollary 15.1 *Coupled behaviors can also be represented by behavior attributes $\{[p_{ij}]_k \mid 1 \le k \le K\}$, then we have the corresponding behavior adjoint matrix:*

$$\begin{aligned} M(\mathbf{B}_c) &:= M(\mathbf{B}) \sum\nolimits_{i1,i2=1}^{I} \sum\nolimits_{j1=1}^{J1} \sum\nolimits_{j2=1}^{J2} f(\theta(\cdot) j1, j2, \eta(\cdot) i1i2) \odot (\mathbf{B}i1j1, \mathbf{B}i2j2) \times \left(\vec{B}_{i1j1}^{T} \vec{B}_{i2j2}\right) \\ &= M(\mathbf{B}) \sum\nolimits_{i1,i2=1}^{I} \sum\nolimits_{j1=1}^{J1} \sum\nolimits_{j2=1}^{J2} f(\theta(\cdot) j1, j2, \eta(\cdot) i1i2) \odot (\mathbf{B}i1j1, \mathbf{B}i2j2) \times ([mpi1i2j1j2]k1k2)K \times K, \end{aligned} \tag{15.4}$$

where Ei_1 and Ei_2 refer to two distinct actors, $\vec{B}_{i1j1}^{T} = \left([p_{i1j1}]_{k1s}\right)_{K \times 1}$ and $\vec{B}_{i2j2} = ([p_{i2j2}]_{sk2})_{1 \times K}$ refer to two distinct behavior vectors with corresponding behavior attributes; $[mp_{i1i2j1j2}]_{k1k2} = [p_{i1j1}]_{k11} \cdot [p_{i2j2}]_{1k2}$ is the (k_1, k_2) element of the matrix multiplication $\vec{B}_{i1j1}^{T} \vec{B}_{i2j2}$; $\sum_{i1,i2=1}^{I} \sum_{j1=1}^{J1} \sum_{j2=1}^{J2} \odot$ means the subsequent behavior adjoint matrix of M(B) is $\vec{B}_{i1j1}^{T} \vec{B}_{i2j2}$ coupled with $f(\theta(\cdot)_{j1,j2}, \eta(\cdot)_{i1i2})$, and so on, with non-determinism; and the following constraints hold: $(i_1 \neq i_2) \vee (j_1 \neq j_2) \vee (k_1 \neq k_2) \wedge (1 \le i_1, i_2 \le I) \wedge (1 \le j_1 \le J_1) \wedge (1 \le j_2 \le J_2) \wedge (1 \le k_1, k_2 \le K)$.

Definition 6.2 (Coupled Behavior Analysis (CBA)) *The analysis of coupled behaviors (CBA problem for short) [40] is to build the objective function $g(\cdot)$ under the condition that behaviors are coupled with each other by coupling function $f(\cdot)$, and to satisfy the following conditions:*

$$f(\cdot) ::= f(\vartheta(\cdot), \eta(\cdot)), \tag{15.5}$$

$$g(\cdot) | (f(\cdot) \ge f_0) \ge g_0. \tag{15.6}$$

The CBA problem is widespread and is applicable to any behavior-oriented application such as intelligent transport systems and community behavior analysis in social media. In [40], an example of using coupled hidden Markov model (CHMM) is reported to model abnormal group-based financial trading behaviors (pool manipulation) in stock markets. As shown in Fig. 15.2, CHMM captures those abnormal group-based investment behaviors that demonstrate exceptional performance in recall and also make abnormal returns in the market, compared with the HMM models built for buy quotes (B-HMM), sell quotes (S-HMM), and trades (T-HMM), respectively, and IHMM which simply adds B-HMM, S-HMM, and T-HMM without considering the couplings.

The above statement shows that coupled behavior analysis (the CBA problem) is a typical non-IIDness learning problem in which the intra-couplings and inter-couplings between behaviors cater for the behavior dependency. The CHMM-based case study further caters for certain heterogeneity between behavior properties. The extension of this approach and the exploration of other effective approaches for CBA show that there are very promising opportunities for the deep analysis of behavioral and social non-IIDness.

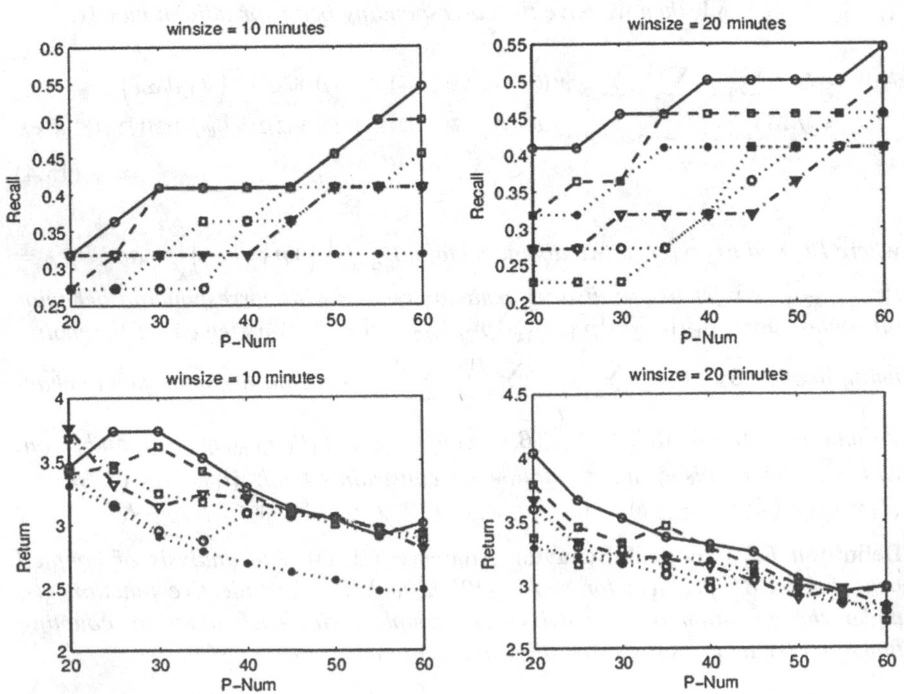

Fig. 15.2 Recall and abnormal return of HMM-based CBA modeling

15.6.2 Coupled Item Recommendation

As discussed in Sect. 15.4.2, the classic collaborative filtering algorithms ignore or only partially consider the couplings between item properties, user properties, and item–user interactions. Here, we present a coupled item-based CF by explicitly considering both intra-coupling and inter-coupling between item attributes and aggregating them in terms of the coupled object similarity (COS) proposed in [37]. Details can be found in [44].

The coupled item similarity (CIS) between categorical items X and Y is defined as follows:

$$\text{CIS}(X,Y) = \sum\nolimits_{j=1}^{n} \delta_j^A(X_j, Y_j), \tag{15.7}$$

where X_j and Y_j are the values of item feature j for X and Y, respectively, and δ_j^A is the coupled attribute value similarity (CAVS).

The CAVS is further described by the *intra-coupled attribute value similarity (IaAVS)* which measures the item feature value similarity by considering the frequency of feature value occurrence within an item feature, and the *inter-coupled attribute value similarity (IeAVS)* which measures the item feature value similarity by taking the item feature dependency aggregation into account.

For item feature j, IaAVS $\delta_j^{Ia}(X_j, Y_j)$ is calculated as per [37, Eq. 4.2], and IeAVS $\delta_j^{Ie}(X_j, Y_j)$ is calculated as per [37, Eq. 4.7]. Accordingly, CAVS δ_j^A between item attribute values X_j and Y_j of item feature j is as follows:

$$\delta_j^A(X_j, Y_j) = \delta_j^{Ia}(X_j, Y_j) \cdot \delta_j^{Ie}(X_j, Y_j). \tag{15.8}$$

Taking the K-modes clustering algorithm as an example, we create a coupled K-modes (CK-modes). Let S be a cluster generated by the previous partition of the K-modes algorithm. There are M items described by categorical item features $\{a_{j1}, a_{j2}, \ldots, a_{jl}\}$ belonging to the cluster S. A mode of the cluster S is an item vector $Q = [q_1, q_2, \ldots, q_l]$ to maximize the sum of the similarity between each element of S and Q. The mode of itemset S with M items is a vector $Q = [q_1, q_2, \ldots, q_l]$ that maximizes

$$\text{Sim}(Q,S) = \sum\nolimits_{i=1}^{M} CIS(Si, Q). \tag{15.9}$$

Within the CK-modes model, the item-based collaborative filtering is adjusted to generate the prediction on item o_i for an active user u. The prediction $P_{u,oi}$ on item o_i for active user u is computed by the following formula:

$$P_{u,oi} = \begin{cases} \dfrac{\sum_{\forall N\, j \in N} (\mathrm{Sim}_{oi,N\,j} * Ru, N\,j)}{\sum_{\forall N\, j \in N} (|\mathrm{Sim}_{oi,N\,j}|)}, & \sum (|\mathrm{Sim}_{oi,N\,j}|) > 0 \\ \overline{r}_u & \sum (|\mathrm{Sim}_{oi,N\,j}|) = 0 \end{cases} \tag{15.10}$$

where N is the intersection of items rated by the active user u and items grouped by the CK-modes algorithm, R_u, N_j represents the rating on item N_j given by the user u, Simo_i,N_j is the coupled item similarity between items o_i and N_j, and $\overline{r}u$ is the average of the active user's ratings.

We evaluate CK-modes against several widely discussed algorithms in recommender systems, including the user-based collaborative filtering algorithm [45], item-based collaborative filtering algorithm [46], and *CLUSTKNN* [47] on the MovieLens data. Figure 15.3 shows the throughputs of all algorithms. Here, throughput represents the number of recommendations generated per second. The user-based recommendation algorithm scans the whole user–item matrix R, and its throughput does not change with the number of clusters. However, the throughput of the item-based recommendation algorithm varies with the number of neighbors selected for prediction. We plot the throughput of the item-based recommendation algorithm by setting the number of neighbors as 30, since this generates the best quality prediction.

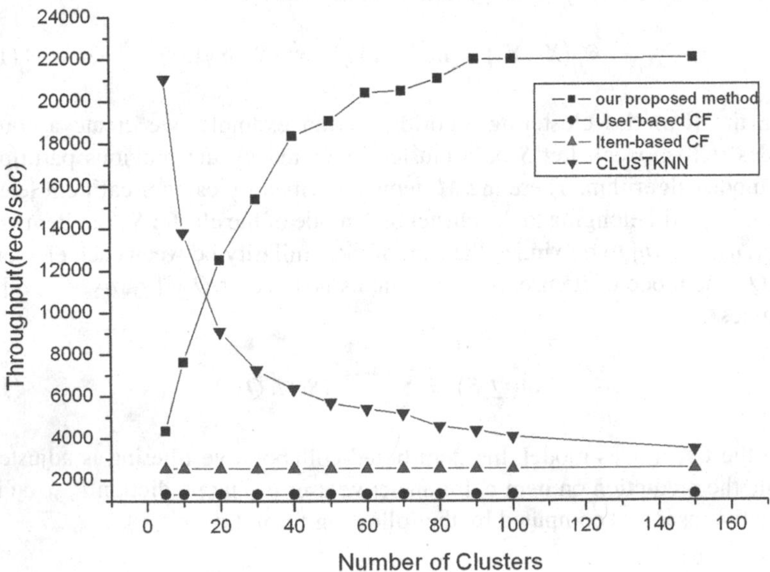

Fig. 15.3 Throughput of the selected recommendation algorithms

15.6.3 Term Coupling-Based Document Analysis

In classic document analysis, typical algorithms such as the bag-of-words model [48] ignore the semantic relations between terms, leading to low learning performance. In [49], a new document clustering framework is proposed which incorporates the *intra-term* coupling between terms within a document, the *inter-term* coupling between terms from different documents, and the aggregative term coupling by combining intra-term and inter-term couplings.

Terms t_i and t_j are *intra-coupled* if they co-occur in at least one document $d_x (d_x \in D)$. The co-occurrence relation between terms t_i and t_j across document base D is quantified as

$$\mathrm{CoR}(t_i, t_j) = \frac{1}{|H|} \sum\nolimits_{x \in H} \frac{W_{xi} w_{xj}}{W_{xi} + w_{xj} - W_{xi} w_{xj}}, \tag{15.11}$$

where w_{xi} and w_{xj} represent the tf–idf weights of t_i and t_j in d_x, respectively, and $|H|$ denotes the number of elements in $H = \{X | (W_{xi} \neq 0) \vee (W_{xj} \neq 0)\}$. If $H = \emptyset$, we define $\mathrm{CoR}(t_i,t_j) = 0$.

We further define the *intra-term coupling* in terms of conditional probability by normalizing the relationship between t_i and t_j, i.e., $\mathrm{CoR}(t_i, t_j)$, to [0, 1] with respect to the total amount of relation between t_i and the other terms. The intrarelation CoR (t_i, t_j) reflects the probability of term t_j co-occurring with term t_i in a document. The *intra-term coupling* $\mathrm{IaR}(t_i, t_j)$ between t_i and t_j is

$$\mathrm{IaR}(t_i, t_j) = \begin{cases} 1 & i = j, \\ \dfrac{\mathrm{CoR}(ti, tj)}{\sum\nolimits_{i=1, i \neq j}^{n} \mathrm{CoR}(ti, tj)} & i \neq j, \end{cases} \tag{15.12}$$

where $\mathrm{CoR}(t_i, t_j)$ is the co-occurrence relationship between terms t_i and t_j. For all the terms t_i $(i \neq j)$, we have $\mathrm{IaR}(t_i, t_j) \geq 0$ and $\sum\nolimits_{i=1, i \neq j}^{n} \mathrm{IaR}(ti, tj) = 1$. $\mathrm{IaR}(t_i, t_j)$ is usually not symmetrical.

The *inter-term coupling* is determined by the context, namely, the co-occurrences between a term and all other terms across the entire document set. Terms *ti* and *tj* are *inter-coupled* if there exists at least one term *tk* such that both $\mathrm{IaR}(ti, tk) > 0$ and $\mathrm{IaR}(tj, tk) > 0$ hold. The term between them, *tk*, is called the *link term*. The *relative inter-term coupling* between terms *ti* and *tj* linked by the term *tk* is formalized as

$$R\_\mathrm{IeR}(ti, tj|tk) = \min(\mathrm{IaR}(ti, tk), \quad \mathrm{IaR}(tj, tk)), \tag{15.13}$$

where $\mathrm{IaR}(ti, tk)$ and $\mathrm{IaR}(tj, tk)$ are the intrarelations between *ti* and *tk*, *tk* and *tj*, respectively. The *inter-term coupling* between *ti* and *tj* is defined by their interactions with all the link terms, formalized as

$$\text{IeR}(ti, tj) = \begin{cases} 0, & i = j \\ \frac{1}{|L|} \sum_{\forall tk \in L} R\_\text{IeR}(ti, tj|tk), & i \neq j, \end{cases} \tag{15.14}$$

where $|L|$ denotes the number of link terms in $L = \{tk|(\text{IaR}(tk, ti) > 0) \wedge (\text{IaR}(tk, tj) > 0)\}$ and $R\_\text{IeR}(ti, tj\,|tk)$ is the relative interrelation between *ti* and *tj* linked by *tk*. If $L = \emptyset$, we define $\text{IeR}(ti, tj) = 0$. The value of $\text{IeR}(ti, tj)$ falls in [0, 1]. When there is no link term for *ti* and *tj*, we regard $\text{IeR}(ti, tj) = 0$.

The overall *term–term coupling* $\text{CR}(ti, tj)$ (viz., the similarity between terms) across all documents is then measured by aggregating the intra-term coupling and the inter-term coupling.

$$\text{CR}(ti, tj) = \begin{cases} 1 & i = j, \\ \alpha \cdot \text{IaR}(ti, tj) + (1 - \alpha) \cdot \text{IeR}(ti, tj), & i \neq j, \end{cases} \tag{15.15}$$

where $\alpha \in [0, 1]$ is the parameter that decides the weight of intrarelation and $\text{IaR}(ti, tj)$ and $\text{IeR}(ti, tj)$ are the respective intrarelation and interrelation between terms *ti* and *tj*.

The term–term coupling-based similarity can then be applied to different document clustering models to cluster documents. Table 15.1 illustrates the Purity, Rand index (RI), $F1$-measure, and normalized mutual information (NMI) of the clustering results by coupled term–term relation model (CRM) for the spherical k-means algorithm incorporated with $\text{CR}(ti, tj)$ on data sets Newsgroups and WebKB, compared to the classic bag-of-words (BOW) model and the GVSM model.

15.7 Summary

In this chapter, we have presented a high-level picture of the non-IIDness learning problem for handling strong couplings and heterogeneity in complex behavioral and social applications. Such problems cannot be tackled well by most extant methods and systems since they overlook or abstract the couplings and heterogeneity by adopting the strong assumption of IIDness (independent and identical distribution).

We have discussed the challenges of analyzing complex behavioral and social applications, the issues surrounding the extant IIDness-based learning approaches in classic behavior analysis, social media and recommendation systems, social network analysis, and the concepts of non-IIDness learning. Several exemplar techniques have been discussed for non-IIDness learning, including coupled behavior analysis for analyzing group behaviors, coupled item recommendation for considering the couplings between items, and term–term coupling-based document analysis by considering the semantic relation between terms across documents.

The concept and ideas of non-IIDness learning allow us to comprehensively, systematically, and deeply explore the couplings between values, attributes,

Table 15.1 Document clustering results [49]

Date	Purity			RI			Fl-measure			NMI		
Sets	*BOW*	*GVSM*	*CRM*	*BOW*	*GVSM*	*CRM*	*BOW*	*GVSM*	*CRM*	*BOW*	*GVSM*	*CRM*
D1	0.79	0.82	**0.88**	0.49	0.57	**0.62**	0.48	0.58	**0.61**	0.32	0.41	**0.44**
D2	0.80	0.80	**0.84**	0.49	0.58	**0.60**	0.62	0.66	**0.67**	0.44	0.48	**0.44**
D3	0.78	0.79	**0.83**	0.27	0.29	**0.36**	0.41	0.43	**0.50**	0.37	0.42	**0.39**
D4	0.82	0.81	**0.85**	0.63	0.65	**0.67**	0.66	0.65	**0.65**	0.32	0.35	**0.37**

objects, methods, and patterns, and the heterogeneity residing in value matching, value frequency, attribute distribution, attribute co-occurrence, objects, methods, and patterns.

References

1. Cao, L.: In-depth behavior understanding and use: the behavior informatics approach. Inform. Sci. **180**, 3067–3085 (2010)
2. Cao, L., Yu, P.S. (eds.): Behavior Computing: Modeling, Analysis, Mining and Decision. Springer, London (2012)
3. Liu, H., Salerno, J., Young, M.J. (eds.): Social Computing, Behavioral Modeling, and Prediction. Springer, Berlin (2008)
4. Steinwart, I., Christmann, A.: Fast learning from non-i.i.d. observations. In: Bengio, Y., Schuurmans, D., Lafferty, J., Williams, C.K.I., Culotta, A. (eds.) Advances in Neural Information Processing Systems, vol. 22, pp. 1768–1776. MIT Press (2009)
5. Zhang, X., Song, L., Gretton, A., Smola, A.: Kernel measures of independence for non-iid data. In: Proceedings of NIPS2009, pp. 1937–1944. MIT Press (2009)
6. Pan, Z., Xiao, Q.: Least-square regularized regression with non-iid sampling. J. Statist. Plan Inference **139**, 3579–3587 (2009)
7. Guo, Z., Shi, L.: Classification with non-iid Sampling. Math. Comput. Model. **54**, 1347–1364 (2011)
8. Cinbis, R., Verbeek, J., Schmid, C.: Image categorization using Fisher kernels of non-iid image models. In: Proceedings of CVPR2012, pp. 2184–2191. IEEE Press (2012)
9. Mohri, M., Rostamizadeh, A.: Rademacher complexity bounds for non-i.i.d. processes. In: Bengio, Y., Schuurmans, D., Lafferty, J., Williams, C.K.I., Culotta, A. (eds.) Advances in Neural Information Processing Systems, pp. 1097–1104. MIT Press (2009)
10. Mohri, M., Rostamizadeh, A.: Stability bounds for stationary ϕ-mixing and β-mixing processes. J. Mach. Learn. Res. **11**, 789–814 (2010)
11. Chazottes, J., Collet, P., Kulske, C., Redig, F.: Concentration inequalities for random fields via coupling. Probab. Theory Relat. Fields **137**, 201–225 (2007)
12. Kontorovich, L., Ramanan, K.: Concentration inequalities for dependent random variables via the Martingale method. Ann. Probab. **36**, 2126–2158 (2008)
13. Modha, D., Masry, E.: On the consistency in nonparametric estimation under mixing assumptions. IEEE Trans. Inform. Theory **44**, 117–133 (1998)
14. Ralaivola, L., Szafranski, M., Stempfel, G.: Chromatic PAC-Bayes bounds for non-IID data: applications to ranking and stationary β-mixing processes. J. Mach. Learn. Res. **11**, 1927–1956 (2010)
15. Zhou, Z., Sun, Y., Li, Y.: Multi-instance learning by treating instances as non-I.I.D. samples. In: Proceedings of ICML2009, pp. 1249–1256. Omnipress (2009)
16. Dundar, M., Krishnapuram, B., Bi, J., Rao, R.: Learning classifiers when the training data is not IID. In: Proceedings of IJCAI2007, pp. 756–761. AAAI Press (2007)
17. Tillman, R.: Structure learning with independent non-identically distributed data. In: Proceedings of ICML2009, pp. 1041–1048. Omnipress (2009)
18. Ping, W., Xu, Y., Ren, K., Chi, C., Shen, F.: Non-I.I.D. multi-instance dimensionality reduction by learning a maximum bag margin subspace. In: Proceedings of 24th AAAI Conference on Artificial Intelligence, pp. 551–556. AAAI Press (2010)
19. Song, L., Smola, A., Gretton, A., Bedo, J., Borgwardt, K.: Feature selection via dependence maximization. J. Mach. Learn. Res. **13**, 1393–1434 (2012)
20. Wang, C., Cao, L.: Modeling and analysis of social activity process. In: Cao, L., Yu, P.S. (eds.) Behavior Computing, pp. 21–35. Springer, Berlin (2012)

21. Cao, L., Dai, R.: Open Complex Intelligent Systems. Post & Telecom, Beijing (2008)
22. Cao, L., Dai, R., Zhou, M.: Metasynthesis: M-space, M-interaction and M-computing for open complex giant systems. IEEE Trans. Syst. Man. Cybern. Part A **39**, 1007–1021 (2009)
23. Cao, L., Zhang, C., Zhou, M.: Engineering open complex agent systems: a case study. IEEE Trans. Syst. Man. Cybern. Part C Appl. Rev. **38**, 483–496 (2008)
24. Cao, L., Zhang, C., Dai, R.: Organization-oriented analysis of open complex agent systems. Int. J. Intell. Cont. Syst. **10**, 114–122 (2005)
25. Cao, L., Luo, D., Zhang, C.: Ubiquitous intelligence in agent mining. In: Proceedings of ADMI, pp. 23–35. Springer (2009)
26. Cao, L., Yu, P.S., Zhao, Y., Zhang, C.: Domain Driven Data Mining. Springer, Berlin (2010)
27. Cao, L.: Domain driven data mining: challenges and prospects. IEEE Trans. Knowl. Data. Eng. **22**, 755–769 (2010)
28. Cao, L.: Combined mining: analyzing object and pattern relations for discovering and constructing complex yet actionable patterns. WIREs Data Min. Knowl. Disc. **3**, 140–155 (2013)
29. Fishbein, M., Ajzen, I.: Predicting and Changing Behavior: The Reasoned Action Approach. Psychology Press, New York (2009)
30. Fisher, W., Piazza, C., Roane, H. (eds.): Handbook of Applied Behavior Analysis. The Guilford Press, New York (2011)
31. Burton, E., Shah, S.: Behavioral Finance: Understanding the Social, Cognitive, and Economic Debates. Wiley, New York (2013)
32. Zhao,Y., Zhang, H., Cao, L., Zhang, C., Bohlscheid, H.: Mining both positive and negative impact-oriented sequential rules from transactional data. In: Proceedings of PAKDD 2009, pp. 656–663. Springer (2009)
33. Zhao, Y., Zhang, H., Wu, S., Pei J., Cao, L., Zhang, C., Bohlscheid, H.: Debt detection in social security by sequence classification using both positive and negative patterns. In: Proceedings of ECML-PKDD2009. Lecture Notes in Artificial Intelligence, vol. 5782, pp. 648–663. Springer (2009)
34. Zheng, Z., Zhao, Y., Zuo, Z., Cao, L., Zhang, H., Zhao, Y., Zhang, C.: An efficient GA-based algorithm for mining negative sequential patterns. In: Proceedings of PAKDD 2010, pp. 262–273. Springer (2010)
35. Dong, X., Zhao, Y., Cao, L., Zhao, Y., Zhang, C., Li, J., Wei, W., Ou, Y.: e-NSP: efficient negative sequential pattern mining based on identified positive patterns without database rescanning. In: Proceedings of CIKM 2011, pp. 825–830. ACM Press (2011)
36. Breese, J., Heckerman, D., Kadie, C.: Empirical analysis of predictive algorithms for collaborative filtering. In: UAI-98, pp. 43–52. AAAI Press (1998)
37. Wang, C., Cao, L. et al.: Coupled nominal similarity in unsupervised learning. In: Proceedings of CIKM 2011, pp. 973–978. ACM Press (2011)
38. Scott, J.: Social Network Analysis: A Handbook. Sage, Newbury Park (1991)
39. Kadushin, C.: Understanding Social Networks: Theories, Concepts, and Findings. Oxford University Press, Oxford (2012)
40. Cao, L., Ou, Y., Yu, P.S.: Coupled behavior analysis with applications. IEEE Trans. Knowl. Data Eng. **24**, 1378–1392 (2012)
41. Wang, C., She, Z., Cao, L.: Coupled attribute analysis on numerical data. In: Proceedings of IJCAI 2013. AAAI Press (2013)
42. Wang, C., She, Z., Cao, L.: Coupled clustering ensemble: incorporating coupling relationships both between base clusterings and objects. In: Proceedings ICDE 2013, pp. 374–385. IEEE Press (2013)
43. Li, J., Wang, C., Cao, L., Yu, P.: Efficient selection of globally optimal rules on large imbalanced data based on rule coverage relationship analysis. In: Proceedings of SDM 2013. SIAM/Omni Press (2013)
44. Yu, Y., Wang, C., Gao, Y., Cao, L., Sun, J.: A coupled clustering approach for items recommendation. In: Proceedings of PAKDD 2013, pp. 365–376. Springer (2013)

45. Resnick, P., Iacovou, N., Suchak, M., Bergstrom, P., Riedl, J.: GroupLens: an open architecture for collaborative filtering of netnews. In: CSCW Proceedings, pp. 175–186. Morgan Kaufmann (1994)
46. Sarwar, B., Karypis, G., Konstan, J., Riedl, J.: Item-based collaborative filtering recommendation algorithms. In: Proceedings of WWW 2001, pp. 285–295. ACM Press (2001)
47. Al Mamunur Rashid, S.K.L., Karypis, G., Riedl, J.: ClustKNN: a highly scalable hybrid model- and memory-based CF algorithm. In: Proceedings of WebKDD 2006. ACM Press (2006)
48. Huang, A., Milne, D., Frank, E., Witten, I.: Clustering documents using a Wikipedia-based concept representation. Adv. Knowl. Disc. Data Min. **5476**, 628–636 (2009)
49. Cheng, X., Miao, D., Wang, C., Cao, L.: Coupled term–term relation analysis for document clustering. In: Proceedings of IJCNN 2013. IEEE Press (2013)

Chapter 16
Opportunities and Prospects

16.1 About Open Complex System Studies

The world is becoming increasingly connected, both loosely and tightly, in terms of explicit or implicit coupling relationships. However large-scale a system is, scale may be less important in the new data and information processing technology evolution than other system complexities, such as invisible heterogeneity, coupling relationships, human involvement, and ubiquitous intelligence [20] in particular.

- Open complex systems and problems are ubiquitous [1–4], appearing in every corner of our daily personal, business, social, and entertainment worlds.
- The challenges and characteristics of complex systems involve many aspects [1, 5–9, 21] of data, business, domain, organization, society, behavior, sociocultural and economic factors which cannot be avoided in attempts to understand a complex system. Simplifying a problem by removing the surrounding contextual factors and systems will not deliver a genuine understanding of the underlying problem; rather, it will deliver outcomes that are dependent on many people seeing only one aspect of the problem instead of the whole picture. This is somehow similar to many blind people work together to build the whole picture of an elephant.
- Typical theories such as system of system, hybrid intelligent systems, and ensemble systems are helpful for handling simple systems. Using a systematic methodology, both the parts and the whole in complex systems can and should be studied, with the personality of parts and the connections between parts being respected. This combines the strengths of reductionism, which focuses on parts, and those of holism, which favors attention on the whole [6].
- Human elements are inbuilt in many complex systems and act as system constituents or problem-solving components; the intelligence contributed to a system by human beings ranges from empirical knowledge and understanding to imaginary and creative thinking, which often plays a more important role [5–9].

L. Cao, *Metasynthetic Computing and Engineering of Complex Systems*,
Advanced Information and Knowledge Processing,
DOI 10.1007/978-1-4471-6551-4_16

- Diversified coupling relationships [10] often appear in an open complex system for a range of reasons, in different forms, and playing a variety of roles; extracting, representing, and learning such couplings are thus very important for building a genuine understanding of the relationship between parts and the whole.
- Sociality [11] is becoming increasingly important in recognizing open complex systems. The interactions between individuals, group interactions, convergence and divergence, hierarchy and structures, reasoning and representation, positive and negative impact on one another, trust and conflict, evolution and dynamics, etc., are becoming more and more important in exploring the societal characteristics of a group, a community and a population. Sociality is also closely related to other important system characteristics including economic, behavioral, cultural, organizational, and environmental aspects [1, 9, 19].
- To disclose the native features and complexities [3, 12, 13] in an open complex system, interdisciplinary research is necessary; this may naturally involve social science, behavioral science, management systems, systems science, mathematics and statistics, computing science, and data science. Computing theories including organizational computing, social computing, behavior computing, service-oriented computing, cloud computing, distributed computing, and parallel computing may also be needed to address the various challenges.
- Data plays an increasingly important role in exploring complex systems, and discovering data intelligence [14] is becoming the next IT revolution and major focus. Data science and engineering are thus becoming fundamental; however, the classic data analysis and learning theories, tools, and systems were not built for tackling system characteristics such as those in open complex systems—this requires revolutionary theoretical breakthroughs and technical innovation in creating data science and technology.
- Understanding the invisibility [10] in open complex systems is becoming important for discovering unknown objective and subjective factors, structures, coupling relationships, dynamics, patterns, and effects. Invisibility may play a major role in forming sophisticated system complexities. The way to explore open complex giant systems is to discover those unknown facts of which we are as yet unaware. Accordingly, learning subjective and objective invisibility in complex systems becomes a major challenge.

16.2 About Metasynthetic Computing and Engineering

Metasynthetic computing and engineering focuses on addressing system complexities in open complex systems and problems [18] which are not limited to software engineering. How to understand, thus compute, open complex problems is a critical challenge, and *metasynthetic computing and engineering* is thus an area that warrants further development.

- To understand a complex problem, we may need to build a problem-solving system that can cater to different stakeholders, including domain experts and end users; system modules, networks, and communication to support cooperation, coordination, and negotiation; outside resources to be accessed on demand; facilities to support the access to and analysis of external resources; tools to enable the involvement of domain experts; knowledge and organizational and social intelligence; and methodologies and tools to support iterative interactions between stakeholders before a firm conclusion is reached. Such a system needs to synthesize and metasynthesize all relevant intelligence, components, resources, and tools that are directly or indirectly related to the problem.
- For open complex problem-solving, group expert interaction and group/social recognition are important. Accordingly, a problem-solving system needs to support cooperation, coordination, negotiation, conflict resolution, and consensus building for all human stakeholders involved in the problem-solving process. This demands the corresponding methodologies, infrastructures, system modules, and interfaces to enable the team-based problem-solving process. In [1, 9, 11], methodologies, systems, and corresponding supporting tools were discussed for enabling the metasynthesis of human intelligence through a human-involved complex problem-solving process.
- Methodologies, systems, and tools need to be studied to support human-centered human–machine cooperation [11]. Such facilities should enable seamless human–machine interaction, user modeling, human cognitive process understanding, social interactions between domain experts, retrieval and analysis of corresponding internal and external data sources relevant to the underlying problem, evaluation of periodical outcomes, adjustment of further process and cooperation, etc.
- The methodologies and systems may take a case-by-case approach [5, 7, 15] and focus on understanding, computing, and engineering those open complex systems closely related to our living, business, and social worlds, especially those systems which are easier to understand and explore, such as artificial systems. Experience and lessons accumulated in these exercises should be evaluated in terms of subjective and objective metrics, with the system, modules, facilities, and outcomes adjusted during the cooperation process.
- A critical challenge facing open complex problem-solving is to understand and build effective cognitive processes, negotiation methods, and working mechanisms for individuals and groups to fully take advantage of individual and social cognitive intelligence during the problem-solving process [16]. This requires cooperation and conflict resolution between domain experts from different disciplines and background who may have varied objectives and expectations; thus, handling conflicts of interest is a significant challenge to be addressed in the design and the system.
- Integration and synthesis of ubiquitous intelligence [14] in problem-solving is necessary for open complex problem-solving. A system needs to be powerful and flexible to incorporate heterogeneous intelligence on demand in different

forms, at different time points, and for corresponding purposes. This is certainly a major challenge to existing methodologies and systems.

- Involving human intelligence, especially imaginary and creative thinking, in open complex problem-solving is necessary for open complex systems [5–7]. The current engineering and computing systems are not good at this. Methodologies and tools to enable creative and innovative problem-solving by individuals and group domain experts in the problem-solving process and system are important.
- The use of the Internet and network theories, tools, and services in open complex problem-solving systems is becoming increasingly important [17]. This requires seamless and instant connections to external resources, networks, experts, and services on demand for information retrieval, sentiment and opinion analysis, cloud computing, crowdsourcing, recommendation, networking, etc.
- The integration of private and public resources and facilities may be necessary, as resources belonging to a particular private system or organization are insufficient for building a full understanding of the nature of a problem or generating a complete picture about what is unknown. In data analytics, for instance, public data from social media and socioeconomic data should be combined with private data from call centers, business processes, and management systems to produce a multifaceted comprehensive picture that will avoid biased findings from local and isolated sources. The implementation of these strategies will, as suggested by our analogy of blind people recognizing an elephant earlier in this book, enable us to see "what the whole elephant really looks like".

References

1. Cao, L., Dai, R.: Open Complex Intelligent Systems (in Chinese). Post & Telecom Press, Beijing PRC (2008)
2. Chu, D.: Complexity: against systems. Theory Biosci. **130**, 229–245 (2011)
3. Miller, J.H., Page, S.E.: Complex Adaptive Systems: An Introduction to Computational Models of Social Life. Princeton University Press, Princeton (2007)
4. Meadows, D.H.: Thinking in Systems: A Primer. Chelsea Green Publishing, White River Junction (2008)
5. Qian, X., Yu, J., Dai, R.: A new scientific field–open complex giant systems and the methodology. Chin. J. Nat. **13**(1), 3–10 (1990)
6. Qian, X.: Revisiting issues on open complex giant systems. Pattern Recogn. Artif. Intell. **4**(1), 5–8 (1991)
7. Qian, X.: Building systematology (in Chinese). Shanghai Jiaotong University Press, Shanghai PRC (2007)
8. Dai, R., Wang, J., Tian, J.: Metasynthesis of intelligent systems (in Chinese). Zhejiang Science & Technology Press, Hangzhou (1995)
9. Cao, L., Ruwei, D., Mengchu, Z.: Metasynthesis: M-space, M-interaction and M-computing for open complex giant systems. IEEE Trans. Syst. Man Cybern. Part A **39**(5), 1007–1021 (2009)
10. Cao, L.: Coupling learning of complex interactions. J. Info. Process. Manag. **51**(2), 167–186 (2015)

11. Dai, R., Li, Y., Li, Q.: Social Intelligence and Metasynthetic System (in Chinese). Post & Telecom, Beijing PRC (2013)
12. Mitchell, M.: Complexity: A Guided Tour. Oxford University Press, New York (2011)
13. Page, S.E.: Diversity and Complexity (Primers in Complex Systems). Princeton University Press, Princeton (2010)
14. Cao, L., Yu, P., Zhang, C., Zhao, Y.: Domain Driven Data Mining. Springer, New York (2010)
15. Cao, L., Zhang, C., Zhou, M.: Engineering open complex agent systems: a case study. IEEE Trans. Syst. Man. Cybern. Part C. Appl. Rev. **38**, 483–496 (2008)
16. Cui, X., Dai, R.: A human-centered intelligent system framework: meta-synthetic engineering. Int. J. Intell. Info. Database Syst. **2**(1) 2(1), 82–105 (2008)
17. Dai, R.W., Cao, L.B.: Internet—an open complex giant system. Science in China (Series E). Sci. China Ser. E. **33**(4), 289–296 (in Chinese) (2002)
18. Dai, R.: Qualitative-to-quantitative metasynthetic engineering. Pattern Recogn. Artif. Intell. **6** (2), 60–65 (1993)
19. Cao, L.: Non-IIDness learning in behavioral and social data. Comput. J. **57**(9), 1358–1370 (2014)
20. Cao, L., Luo, D., Zhang, C.: Ubiquitous intelligence in agent mining. ADMI **5680**, 23–35 (2009)
21. Waldrop, M.M.: Complexity: The Emerging Science at the Edge of Order and Chaos. Simon & Schuster, New York (1992)

Index

L. Cao, *Metasynthetic Computing and Engineering of Complex Systems*,
Advanced Information and Knowledge Processing,
DOI 10.1007/978-1-4471-6551-4

Printed in the United States
By Bookmasters